Water: A Critical Introduction

Critical Introductions to Geography

Critical Introductions to Geography is a series of textbooks for undergraduate courses covering the key geographical sub-disciplines and providing broad and introductory treatment with a critical edge. They are designed for the North American and international market and take a lively and engaging approach with a distinct geographical voice that distinguishes them from more traditional and outdated texts.

Prospective authors interested in the series should contact the series editor:

John Paul Jones III
School of Geography and Development
University of Arizona
jpjones@arizona.edu

Published

Water: A Critical Introduction
Katie Meehan, Naho Mirumachi, Alex Loftus, and Majed Akhter

Environment and Society: A Critical Introduction, 3e
Paul Robbins, John Hintz, Sarah A. Moore

Political Geography: A Critical Introduction
Sara Smith

Political Ecology: A Critical Introduction, 3e
Paul Robbins

Economic Geography: A Critical Introduction
Trevor J. Barnes, Brett Christophers

Health Geographies: A Critical Introduction
*Tim Brown, Gavin J. Andrews, Steven Cummins, Beth Greenhough,
Daniel Lewis, Andrew Power*

Urban Geography: A Critical Introduction
Andrew E.G. Jonas, Eugene McCann, Mary Thomas

Geographic Thought: A Critical Introduction
Tim Cresswell

Mapping: A Critical Introduction to Cartography and GIS
Jeremy W. Crampton

Research Methods in Geography: A Critical Introduction
Basil Gomez, John Paul Jones III

Geographies of Media and Communication
Paul C. Adams

Social Geography: A Critical Introduction
Vincent J. Del Casino Jr.

Geographies of Globalization: A Critical Introduction
Andrew Herod

Forthcoming

Energy, Society, and Environment: A Critical Introduction
Dustin Mulvaney

Geographic Thought: A Critical Introduction, 2e
Tim Cresswell

Introduction to Cultural Geography: A Critical Approach
Jamie Winders, Declan Cullen

Water

A Critical Introduction

Katie Meehan, Naho Mirumachi,
Alex Loftus, and Majed Akhter

WILEY Blackwell

This edition first published 2023

Registered Offices
John Wiley & Sons, Inc., 111 River Street, Hoboken, NJ 07030, USA
John Wiley & Sons Ltd, The Atrium, Southern Gate, Chichester, West Sussex, PO19 8SQ, UK

For details of our global editorial offices, customer services, and more information about Wiley products visit us at www.wiley.com.

Library of Congress Cataloging-in-Publication Data
Names: Meehan, Katie, author. | Mirumachi, Naho, author. | Loftus, Alex, author. | Akhter, Majed, author.
Title: Water : a critical introduction / Katie Meehan, Naho Mirumachi, Alex Loftus, and Majed Akhter.
Description: Hoboken, NJ : Wiley, [2023] | Series: Critical introductions to geography | Includes index.
Identifiers: LCCN 2022052847 (print) | LCCN 2022052848 (ebook) | ISBN 9781119315216 (paperback) | ISBN 9781119315193 (adobe pdf) | ISBN 9781119315162 (epub)
Subjects: LCSH: Water–Philosophy. | Water-supply. | Water and civilization.
Classification: LCC GB671 .M44 2023 (print) | LCC GB671 (ebook) | DDC 553.7–dc23/eng20230123
LC record available at https://lccn.loc.gov/2022052847
LC ebook record available at https://lccn.loc.gov/2022052848

Cover Design: Wiley
Cover Image: Dam at end of Fassa Valley in Dolomites © devid75/Getty Images

Set in 11/13.5pt Minion by Straive, Pondicherry, India

Contents

Contents

List of Figures

List of Tables

List of Boxes

Acknowledgments

We hope you read this book as a love letter to the power of critique. The references section, itself lengthy as a chapter, captures just a small sample of the muscular, insightful, and brilliant critical scholarship on water – including original research and stories of our own. In the places we research and work, we are grateful to our many collaborators, informants, and participants, who have deeply shaped and sharpened our worldviews. May the works cited here inspire questions, discoveries, and more skeptical superheroes.

This book is the product of four people with very different backgrounds, interests, and life experiences. What unites us is a curiosity about the world, a relentless drive to question, and a responsibility to the people, places, and waters we have met along the way. We hope that this book serves as a launch pad for readers to do the same and see the world's waters around them a little differently.

We are fortunate to work together at King's College London in the Department of Geography, a truly inspiring and collegial environment that supported the production of this book. The book arrives amid exciting developments with King's Water Centre, and we are thankful for our new interdisciplinary community that has formed around grand challenges of water sustainability. We are grateful to the Department of Geography Research Impact Fund and the PLuS Alliance (Arizona State University, King's College London, and UNSW Sydney) for providing funds that were crucial to making the book more accessible and engaging for our readers. A two-month fellowship from the University of Arizona's Agnese Nelms Haury Program in Environment and Social Justice helped to shape early content and structure.

It takes a village to make a book. The series editor, John Paul Jones III, believed in this project (and us) from the start, and we are grateful for his vision, patience, and skilled editorial pen – literally, scribbled in the margins of scanned pages of text. At Wiley, editors Charlie Hamlyn, Clelia Petracca, and Justin Vaughn shared the faith and created a structure for this project to flourish and manifest. We also thank Verity Stuart and others at Wiley who helped bring this book to publication. Philip Stickler did wonders with the maps and cartographic design.

A special thanks is due to Jen McCormack, our copyeditor, who provided superb editorial support, poring over the text with a keen eye and unwavering fidelity to critique. There was no hiding from Jen's quality control: she pushed us to improve our craft and write to the student at the back of the classroom. All faults and errors remain ours.

Peer review is the often-invisible and ever-vital reproductive labor of our praxis. We are grateful to the anonymous reviewer for their close reading, incisive comments, and big picture thinking. Their excellent comments were joined by three outstanding student reviewers – Lucy Everitt, Diana Kim, and Tamara Sbeih – who provided constructive and candid feedback. It was a privilege to be closely read by such brilliant people.

Writing this book involved a surprising degree of joy and intellectual freedom. Our classrooms provided a supportive space for thought experimentation. Special thanks is owed to University of Oregon students from Katie's International Water Policy class – all nine years of you! – and members of the Salon (Shiloh Deitz, Fiona De Los Ríos, Lourdes Ginart, Olivia Molden, and Kate Shields) who generously read the book proposal and commented on early material. At King's College London, we thank students in our current and past modules, especially Water and Development, Environmental Justice, and Water Sustainability.

The writing of this book coincided with COVID-19 pandemic upheavals and major structural transformations (and struggles) within British higher education. Chapters were frantically written between "pingdemics" and picket lines, including many weeks of industrial action over pension dispute, fair pay and working conditions, gender and racial equity – all fights against the devaluation of our profession and the erosion of education.

A healthy sense of outrage is always a good place to start writing a critical introduction. Many people supported us along the way – cooking meals, adjusting plans, telling us when things got boring – and we would especially like to thank our friends and families, who have been our best (or at least, most honest) critics and confidants.

This book is dedicated to our students, past and present, who continue to create new worlds and teach us.

Part 1

Foundations

Chapter 1

The Hydrosocial Cycle

The Month of Big Rains

May is the last month of the school year in Tucson, Arizona, USA. At Manzo Elementary School, students lead us on a tour of their school gardens.[1] Under the desert sun, shade from fruit trees and the cool touch of goldfish ponds offer respite in this arid city. We stop near a rain tank (Figure 1.1). Chickens peck at bugs and sip water collected from the roof. A desert tortoise peeks out from under an agave plant. At Manzo, students learn science in the garden classrooms. They also develop community-building skills of empathy, leadership, mental wellbeing, and the care work involved in cultivation (Lohr et al. 2022). Manzo students are architects of life, transforming rain into vegetables, flowers, fruit, and eggs. At their weekly market, they sell garden products, manage customers, plan budgets, fix systems, and feed their neighborhood (Figure 1.2).

---

[1] The Manzo project is part of the Community and School Gardens Project (CSGP), a joint effort between the University of Arizona and the local Tucson Unified School District in over 70 Tucson-area schools: https://schoolgardens.arizona.edu/

Water: A Critical Introduction, First Edition. Katie Meehan, Naho Mirumachi, Alex Loftus, and Majed Akhter.
© 2023 John Wiley & Sons Ltd. Published 2023 by John Wiley & Sons Ltd.

Figure 1.1　The school gardens at Manzo Elementary School in Tucson, Arizona.
Source: Courtesy of Community and School Gardens Program.

Figure 1.2　The Manzo farmer's market. Here, student leaders explain how they grow
vegetables with harvested rainwater and sell their products in a community market.
Source: Katie Meehan (author).

Can a desert support life? Outsiders tend to assume that a desert is a wasteland – a site of scarcity, a harsh landscape devoid of water and therefore life.[2] But in the Sonoran Desert, the Manzo students show us how life is infused in every raindrop. Plant flowering coincides with the North American monsoon season of July through September. Thunderhead clouds build pressure and water droplets, then break in dramatic displays of thunder, lightning, and heavy evening rain. Water floods streets and arroyos – trapping cars and washing out paved roads – and recharges rivers and aquifers. Plants bloom, sprout, seed, and germinate in a few crucial weeks. Most of the crops at Manzo are rainfed. Water from the municipal piped network (brought hundreds of miles from the Colorado River) is a backup source.

Water at Manzo is an example of the **hydrosocial cycle**, the view of water as inseparable from society. A hydrosocial approach argues that water is fundamentally *relational* (Loftus 2007). Water is the product of social, spatial, and ecological relations – a point of view that positions us (people) as *internal* to the production of the thing we call "water." The hydrosocial cycle asks questions like: How is water produced? Where is it sourced from and to whom does it flow? What work does water do? And what conditions does a water cycle create?

The hydrosocial thesis comes into focus at the end of the Manzo school tour, as we pause at a colorful mural (Figure 1.3). Our student guides explain that the mural is the traditional Tohono O'odham calendar for weather, agriculture, and ecological knowledge. Experts in dryland agriculture, the Tohono O'odham are Indigenous people of the Sonoran Desert (including Tucson), residing primarily in what is present-day Arizona (USA) and Sonora (Mexico).

Each month marks a water-related event or task. In April, cacti and flora bloom in spectacular colors, following a season of slow winter rains. May is the ideal time to collect beans from mesquite trees, which are dried and ground into flour. In June, the saguaro cactus called *ha:san* in Tohono O'odham bears fruit called *baidaj* which ripens in scorching temperatures. June is also the Tohono O'odham new year, called *ha:san baidaj* (or *bak*) *masad* (NAAF 2021). This celebration connects Tohono O'odham lifeways or *himdag* to the harvest of sweet, fuchsia-colored *baidaj*. By July – the month of big rains – the North American monsoon cracks open, unleashing torrents of hard rain and thunderclaps across a thirsty desert landscape. At its heart, the O'odham calendar depicts a *situated* worldview of water and society – the opposite of what scholar Donna Haraway (1988) calls the "view from nowhere" that characterizes modern science. Tohono O'odham Nation citizen and agriculturalist Nacho Littleagle Flores (CSGP 2022) explains how the calendar sustains O'odham culture, identity, and language, and incorporates biogeography, seasonal weather, human labor, and the intimate relations of water.

---

[2] Like any landscape, deserts are not innocent. As geographer Natalie Koch (2021, p. 87) argues, "[E]nvironmental imaginaries about deserts are geopolitics imaginaries, actively constituting and constituted by relationalities, identities, and potentialities across time and space."

Figure 1.3   The O'odham calendar at Manzo Elementary School. Source: Katie Meehan (author).

A hydrosocial approach opens the sluice to a whole array of radical possibilities. In contrast to the hydrologic cycle, which "naturalizes" the nature and behavior of H_2O, the hydrosocial cycle challenges us to ask how "nature" – like a flooded field, a broken dike, a submerged city, a parched town, a thirsty household – comes to be. Why is Jakarta sinking? Why did New Orleans flood when Hurricane Katrina struck the US Gulf Coast in 2005? Why is the Middle East touted as the hot spot for water scarcity? What made "Day Zero" in South Africa such a terrible crisis? What explains the global rise in large dams? Who benefits from clean, safe piped water – and who does not? *Why?* Why is our world this way? And what can we do about it?

The Hydrosocial Cycle

Imagine water in action. What do you see? Nearly every science textbook and school lesson begin with the classic image of the hydrologic cycle: a sweeping visual trace of water's planetary travels through clouds, oceans, lakes, rivers, aquifers, trees, and occasionally a crop field or town. The hydrologic cycle is a cornerstone of water science and expert knowledge. In most textbook versions, water moves seamlessly against a temperate backdrop – a hint of its Northern origins (Linton 2008) – and flows without friction through different sites and states of being.

Water in the hydrologic cycle obeys a supposed "natural" rhythm and logic, neatly illustrated by arrows, names, and occasionally numbers. This water spends a long time underground, and comparatively, just seconds in the upper reaches of the atmosphere. Fueled by energy from the sun, water in the hydrologic cycle flows like a machine: a predictable substance that quietly follows the laws of physics and nature. Precipitation, infiltration, evaporation – these states of water are "scientific" and devoid of human influence or touch. Our task, as students and viewers, is to take notes. And then take a test.

Of course, water does obey rules. Rain falls, according to gravity and physics, even in the Arizona desert. But as the Manzo students remind us, water is more than a simplified scientific representation – which, even on its best days, captures knowledge about water that is important but partial, contingent, and produced (Haraway 1988). Critical scholars have shown us how the very categories of "nature," "technology," "wilderness," and "culture" are not stable and pre-given, but contingent products of human minds, social conventions, colonial histories, state institutions, and positions of privilege (Cronon 1996; Jasanoff 2004; Latour 1993; Ottinger et al. 2016). This critique is true of water. "Our starting point is that the hydrologic cycle is not merely a neutral scientific concept," argue Jamie Linton and Jessica Budds (2014, p. 171), "but can be regarded as a social construct with political consequences." This idea – that knowledge is produced, and no environment is apolitical (Robbins 2019) – anchors the journeys we take in this book.

In the mid-nineteenth century, for example, the US West and British Punjab regions were punctuated by large dams and massive irrigation projects of "desert reclamation" – a topic we explore in more depth in Chapters 2 and 6. These infrastructures were made possible by hydrologic studies and "truths" established by western science. This intellectual position was backed by the foreign capital and development muscle of American and British colonial rule – a confluence of science, capital, and power called the **technozone** (Akhter and Ormerod 2015). Experts deemed arid environments as "deficient" landscapes in need of development intervention to maximize their full potential as productive landscapes (Koch 2021). Drylands, the message went, must be tamed, properly managed by experts, and "scarce water" should not be wasted. In short, technozone thinking produced a scientific idea of water in desert regions that went hand in glove with large-scale infrastructure and development interventions. As we will analyze, this is not "neutral" knowledge but a political worldview.

The hydrologic cycle is a relatively recent invention. Jamie Linton (2010) explains how the hydrologic cycle emerged during an early twentieth-century struggle among scientists to define hydrology as a "pure natural science" and legitimate discipline, backed by quantitative force. In 1931, Robert E. Horton created the first scientific depiction of the hydrologic cycle, published in his landmark article and announced in a public address, launching the field of hydrology.[3] "Hydrology is described as having origins in ancient philosophy" – a narrative promoted by Horton that supports the modernist idea that the water cycle was "just sitting there" awaiting discovery and simply needed a new discipline to illuminate it (Linton 2010). Taking a critical approach to history, Linton (2010, p. 109) excavates hydrology's origins "with the quantitative, basin-scale studies of French and English proto-hydrologists in the seventeenth century." Through this "new" scientific representation of water, Linton argues (2010, p. 105), the hydrologic cycle was "an intellectual move that allows us to quantify water and abstract it from cultural contexts that otherwise define its social nature(s)."

Horton was no stranger to these ambitions. In his hand-drawn version of the hydrologic cycle, water follows a precise order and quantitative logic. Any relations are severed: humans reside somewhere "external" or outside of water. Indeed, Horton's water cycle does not feature people at all! The effect of this representation was to "naturalize" water's circulation – as timeless, placeless, and devoid of human influence (Linton and Budds 2014; Schmidt 2014; Swyngedouw 2004). A seemingly innocent diagram, the hydrologic cycle has had major implications for how we understand people and nature:

> Because it is understood as the natural circulation of water on earth, the only possible way that people can involve themselves in the hydrologic cycle is to *alter* it, thus inevitably producing an antagonistic kind of relationship. Instead of allowing for the increasingly hybrid (socio-hydrological) nature of the circulation of water, the hydrologic cycle conditions an understanding that keeps water and people in separate, externally related spheres. (Linton 2010, p. 106)

A scientific field was born. Water, Horton argued, deserves a separate field of inquiry called hydrology, constituted by a certified body of experts (known as "hydrologists") who specialize in the "science of water" and bring technical knowledge and authority over its dynamics (Linton 2010, p. 171). This new framing of water dovetailed with national development agendas – think of the US West and British Punjab examples – and the restless movements of global capital, looking to invest in new infrastructure projects. By the mid-twentieth century, Linton (2010, p. 106) describes, "[T]he hydrologic cycle was quickly taken up by planning agencies of the US federal government as a means of envisioning the nation's water resources and rendering them to a 'calculable coherence' to use Heidegger's term." From Mexico to Pakistan, the science of hydrology supported national development agendas – cue the big dams (Chapter 6).

The hydrologic cycle is undoubtedly a major achievement. But this book is guided by a different notion: the hydrosocial cycle, the idea that water is inseparable from society and shapes – and is shaped by – our lives, places, practices, and geometries of

---

[3] Appropriately, the article was called "The Field, Scope, and Status of the Science of Hydrology" and was published in the flagship journal of the American Geophysical Union (Linton 2008, 2010).

power (Linton and Budds 2014). The hydrosocial cycle is a **heuristic**, a tool for thinking about relations that might otherwise remain hidden in mainstream accounts of water. As a tool, the hydrosocial cycle directs our inquiry into the very production of water flows, facts, narratives, and ideals. The hydrosocial cycle queries assumptions, challenges mythologies, and questions authority, even as it traces the material flows of water. We can ask important questions about the state of the world:

- What is water? How do social groups differentially construct water? How does water's **materiality** – its material properties – shape the ways people know and manage water?
- Where is water? Who experiences its scarcity or (over)abundance? Why? What does the spatiality of water reveal about its social and ecological relations?
- Whose knowledge about water counts? Under what circumstances? How do these knowledges articulate or refract vested interests or structures of power?
- How is water produced? What conditions make water "scarce" or "plentiful"? Who (or what) makes these conditions? What does water reveal about broader trends, politics, or power? Why is a drought (or a flood) never just a drought (or flood)?
- What causes water injustices? Why do they still occur, despite major advances in technology and management?
- What is the future of water? How can we manage water for more just and sustainable futures?

Several key elements of hydrosocial thinking bracket this book. First, the hydrosocial cycle forces us into a **relational** state of mind. Why is this important? Water is fundamentally relational, the product of social, spatial, and ecological relations – a point of view that positions us (people) as fundamental to the production of water (Loftus 2007). For example, a relational view asks how the lack of universal piped water provision in Durban, South Africa, and San Francisco, California, are *manufactured* crises generated by social institutions, ideologies, and power relations (Deitz and Meehan 2019; Loftus 2007, 2009; Meehan et al. 2021). A relational point of view focuses our analysis on the *conditions* of water and its production – and how we, as people, are part of that production.

Second, while the relational aspects of water transcend space and time, an understanding of the hydrosocial cycle is necessarily attuned to **place**. A place-based perspective provides a sharper view into power and the production of spatial and social difference (Massey 2005), including racial, ethnic, classed, caste, and gender-based lines of difference and intersection. Consider the hydrosocial cycle of Tijuana, Mexico, a vibrant coastal desert city on the Mexico–United States border (Figure 1.4). Often stereotyped by images of narco-violence, NAFTA, and Nortec music, on closer look, Tijuana is a city rooted in struggles over water (Meehan 2014).

Water moves unevenly through Tijuana, shaped as much by infrastructure, power, and money as by the energy from the sun. Figure 1.4 depicts the hydrologic and political production of water in Tijuana (Meehan 2010). At the city's edge, reservoirs store

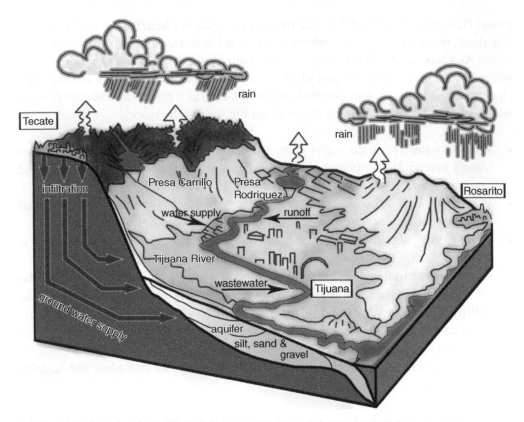

Figure 1.4   The Tijuana hydrosocial cycle. Water moves unevenly through the Tijuana urban region, fueled as much by "natural" forces (gravity, energy from the sun, ecological functions) as by "social" dynamics and institutions (law, science, urban planning). A hydrosocial perspective invites us to ask: what makes these conditions? Source: Katie Meehan, with help from Josh Gobel.

a precious water supply imported at great distances from the transboundary Colorado River. The reservoir supply is governed by a century-old international treaty and legal regimes that favor "beneficial use" for elite parties over long-term sustainability (Chapter 3). Tijuana has other vital inputs and outputs of water. Groundwater is a secondary source for Tijuana's industry and regional agriculture. The winter rains from the Pacific Ocean are a complicated water source. Big storms trigger street flooding and sewer overflows, causing problems for residents – especially those living in shacks in Tijuana's denuded canyons – and life in the Tijuana River estuary, a complex and beautiful ecosystem at the heart of the San Diego–Tijuana region. Rather than assume a "universal" movement of water through space, the Tijuana water cycle reflects the intersecting global and local dynamics of water in place.

Third, hydrosocial thinking attunes us to a critical and generative reading of the *politics* of water, by focusing on its **production**. "While the hydrologic cycle has the effect of separating water from its social context," argue Linton and Budds (2014,

p. 170), "the hydrosocial cycle deliberately attends to water's social and political nature." What does it mean to say that water is "produced"? A hydrosocial perspective does not dispute the existence of nature, reality, or established facts – boring! And by "critical" we do not imply "nihilistic" or "unproductive" or "critical for critique's sake" – doubly boring! Rather, a critical approach to water moves us past a reading of "external" human influence on the environment – because water is already relational, and therefore always political – and unlocks an urgent set of questions: how and why a waterscape is produced, and with what implications, where, and for whom?

In the following pages, we put our heuristic – the hydrosocial cycle – to work. Just like the original hydrologic cycle illustrations of gorgeous swirling vapors and globe-trotting reach, this book will follow water through different biomes, sites, controversies, and dimensions. We will discuss toilets, treaties, food crops, market logics, big dams, Pinochet, Pakistan, and sex. To build our approach, the remainder of this chapter presents four core arguments to structure this text. Think of them as the four joists that underpin the foundation of this book. They are:

1. Knowledge is power.
2. Scarcity is made.
3. Water is life.
4. Camp is everywhere.

Box 1.1 Make Your Own Hydrosocial Cycle

How does water flow in your neighborhood or region? Representations are a form of **visual discourse**: a system of order and power that uses imagery (not just words) to express a worldview. As observers and participants in the world, we – the authors of this book, and you, the reader – also produce discourse.

In a class assignment, students at the University of Oregon first listened to a news podcast story ("Valley of Contrasts" by journalist Antonia Cereijido) about water in Coachella Valley, California, the site of a retirement community and famous music festival. "Coachella is divided into two parts: the west side and the east side," Cereijido (2017) explains. "While the westsiders have pools, golf courses, and sprawling lawns – all which require a lot of water – there are parts of the east (such as mobile home parks) with up to ten times the safe level of arsenic in their water." Working in small groups, Oregon students listened to the podcast and teamed up to illustrate their version of how water flows through the valley (Figure 1.5). Just like the Horton water cycle, their illustrations are not "neutral" depictions, but a visual analysis and argument.

We invite you to put down this book and pick up a pen or pencil. Find a flip-chart, a notebook, a whiteboard, a tablet, or even a sidewalk. Think of a place

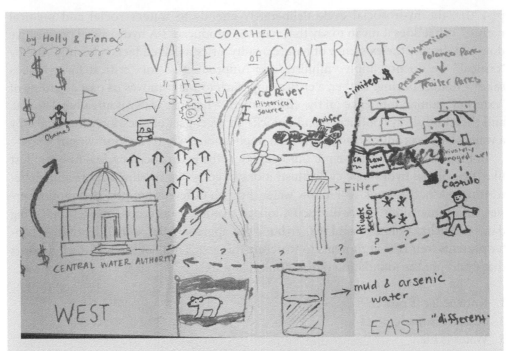

Figure 1.5 Hydrosocial flows in the Coachella Valley. Illustration by students (Fiona De Los Ríos and Holly Moulton) at the University of Oregon. After listening to the "Valley of Contrasts" podcast, students discussed the story in class and created their own interpretation of the Coachella hydrosocial cycle. Source: Katie Meehan (author).

like your hometown, or perhaps, a place cited in current events or a lecture. How does water flow in this place? Who and what are its sources and destinations? What shapes these physical and material circulations? What landscape features are critical or important? What impedes, diverts, or transforms water's flow? With what effects or results? Does water obey the rules of physics – like the alchemy of rainfall, the mechanics of evaporation – or are there other, perhaps more social elements and forces at work? Draw and discuss your findings. Every image will be unique, because (i) water has certain universal properties but its flow is dependent on places, which are unique; and (ii) drawings will depend on you – the artist and analyst. Ask yourself: What does a hydrosocial cycle allow us to "see" about water and society?

Knowledge is Power

A key tenet of the hydrosocial cycle is that knowledge is power. In other words, scientific knowledge, expertise, and authority shape how water is understood, managed, and legitimized. In La Ligua basin in Chile, Jessica Budds (2009a) explores how routine groundwater assessments by hydrologists became fraught politicized tools used by the water authority, the National Water Directorate (DGA). These assessments shape water rights and access for farmers in uneven ways. Far from science playing a neutral role in water management, the story of La Ligua reveals how social power saturates the nexus of science and decision-making, influencing which (and whose) ideas are adopted into practice.

Knowledge is not innocent or neutral but is a human construction that shapes (and is shaped by) the world. Drawing on the field of science and technology studies, Sheila Jasanoff (2004) argues that scientific knowledge is **coproduced** by social practice and spatial orders. In short, science and politics are a two-way street, as the world seeps into science from the outset of knowledge creation.

François Molle (2008) expands on this approach to explain why certain flagship ideas manifest as practice in international water policy. Molle examines the emergence of Integrated Water Resource Management (best known by its acronym, **IWRM**), a prominent concept that promotes a coordinated and cross-sectoral approach to managing water. While IWRM was welcomed as an antidote to the perceived "chaos" of uncoordinated water management, Molle argues that IWRM, like any hegemonic concept, tends to obscure the political nature of natural resource management. "Ideas are never neutral and reflect the particular societal settings in which they emerge, the world views and interests of those who have the power to set the terms of the debate, to legitimate particular options and discard others, and to include or exclude particular social groups" (Molle 2008, p. 131). For Molle, the global IWRM bandwagon operates not due to "genius" insight but because of the social conditions which brought IWRM into existence in the first place. For us, a critical focus sparks even more questions: Whose water knowledge counts? Why? To what effect or end?

A productive way forward is to consider how expert or scientific **discourse** about water is more than a set of constructed facts or words. Scientific discourse reflects a categorization and system of power. Social power shapes our realities, knowledge, and claims about water – it even produces the ways in which truth is *made true*, what philosopher Michel Foucault (2020 [1975]) calls "regimes of truth." Rutgerd Boelens draws on this theory to explore the rationalities in different kinds of Andean water knowledge, including "scientific" knowledge and other knowledge systems categorized as "local" or less salubrious to water management. From the vantage point of Western science, the pecking order of knowledge sits in a hierarchy:

> Rules, rights, and duties attached to water flows and hydraulic infrastructure are closely linked to systems of meanings, symbols, and values, involving institutions and networks of human, non-human, and supernatural actors and power that influence water control. This domain – often erroneously associated with only "social" and not with, for example, technology – is essentialized in romantic representations and contested or ignored in natural sciences. (Boelens 2014, p. 240)

Discourses have the effect of designating "legitimate" knowledge, truths, and frames of reference (Boelens 2014, p. 235) at the expense of creating (subordinate) categories of "local" (or vernacular) knowledge (Klenk et al. 2017). In this way, the hydrologic cycle works to (i) separate "legitimate" forms of water knowledge, rights, and access from "illegitimate" forms; (ii) naturalize policy models as scientific and reinforce elite and state control over water resources (Boelens and Vos 2012). Struggles over water are not limited to physical allocation, they include struggling over truth regimes and defining the very order of things (Boelens 2014, p. 235).

How do ideas about water (and people) travel? Who comes up with influential policy principles and what makes them stick? Why are some policy models so seductive, pervasive, and powerful? Our knowledge about water is profoundly shaped by **expert networks** and hegemonic policy narratives (Conca 2005). A good idea is not enough, argues Ken Conca (2005), an idea needs a networked cadre of elites, organizations, and institutional structures that grease the tracks for policy ideas to circulate globally, gain authority, and reproduce in settings beyond their origins. Michael Goldman (2007) illustrates this argument with his account of how the World Bank seized the "pro-poor" narrative of "water for all" and mobilized it into a policy prescription for water privatization. Goldman (2007, p. 788) shows how this policy idea stretched beyond the Bank, as it "requires active participation and contributions from actors in corporations, NGOs, think tanks, state agencies, and the media, across the global North and South." In this case, the expert networks converged to create a "global consensus" on reforming water "for all," with privatization as its answer.

Mary Galvin (2015) offers an equally compelling analysis of community-led total sanitation (CLTS) – a prominent technique in the water, sanitation, and hygiene (**WaSH**) sector, implemented in over 56 countries. In the appropriately titled article "Talking Shit," Galvin investigates the ideology that underpins CLTS and chronicles the expert networks that enabled its "thrilling success" in the WaSH sector.

> What is particularly distinctive about CLTS is that it forces participants to confront their "shit" by using this word, visiting places where people openly defecate and tracing the fecal to oral transmission route to the glass of water on the table. (Galvin 2015, p. 10)

We discuss further how disgust and shame-based methods create social damage in Chapter 8 (see also Brewis et al. 2019a). Galvin, meanwhile, dissects organizational pathways taken by UNICEF, WaterAid, and the UN Special Rapporteur on the Right to Water and Sanitation that have put CLTS in motion. "Communities may be driving," argues Galvin (2015, p. 17), "but the roads have been built by these organizations."

Of course, in our uneven world, not all knowledge gets to *be* mobile or count equally. As Diné geographer Andrew Curley (2019a, 2021b) argues, quantification is central to the logic and mechanics of water law in the US West – a logic informed by the hydrologic sciences. In the Colorado River basin, US water law literally "divides up the river" into segmented, quantified units – divorced from their context and place, stripped of time and kin (human and non-human), and pegged to settler-defined geographical units. This logic stands in contrast to the worldview of water held by many Indigenous communities (see Further Reading section). Curley (2021b, p. 21) argues that Indian water settlements, a type of legal agreement between Native Nations and the US federal government, are forms of colonial enclosure, "built on a lineage of law that replaces and perpetuates settler-colonial dispossession." In reproducing law, we reproduce these hegemonic systems of knowledge.

In sum, a critical approach to water does not take knowledge at face value, but asks: How is knowledge about water produced? By whom? What kinds of knowledge are designated as "legitimate" or gain authority – and which do not? Why? What work does knowledge do?

Scarcity is Made

For desert cities, like Tucson and Tijuana, the notion that water is a "scarce" resource is a common refrain in many influential documents, textbooks, and policy principles. But what if, following Erik Swyngedouw (2004, 2009), we started with the idea that scarcity is relational and constructed? That a lack of available water – or clean, accessible, secure, safe water – is not an inherent feature or pre-given reality, but the *outcome* of uneven conditions, logics, and practices? How do we account for the fact that water – the molecule H_2O – is one of the most abundant elements on planet Earth and yet out of reach for so many? What explains resource scarcity?

The narratives of many "scarcity" debates can be traced to Thomas Malthus, an English economist and demographer in the late eighteenth century, whose ideas influenced generations of key thinkers, including Charles Darwin. In his book *An Essay on the Principle of Population*, first published in 1798, Malthus (1992) introduced the concept of population growth causing environmental degradation. Malthus predicted that the human population would outgrow the available food (and water) supply, using the artfully simple logic – he called it "logical empiricism": that (i) people reproduce geometrically (exponentially) and yet, (ii) food supply reproduces in an arithmetic (linear) progression. Too many people, not enough food.

Malthus didn't stop there. Indeed, he argued the human population will expand to the limits of subsistence and only through techniques of "vice" (including war and violence), "misery" (including famine, illness, and drought), and "moral restraint" (i.e. abstinence and Protestant morality) could the world check excessive population growth and avoid environmental destruction. Welfare or charity (embodied by the Poor Laws) was a useless exercise, creating more "dependency" by subjects on the state. Malthus's own words offer a window into his worldviews on race, gender, and class:[4]

> The Poor Laws of England tend to depress the general condition of the poor . . . they may be said, therefore, to create the poor which they maintain. (p. 100)
>
> It can scarcely be doubted that, in modern Europe, a much larger proportion of women pass a considerable part of their lives in the exercise of virtue than in past times and among uncivilized nations. (pp. 43–44)
>
> In some of the southern countries where every impulse may be almost immediately indulged, the passion sinks into mere animal desire, is soon weakened and extinguished by excess. (p. 212)

With the winds of privilege at his back, Malthus's ideas about people and the environment spread like wildfire. His ideas are found, for example, in popular World Bank claims that the "world is running out of freshwater." They lurk in "population bomb" arguments and other accounts that smack of environmental determinism (see Robbins 2019 for explanation and critique).

What does this "scarcity" argument overlook or leave out? Let's go back to Tijuana. Water in Tijuana can easily be labeled as "scarce" – but this claim requires surgical attention and critical analysis. Tijuana is a desert city, but the provision of water to homes and businesses is mediated by infrastructure, social institutions, law and legal status, and money. For example, *maquiladoras* (export-oriented manufacturing plants) are thirsty customers that never run out of municipal water; yet informal housing settlements (*colonias*) are routinely denied piped water service and sewerage based on their tenure status, and unhoused (homeless) people are reliant on precarious or polluted water sources (Meehan 2013).

For the beneficiaries of Colorado River water, including Tijuana, scarcity is felt unevenly (see Figure 6.3 in Chapter 6 for the map). In Southern California, the Imperial Valley is one of the system's major recipients; the valley gobbles the bulk of regional water allocation rights to grow sod for lawns, parks, and sports fields. The service districts of Los Angeles and San Diego are also well watered. Tijuana, as a major metropolitan area and economic engine, sucks up the bulk of Mexico's allocated Colorado River water through a complex pipeline system that flows west over the Sierra Madres mountains and delivers water to its reservoirs, after which it is unevenly distributed to city residents. Some users can afford to pump groundwater, at a considerable cost. South of the international border, in the Colorado River Delta region of Mexico, small-scale and subsistence farmers eked out an agricultural livelihood on the escaped flows from irrigation system leaks (at least, before the Imperial Valley engineers sealed the leaks). In parts of the Delta, Indigenous users have been marginalized in terms of water access and rights allocation (Muehlmann 2013). Now, the mighty

---

[4] For a brilliant and blistering critique of Malthusian thinking, see Harvey (1974). For key updates, see Robbins (2019) and Robbins and Smith (2017).

flows of the Colorado River often never make it to the sea, ending in a sad trickle in a desiccated wetland.

This story asks a deeper question: What produces water scarcity? Our example does not deny the reality of aridity, or the fact that deserts receive less rain than their temperate counterparts, or the fact that the Tijuana metro region is urbanizing rapidly and placing new pressures on the existing water supply.

A critical approach invites us to ask important questions about conditions. Scarcity is *produced*, not ready-made. As Tijuana illustrates, what is "scarce" – and to whom, where, and why, across the Colorado basin – is an outcome of relations. And so, we invite you to put Malthusian thinking in the dustbin. In its place, we invite you to think with the people of Standing Rock, who offer a far more interesting and critical thesis (the third plank of this book): water is life.

Water is Life

Early in 2016, the people of the Standing Rock Sioux Tribe mobilized to prevent a crude oil pipeline called "The Black Snake" from crossing their unceded lands and threatening the Missouri River, a major water source along the tribe's eastern boundary, located in the Upper Midwest region of the United States (Figure 1.6). The Black Snake is the Dakota Access Pipeline (DAPL), a US$3.8 billion, 1 172-mile underground pipeline that runs 570 000 barrels of oil every day across four US states, two major rivers, and over two hundred creeks and streams.

Early on, Indigenous youth organized camps and led protests, galvanized by calls to "ReZpect Our Water" and defend Indigenous sovereignty (Curley 2019a). By late September, more than three hundred Native Nations and allies joined the movement, creating a #NoDAPL movement and planting their flags in solidarity at Oceti Sakowin Camp, the largest of several protest camps (Figure 1.7). *Oceti sakowin* is translated as the seven council fires of the Lakota, Dakota, and Nakota Nations, known by some as the Great Sioux Nation. Not only the name of the largest camp, *oceti sakowin* is also the unity of the *oyate* or people of this geography.[5] The Standing Rock Sioux were no strangers to their roles as Water Protectors. "#NoDAPL [w]as not a departure from so much as it was a continuation – a movement within a larger movement, but also a movement within a moment – of long traditions of Indigenous resistance deeply grounded in place and history" (Estes and Dhillon 2019). At its peak, the #NoDAPL struggles included thousands of Water Protectors, galvanized by the phrase *mni wiconi* or "water is life."

The physical geography of the DAPL is massive. Operative since July 2014, it carries light sweet crude oil from six sites in the Bakken and Three Forks oil-producing regions of North Dakota, through South Dakota and Iowa, to a terminal in southern

---

[5] The name of the camp, Oceti Sakowin, shows the power of bringing Lakota people together, including Standing Rock but also people of Pine Ridge and Rosebud, who share the worldview of *mni wiconi*.

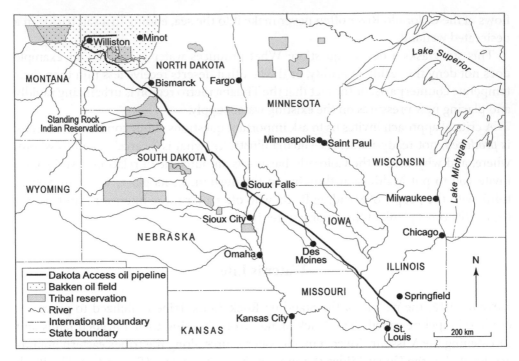

Figure 1.6   The Black Snake. This map illustrates the geography of the Dakota Access Pipeline as it routes oil across the Standing Rock Sioux Reservation, traditional Oceti Sakowin Lands, and the northern Great Plains to a distribution center in Illinois, near the Mississippi River. Source: The Authors, cartography by Philip Stickler.

Figure 1.7   The Oceti Sakowin Camp at Standing Rock. Source: Courtesy of Teresa Montoya.

Illinois, where it links to refineries in the Midwest and Gulf Coast (Mazer et al. 2019, p. 356). The Bakken and Three Forks production areas are part of a deposit that runs all the way north into Montana and the Canadian provinces of Saskatchewan and Manitoba. DAPL fuels a carbon-dependent empire, transnational in finance and commodity flows (Mazer et al. 2019). A rupture of "sweet crude" jeopardizes the drinking water of millions of people and other-than-humans who depend on clean rivers and aquifers for life.

The history of the Black Snake rests on shady grounds. The DAPL violates the Fort Laramie Treaties of 1851 and 1868 (Mazer et al. 2019), treaties made between Lakota, Dakota, Nakota, Arapaho, Crow, and other Native Nations and the US federal government. In neither of these agreements with the US federal government did the Oceti Sakowin cede the land in question surrounding the DAPL route. Since this time, in its rush to seize resources and secure access to land, the United States has repeatedly violated the terms of the Fort Laramie Treaties, including the occupation of sovereign lands, which currently represent a fraction of what is stipulated in the treaty (Mazer et al. 2019, p. 358). Along with #NoDAPL protests, the Standing Rock Sioux Tribe sued the US Army Corps of Engineers, the federal water agency that had approved key permits and authorized DAPL construction.

What does it mean to say that **water is life**? At its core, the concept reflects a relational worldview of water, land, and society – the heart of hydrosocial thinking. In the compilation *Standing with Standing Rock: Voices from the #NoDAPL Movement*, Nick Estes and Jaskiran Dhillon (2019, pp. 2–3) expand on this worldview:

> [W]ater, in general, is not a thing that is quantifiable according to possessive logics. *Mni Sose* is a relative: the *Mni Oyate*, the Water Nation. She is alive. Nothing owns her. Thus, the popular Lakotayapi assertion "*Mni Wiconi*": water is life or, more accurately, water is alive. You do not sell your relative, Water Protectors vow. To be a good relative mandates protecting *Mni Oyate* from the DAPL's inevitable contamination. This is the practice of *Wotakuye* (kinship), a recognition of the place-based, decolonial practice of being in relation to the land and water.

You do not sell your relatives. Relational thinking, in this case, demands a different set of logics, analytics, and practices than a settler colonial mindset. "Land is much more than a site for cultural and spiritual practice, it is also space for social reproduction," explains Andrew Curley (2019a, p. 164), "Without land, life is not possible." Water, too, is part of **social reproduction**: the material social practices, infrastructures, and ecologies that make life possible – for work, play, and freedom. In this way, water is part of the "messy, fleshy" materiality to sustain life, within and beyond the planetary order of **capitalism** (Katz 2001). The notion that "water is life" advances this critique of hegemonic planetary order one step further. Elders from the Yukon First Nations, for example, argue that "Water is a living entity, with the 'person-life' quality of agency referred to as 'spirit'" (Wilson and Inkster 2018, p. 524). "From this perspective water not only enables human life by meeting physical needs, but water *is* life or alive."

In Western understandings – such as Horton's hydrologic cycle – water is often described as a resource, a thing to be abstracted, measured, quantified, valued in commensurable terms, and put to "beneficial" use by law. Rivers running to the sea untapped are considered "wasted" in California water law, for example (Cantor 2017). For some Indigenous communities, such hegemonic thinking runs counter to their fundamental **ontology** of water, as Anishinaabe scholar Deborah McGregor (2014, p. 496) explains:

> Water, in the dominant Western Euro-Canadian context, is conceptualized as a resource, a commodity to be bought and sold. Federal and provincial governments therefore make decisions about water based on a worldview, philosophy and set of values which stands in direct contrast to the views of First Nations people.

Ontology is a word to describe what the world *is* (and how we see it), which is linked to how we *know* the world (**epistemology**). Ontology is collective, not singular. Ontologies are collective systems of identification and classification that serve as models of relationality, points of reference for contrasting worldviews, and wellsprings of identity and knowledge about the world. Put simply, ontology describes what a thing – like water – is and how, why, and where it fits in the order of the universe.

To say that water is life, then, implies an ontology of water that relates internally to society. "Water is life" implies a politics of kinship and respect, and water is "part of extended networks of kinship or kin relations" (Wilson and Inkster 2018, p. 525). An example from a Tlingit salmon ceremony puts these relations into practice. "By returning the fish bones to the water, a salmon ceremony is an act of respect for fish and water," explain Nicole Wilson and Jodi Inkster (2018, p. 525). "Reciprocity is therefore about engaging with water according to protocols to ensure mutual survival. In other words, if you take care of the water, it will take care of you."

> In other words, for Indigenous peoples' water (and land) are understood not simply as a physical asset, but as a way of knowing (epistemology) and being (ontology) embedded in a universe of relations between human and non-human beings that can guide forms of governance and resistance. Through this lens, respect for water can be understood as a politics of kinship that is inextricably linked to Indigenous understandings of water and water governance. (Wilson and Inkster 2018, p. 527)

Conflicts between different regimes of water governance inform **ontological politics**: the friction and lived implications of different worldviews. In the Canadian North, settler ontologies and forms of governance permeate every element of water management, from the co-management boards (shaped by land claims) to the baseline declaration that "water belongs to the Government" in the Yukon Waters Act (Wilson and Inkster 2018). To be clear, the stakes of ontological politics involve erasure and violence. "Although both colonialism and settler colonialism are based on domination by an external power, only settler colonialism seeks to replace Indigenous peoples with a settler society," explain Wilson and Inskter (2018). Settler colonialism is fundamentally about access to territory, in this case, land *and* water. And "invasion" in the context of settler colonialism "is a structure not an event" and therefore never ends (Wolfe 2006).

What lessons can we learn from Standing Rock and other sites of struggle? First, the ontological politics of water is not mere wordplay. Water protectors at Standing Rock endured police attack dogs, pepper spray, concussion grenades, and high-frequency noise devices. Police used high-pressure water cannons on protestors in freezing winter weather. By October of 2016, there had been over 140 arrests. For Estes and Dhillon, state violence against Water Protectors reveals high stakes at play. "Water Protectors became criminal precisely because they were generating and upholding a different kind of law contrary to settler law (one that places relations with nonhumans, the land, and water equal to, or sometimes surpassing, human-made laws), while also reminding the United States of its own obligations to uphold its own treaties – its original agreements – with the Oceti Sakowin" (Estes and Dhillon 2019, p. 2).

Second, Standing Rock shows us a critical praxis of water is alive and resurgent. Far beyond the prairies, the message "Water is Life" rippled across the globe. "No one could have predicted the movement would spread like wildfire across Turtle Island and the world, moving millions to rise up, speak out, and take action," observe Estes and Dhillon (2019, p. 1). "Few could have imagined it would happen in their lifetimes, except for, perhaps, the visionaries themselves who kept the dream alive; and yet it happened, in the isolated, rural geography of dirt roads, farmlands, and the lush shorelines of the *Mni Sose*, the Missouri River." Water protection is bound up with questions of sovereignty, territory, and justice. Whose (and what) future gets to count? For Estes and Dhillon, "The good people of the earth have always been the vanguards of history and radical social change. Such was the case at Standing Rock: everyday people taking control of their lives" (2019, p. 4).

Finally, Standing Rock signals a different kind of **futurity** in the making (Curley and Smith 2022; Daigle 2018). *Mni Wiconi* "simultaneously speaks to the past, present, and future – catapulting us into a moment of critical, radical reflection about the colonial wounds and wounding in the spaces between calls to save planet Earth and the everyday sociopolitcal realities facing Indigenous peoples" (Estes and Dhillon 2019, p. 3). Good relations are born and reborn every day.

"Water *does* connect us," writes Diné anthropologist Teresa Montoya (in Mari Birkett and Montoya 2019). "This understanding goes beyond environmentalist assertions of justice, not because they are wrong but because they are incomplete. It is why, time and time again, our relatives have felt so implored to defend this being." Montoya bends our focus from the individual – a privileged site of Western epistemology – and toward a collective, relational approach to water and society. "When you ask, where do I carry this knowledge," she muses, "I contemplate instead, where does *it* carry us?"

Camp is Everywhere

Standing Rock compels us to ask hard, critical questions – for life, not just for school. By now, you might be asking: What exactly *is* a critical introduction to water? What does a specifically geographic take bring to the water table? And why should we care?

A critical perspective is our entry point to water. Critique is a tricky and beautiful thing – it is a skill, sharpened at the edges by practice, reading, listening, debate, curiosity, doubt, risk, and self-reflection. Critique is a **praxis**, indivisible from what philosopher Hannah Arendt (1998) calls "everyday political action" and our capacity to wrestle with ideas through the active life (*vita activa*). Critique is not only words on a page, but also the practices, knowledge, and action taken by folks with the Standing Rock camp.

We break a few rules in this book. First, we refuse to give you a test. A critical approach to water, as we try to model in this book, is learning for life – not just the exam. On a heating planet, marked by melting glaciers, species extinction, polluted water, state violence, struggling families, and widening social inequality, no exam will be sufficient to meet the task of urgent critical praxis. With this book, we hope to equip you with the tools to learn long after class is dismissed. We will try our best to make it memorable and fun.

Second, we refuse disciplinary silos and easy narratives – we go for the hard story. For us, a critical perspective is a stance rooted in skepticism: a wariness of institutionalized power and seductive narratives that tra-la-la around the globe. Skepticism is not cynicism or a life without hope. Skepticism is a superpower. Skepticism allows us to slow down, see past the hype, and think about the long arc of a narrative: such as ideas that the "law" delivers water justice, or that the "market" is the best allocator of water resources. With these superpowers, we can ask the important questions: How did we get here? What produces a water crisis? Who (or where) benefits from things like water treaties, infrastructures, or technologies – and who (human and otherwise) does not? Why does water remain "hidden" in our breakfast and dinner? What makes these conditions – and why? And what do we do about it?

Who are "we"? At the heart of this book are four people – teachers, writers, editors, and friends – whose situated knowledges, voices, and unique life experiences shape the ideas and stories presented here. Despite the "we" used in the book – the plural, first-person pronoun – our "we" voice is more like a piece of felt than a singular perspective. In this book, we have felted together our situated insights and areas of expertise, informed by our unique subject positions and a mutual training in geography.[6]

Standing Rock is an invitation to camp. Lessons learned on the Dakota prairies enter our classrooms and everyday lives. Camp is planetary and frames this book.

---

[6] Disciplines are freighted with questions of power and knowledge production, which we actively explore in this book. Geography, for example, carries a deep history of colonial science that still shapes our present-day community and praxis. For us, a geographic approach offers far more than a love of maps; it gives us the tools to explore the "why" and "how" of social and spatial interconnections between water and society.

Camp is a form of struggle – using Arendt's term, a form of situated praxis and critique. In *Standing with Standing Rock*, Teresa Montoya and Tomoki Birkett (Birkett and Montoya 2019, p. 270) illustrate the power of "camp" through a beautiful exchange of letters. Reflecting on water development in the American West, including its extractive history and transhemispheric reach, Montoya recalls a story by LaDonna Bravebull Allard. A camp founder and movement leader, LaDonna Allard discusses a tributary of the Missouri River, not far from the Standing Rock camps.

> In her telling, Allard emphasizes that the true name for the Cannonball River is *Inyan Wakangapi Wakpa*, which means "River That Makes Sacred Stones." This name refers to a once active whirlpool whose movement shaped "large, spherical sandstone formations" in the river's bed. Back in the 1950s, however, the US [Army] Corps of Engineers severed this flow when they flooded the area for the construction of Oahe Dam. The project resulted in a loss of 150 000 acres for the Cheyenne River Indian Reservation. But the greater loss was not quantifiable or limited to one Nation or another. Allard writes, "They killed a portion of our sacred river. I was a young girl when the floods came and desecrated our burial sites and Sundance grounds. Our people are in that water. This river holds the story of my entire life."

Co-author and colleague Tomoki Birkett (p. 274) writes back,

> The movement to protect the water at Standing Rock spirals out in so many directions, through seeping groundwater, as evaporated atmosphere, through ocean currents. As you wrote, having enduring relations, despite losses and through victories, has sustained Indigenous practices of decolonization for hundreds of years. In interviews since the camps at Standing Rock were emptied, leaders of the NoDAPL movement have emphasized this continuity. Holy Elk Lafferty has said, "For me it's been a continuum. It has never stopped. We're all continuing to fight. Now, camp is the globe. Camp is everywhere."

Camp is everywhere. We invite you on this journey. Step with us, into the whirlpool.

Summary and What's Next

This chapter introduced the central thesis of this book: the hydrosocial cycle, the idea that water is inseparable from society, and that water shapes – and is shaped by – social practices and geometries of power. We use the hydrosocial cycle as a tool to reveal hydrosocial relations often hidden in mainstream accounts of water. To build our approach, we introduced four core arguments that structure this text: (i) knowledge is power; (ii) scarcity is made; (iii) water is life; and (iv) camp (praxis) is everywhere. These "planks" provide the foundation for our central argument and you will see them resurface in future chapters. We situated ourselves within this book and extended an invitation to learn for life, not just for the exam. Moving forward, the next chapter takes on hydrosocial relations that have permeated every corner of the globe: the relations of water, power, and empire.

Further Reading

The hydrosocial cycle

Budds, J. (2009). Contested H$_2$O: Science, Policy, and Politics in Water Resources Management in Chile. *Geoforum* 40 (1): 418–430.

Linton, J. (2008). Is the Hydrologic Cycle Sustainable? A Historical-Geographical Critique of a Modern Concept. *Annals of the Association of American Geographers* 98 (3): 630–649.

Linton, J. (2010). *What Is Water? The History of a Modern Abstraction*. Vancouver, BC: UBC Press.

Linton, J. and Budds, J. (2014). The Hydrosocial Cycle: Defining and Mobilizing a Relational-Dialectical Approach to Water. *Geoforum* 57 (1): 170–180.

Schmidt, J.J. (2014). Historicising the Hydrosocial Cycle. *Water Alternatives* 7 (1): 220–234.

Swyngedouw, E. (2009). The Political Economy and Political Ecology of the Hydro-social Cycle. *Journal of Contemporary Water Research & Education* 142 (1): 56–60.

Indigenous sovereignty and water

Arsenault, R., Diver, S., McGregor, D., Witham, A., and Bourassa, C. (2018). Shifting the Framework of Canadian Water Governance through Indigenous Research Methods: Acknowledging the Past with an Eye on the Future. *Water* 10 (1): 49.

Chief, K. (2018). Emerging Voices of Tribal Perspectives in Water Resources. *Journal of Contemporary Water Research & Education* 163 (1): 1–5.

Chief, K., Meadow, A., and Whyte, K. (2016). Engaging Southwestern Tribes in Sustainable Water Resources Topics and Management. *Water* 8 (8): 1–21.

Curley, A. (2019). "Our Winters' Rights": Challenging Colonial Water Law. *Global Environmental Politics* 19 (3): 57–76.

Curley, A. (2021). Unsettling Indian Water Settlements: The Little Colorado River, the San Juan River, and Colonial Enclosures. *Antipode* 53 (3): 705–723.

Daigle, M. (2018). Resurging through Kishiichiwan: The Spatial Politics of Indigenous Water Relations. *Decolonization: Indigeneity, Education & Society* 7 (1): 159–172.

Gergan, M.D. (2017). Living with Earthquakes and Angry Deities at the Himalayan Borderlands. *Annals of the American Association of Geographers* 107 (2): 490–498.

Estes, N. (2019). *Our History is the Future: Standing Rock Versus the Dakota Access Pipeline and the Long Tradition of Indigenous Resistance*. London: Verso.

Estes, N. and Dhillon, J. ed. (2019). *Standing with Standing Rock: Voices from the #NoDAPL Movement*. Minneapolis, MN: University of Minnesota Press.

McGregor, D. (2014). Traditional Knowledge and Water Governance: The Ethic of Responsibility. *AlterNative: An International Journal of Indigenous Peoples* 10 (5): 493–507.

Poelina, A., Taylor, K.S., and Perdrisat, I. (2019). Martuwarra Fitzroy River Council: An Indigenous Cultural Approach to Collaborative Water Governance. *Australasian Journal of Environmental Management* 26 (3): 236–254.

Poelina, A. (producer) and McDuffie, M. (director). (2021). *A Voice for Martuwarra.* Madjulla, Inc. Broome, Australia. https://vimeo.com/424782302 (accessed 12 July 2022).

Prieto, M. (2021). Indigenous Resurgence, Identity Politics, and the Anticommodification of Nature: The Chilean Water Market and the Atacameño People. *Annals of the American Association of Geographers* 12 (2): 487–504.

Wilson, N.J. (2014). Indigenous Water Governance: Insights from the Hydrosocial Relations of the Koyukon Athabascan Village of Ruby, *Alaska. Geoforum* 57 (1): 1–11.

Wilson, N.J. and Inkster, J. (2018). Respecting Water: Indigenous Water Governance, Ontologies, and the Politics of Kinship on the Ground. *Environment and Planning E: Nature and Space* 1 (4): 516–538.

Wilson, N.J., Montoya, T., Arsenault, R., and Curley, A. (2021). Governing Water Insecurity: Navigating Indigenous Water Rights and Regulatory Politics in Settler Colonial States. *Water International* 46 (6): 783–801.

Expert networks

Conca, K. (2005). *Governing Water: Contentious Transnational Politics and Global Institution Building.* Cambridge, MA: The MIT Press (see chapter 5 "Expert Networks: The Elusive Quest for Integrated Water Resources Management").

Galvin, M. (2015). Talking Shit: Is Community-Led Total Sanitation a Radical and Revolutionary Approach to Sanitation? *WIREs: Water* 2 (1): 9–20.

Goldman, M. (2007). How "Water for All!" Policy became Hegemonic: The Power of the World Bank and its Transnational Policy Networks. *Geoforum* 38 (5): 786–800.

Molle, F. (2008). Nirvana Concepts, Narratives and Policy Models: Insights from the Water Sector. *Water Alternatives* 1 (1): 131–156.

On drylands and deserts

Abbey, E. (1968). Water. In: *Desert Solitaire: A Season in the Wilderness*, 112–127. New York: McGraw-Hill.

Koch, N. (2021). The Political Lives of Deserts. *Annals of the American Association of Geographers* 111 (1): 87–104.

On water justice

Boelens, R., Perreault, T., and Vos, J. ed. (2018). *Water Justice.* Cambridge, UK: Cambridge University Press.

Perreault, T. (2014). What Kind of Governance for What Kind of Equity? Towards a Theorization of Justice in Water Governance. *Water International* 39 (2): 233–245.

Pulido, L. (2016). Flint, Environmental Racism, and Racial Capitalism. *Capitalism Nature Socialism* 27 (3): 1–16.

Ranganathan, M. and Balazs, C. (2015). Water Marginalization at the Urban Fringe: Environmental Justice and Urban Political Ecology across the North-South Divide. *Urban Geography* 36 (3): 403–423.

Sultana, F. (2018). Water Justice: Why It Matters and How to Achieve It. *Water International* 43 (4): 483–493.

Zwarteveen, M.Z., and Boelens, R. (2014). Defining, Researching and Struggling for Water Justice: Some Conceptual Building Blocks for Research and Action. *Water International* 39 (2): 143–158.

Chapter 2

Water and Empire

Stillsuits and Spice

Picture Arrakis. The vast desert planet is the center of Frank Herbert's science fiction epic *Dune*. Arrakis serves as a battleground for various factions of the Empire that dominates the universe. Natural resource geopolitics is a major theme of *Dune*. "Spice" is a hallucinogenic substance that enables inter-planetary travel, trade, and governance. As the archvillain in the story, Baron Harkonnen muses, "He who controls the spice, controls the universe." The message could not be clearer: spice builds empire.

But the geopolitics of *Dune* is shaped more subtly and powerfully by another natural resource: water. The relationship between water and empire – in the fictional world of *Dune* and our own – is intricate, intimate, and multi-dimensional. Arrakis is difficult to conquer because of the extreme aridity and lack of water. The Fremen's adaptation to water scarcity, in part through technologies like stillsuits that recycle the body's sweat, projects a sense of ruggedness and military edge against invaders. Both the Fremen and the colonizers have feverish watery visions for the future of Arrakis. One day, rivers will gush through verdant gardens where there was once a desert. For most fans, power in *Dune* is spice. But the true masterpiece of *Dune* is its complex and nuanced portrait of an imperial society shaped on every level by water and struggles with water.

Water: A Critical Introduction, First Edition. Katie Meehan, Naho Mirumachi, Alex Loftus, and Majed Akhter.
© 2023 John Wiley & Sons Ltd. Published 2023 by John Wiley & Sons Ltd.

In this chapter, we follow Herbert's tracks to examine the relationship between empire and water. Taking forward our critical approach, we argue that the hydrosocial relations of empire have profoundly shaped many different aspects of environmental, social, political, and cultural life. **Empires** are established by states expanding into "new" areas and establishing control over different population groups and resources. Empire implies the conquest, domination, and re-engineering of natural environments. Because water is central to life and economics, imperial states must deal with how water resources are governed and developed. The construction of water infrastructures and technologies – like the stillsuits in *Dune* – is central to how empires are created and managed.

The historic sweep of this chapter connects different centuries and geographies. The unifying thread is that water is both a lens and a medium to empire. Looking at empire through water reveals a vital function of imperial society. Water and water infrastructures – a centerpiece of this chapter – are core to the political, economic, and cultural processes that compose our world. In the context of imperialism and the formation and expansion of the **nation-state** into supposedly "new" geographies, the meaning of water remains situated, reflecting broader techniques, logics, and processes of violence, dispossession, uneven development, and control.

Theories of Empire and Water

The environment, natural resources, and especially water matter in the making of empires. But there are significant differences in *how* water comes to matter. This section explores two major approaches to water – mechanical and dialectical – and argues in favor of the latter. **Mechanical** approaches hold that ecological conditions unidirectionally cause political effects. On the other hand, **dialectical** approaches hold that ecology and politics, and other factors like culture and society, shape each other continuously and interactively. The dialectical approach builds on our relational thinking, introduced in Chapter 1. While mechanical approaches have an appealing simplicity, they are less effective at capturing the complex nature of how water comes to matter politically.

Mechanical approaches

One figure stands out in any discussion of water and empire: Karl Wittfogel (Figure 2.1). In his most famous work, *Oriental Despotism: A Comparative Study of Total Power* (1957), Wittfogel argued how large-scale water control, especially in arid regions, led to political power being concentrated in a strong central state. By the term "oriental despotism" Wittfogel meant the states in Asia that built large infrastructures and controlled the essential resource of water would necessarily become authoritarian, or "despotic."

Figure 2.1 Sketch of Karl Wittfogel by the Hungarian artist Lajos Tihanyi (1926).
Source: Lajos Tihanyi / Wikimedia Commons / Public domain.

As part of a long line of mistaken and self-serving European political thought, Wittfogel assumed there was no private property in Asia. He thought a lack of private property allowed states to establish exclusive control over water by constructing large infrastructure systems and maintaining overbearing bureaucracies to run them. In this view, the state commanded society at large (Wescoat 2000). Wittfogel mechanically linked the state's control of water to political despotism – an attempt to build a "universal" theory. This mechanical cause-effect relationship has been disproved and critiqued for many decades. For example, state-led irrigation did not result in the centralization of imperial power in the Indus Valley Civilization or the terraced rice fields of Bali (Lansing 2009; Wescoat 2000).

Yet, like Malthusian theories of population and scarcity (Chapter 1), Wittfogel's thesis has never faded away. One reason for its durability is China and its long association between empire-building and water control. There is "one indisputable fact about statecraft in China: water is vital to it, and it has always been shaped by patterns of water management, control, and access" (Ball 2017, p. 104). For example, the most famous water infrastructure in China's history is the Grand Canal. Although construction began on this project in 486 BCE, the Canal was not completed until around 1300 CE. Designed to facilitate navigation, this waterway was built by over a million forced laborers during the peak of construction in the early seventh century. From the Hangzou-Shangahi area in the south up toward Beijing in the north, the Grand Canal linked a wet region to a dry one. Flows of money quickly followed. Indeed, the Grand Canal "was not so much a means of moving food for the masses from an area of abundance to one of scarcity as it was a means of moving taxes from the periphery to the capital" (Worster 2011, p. 9).

Large-scale infrastructures like the Grand Canal were a raw display of imperial power. So, it is not hard to see how Wittfogel came to his theories. Nonetheless, the idea of a unitary imperial state developing hydraulic infrastructures to centralize power does not hold up in Chinese history. Large water projects were built without state support, for instance, and the operations of power in China were often decentralized as a policy choice. There were periods of generous support from the central state "that intervened to encourage industry and growth, not least to stimulate development of more remote areas of the empire" (Ball 2017, p. 103).

For better or worse, Wittfogel was an influential thinker. Contemporary scholars still reference and expand on Wittfogel's ideas. Even though his major arguments have been refuted – many times! – the linkages he established between water infrastructures, bureaucracy, and state power mean Wittfogel is an unavoidable entry point.

Dialectical approaches

Environmental and social change, and the role of political power, cannot be reduced to simple causal relationships. We need a different explanation for water and empire. As Chapter 1 explains, the hydrosocial cycle takes a relational approach to water and society – a point of view that positions us (people) as internal to the production of water. Water is inseparable from all aspects of society, including imperial contexts, and relates to ecology, space, politics, and culture. To understand how water resources and infrastructures shape the formation of empire, we ought to move beyond the type of unidirectional causal relationships that Wittfogel was keen to emphasize. Hydrosocial thinking asks us to apply a dialectical approach to query how water comes to matter and why factors such as religion, property, caste, spatial bordering, and race and indigeneity colligate in the imperial context.

What do we mean by a dialectical approach? Imagine we see the relationship between water and empire through a crystal, not a straight line. This crystal would look different depending on how you turn the mineral toward the light. Turning one way, a prism of religion and culture would appear more prominent. But if you turn the crystal another way, a spectrum of political economy and property relations would be refracted. The goal of a dialectical approach is not to reach a static conclusion that applies to all empires, but to appreciate the complexity and multiplicity of ways that water makes empire and empire makes water – to sharpen the crystal. In what follows, we turn the crystal of water and empire over and over, to illuminate different facets of this complex relationship.

Water at the Frontier

Religious and cultural values are central to the ways most people in the world lead their lives. The most famous example of the religious value of water is the sacred status of the Ganga (Ganges) River in North India in Hindu traditions. Similarly, fresh water from the ZamZam spring in Makkah is significant for Muslims by association with the sacred Hajj, or pilgrimage ritual. But religious values – like all cultural attitudes and norms – change over time, and this change can happen with the construction and maintenance of water infrastructures.

To illustrate, consider the expansion of the Mughal Empire into the densely populated region of Bengal between the thirteenth and nineteenth centuries. The region of Bengal is defined by the deltaic plains of the Ganga-Brahmaputra River systems in eastern South Asia. By the early eighteenth century, Bengal was one of the most dynamic and prosperous regions of the world. It was densely populated and the living standards were extremely high. Due to its extraordinary agrarian and trading wealth and population density, Bengal was the jewel in the crown of the Mughal Empire, one of the most geographically expansive and culturally complex empires in the world.

The dynastic empire reigned in some form in South Asia between 1526 and 1857. The Mughals relied on an enormously productive agrarian system, enabled by modified waterscapes and irrigation tanks. However, the interrelation between these water infrastructures, the agrarian economy, and the power of the imperial state was not as simple as Wittfogel observed. The political, hydraulic, and agro-ecological evolution was an evolving process in which religious affiliation generated a seismic shift. East Bengal grew into a settled peasant society reliant on irrigation works and a Muslim-majority region.

The historian Richard Eaton (1993) explained Bengal's evolution in the early modern period as the interactions of several distinct, moving **frontiers**. The most important interactions were between the (i) political, (ii) agro-ecological, (ii) fluvial, and (iv) religious frontiers. The political frontier was represented by the large imperial states expanding outward into Bengal from North India, most notably the Mughal Empire (1575–1765), but also its precursor the Delhi Sultanate (1204–1575). This political frontier represented the major influence of Islamized Turkic armies who had swept into North India from Central Asia, through today's Afghanistan.

The political frontier interacted with another moving frontier: the agro-ecological frontier. Bengal's agrarian frontier "divided the delta's cultivated terrain from the wild forests or marshlands that were as yet unpenetrated, or only lightly penetrated, by plow agriculture and agrarian society" (Eaton 1993, p. xxiii). Settled agriculture allowed the production of massive food surpluses in the Bengal region, where before there was a thickly forested landscape with low population density. Famous for tigers, the Bengal forests were hideouts for people escaping the violence of state formation or the long arm of the law. To tap the rich soils of the Bengal delta for agriculture, the forest needed to be cleared and irrigation infrastructures had to be constructed. Moving this agrarian frontier along in the early modern period was not an easy task and required multitudes of laborers working around the clock. But – contrary to what Wittfogel predicted – this task was not carried out by a highly centralized state bureaucracy. Rather, expanding imperial states "outsourced" this task to a decentralized army of pious pioneers, who cleared the jungles and established mosques.

Bengali oral traditions recount the exploits of one of these pioneers, Khan Jahan (d. 1459). Jahan was a high-ranking officer in the Bengal Sultanate, who commanded the "superior organizational skills and abundant manpower necessary for transforming the regions' formerly thick jungle into rice fields." To push the agrarian frontier deeper into Bengal, "the land had to be embanked along streams in order to keep the saltwater out, the forest had to be cleared, tanks had to be dug for water supply and storage, and huts had to be built for workers," (Eaton 1993, p. 210). Oral traditions credit Jahan with the construction of almost 500 water tanks in the region. This mandate to clear the forest, build irrigation infrastructures, and establish mosques, was repeated across Bengal, over many generations. Muslim pioneers slowly engineered a new type of waterscape that would make Bengal one of the most fertile agrarian regions in the world.

Underlying the movement of this agro-ecological frontier – an impossible feat without the patronage and expansion of the imperial states pushing east from North India – was a third frontier: the fluvial frontier. Rivers in delta regions are unruly. They deposit tons of silt annually, which causes water to routinely jump channels, change course, and assume new routes. Massive amounts of labor and money are applied to keep rivers in place and to sustain heavily populated cities in river deltas: the city of New Orleans, Louisiana (USA), in the Mississippi River Delta is a good example.

Bengal's rivers in the early modern period (as they still do today) underwent a historic drift from west to east. Over the course of centuries, the western portions of the delta lost their channels as they shifted eastward. This meant that "as new river systems gave access to new tracts of land and deposited on them the silt necessary to fertilize their soil, areas formerly covered by dense forest were transformed into rice fields, providing the basis for new agrarian communities" (Eaton 1993, p. xxiv).

This eastward migration of Bengal's deltaic rivers coincided with the eastward push of the political frontier and established the soil conditions for the eastward push of the agrarian frontier. This pushing of the agrarian frontier involved the transformation of the landscape in part by building vast amounts of irrigation infrastructure. The colonization of these forests and the interaction of these moving frontiers had epochal consequences. Indigenous communities of nomadic cultivators and fisherfolk were compelled into a sedentary peasant lifestyle and the food supply increased from the expansion and intensification of rice cultivation. Islam was the first monotheistic and imperially associated religion that these communities encountered.

Coupled with the reality that the pioneers were mostly Muslim, Bengal peasantry experienced a slow process of mass conversion to Islam. This is the final frontier: the religious frontier. The moving religious frontier had dramatic reverberations lasting into the twentieth century, namely, the creation of the eastern wing of Pakistan (today Bangladesh) out of the Muslim-majority regions of British India in 1947. Religious identity is a fluid and dynamic process. Mass conversion does not always happen overnight because people "see the light." Instead, religious affiliation can develop from the dynamic process of imperial expansion that also involves political and ecological factors.

While this is only a snapshot of a centuries-long process that involved the movement and interaction of multiple frontiers in Bengal, the contrast with mechanistic understandings of water and empire cannot be more dramatic. Wittfogel claimed a mechanical causal relationship between the construction of hydraulic infrastructures and the establishment of empire. But in our exploration of irrigation infrastructures in early modern Bengal, several key lessons emerge.

First, states do not act as monoliths. They are rooted in diverse ways with distinct parts of society. The state did not formally enlist and employ an army of water engineers to transform the agro-ecology of Bengal, this task was done by a decentralized and entrepreneurial band of pious pioneers.

Second, ecological conditions are not static. They change because of natural forces (rivers jumping their banks) and social forces (the construction of irrigation tanks and embankments). We cannot assume natural or ecological conditions as a given. Through a dialectical approach, we examine ecological movement with social and political forces, as part of the complex crystal of relations between water and empire.

And third: place matters. Mass conversion to Islam did not happen everywhere the Mughal Empire ruled. In fact, the vast majority of India remained non-Muslim during the long centuries of Muslim rule across the subcontinent. The interplay between the colonization process in Bengal, the movement of the rivers, and the expansion of imperial authority shaped the conversion to Islam in east Bengal.

In the twentieth century, this led to the partition of India and Pakistan along religious lines. Bangladesh fought for independence from Pakistan in 1971 and was downstream to India on the very same deltaic rivers that have shaped Bengal for so long. Today, Bangladesh and India negotiate – at times in a contentious way – over the governance, development, and control of these transboundary rivers (Thomas 2021). These rivers cross borders whose origins lie in the movement and interaction of several types of frontiers since the early modern period. Thus, a dialectical view of water and empire, and political ecological changes occurring over many centuries, allows for a deeper and more nuanced understanding.

Property, Race, and Caste

Let us now shift the crystal, letting the light hit it differently to reveal another aspect of the complex relations that make water and empire. **Property relations** – or who owns what – shape who holds power in society and who reaps the bulk of economic benefits. Property applies not only to real estate and buildings, but also to natural resources like land and water, a theme we return to in Chapter 3. A new empire establishes authority in an area by staking property as one of the first actions. By determining who owns what and who gets to claim the profit from natural resources, empires establish their authority and secure revenue. Staying in South Asia, but moving ahead in time, we can examine a case from one of the most powerful empires in world history, the British Empire (Figure 2.2).

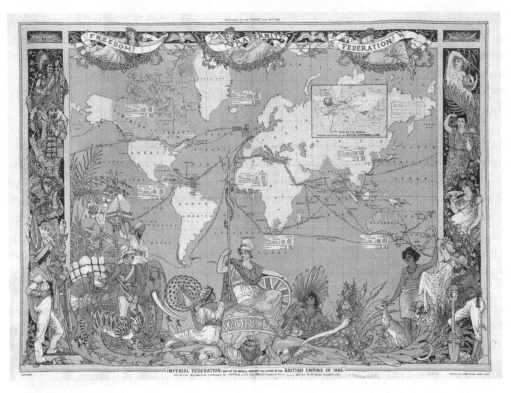

Figure 2.2 The "Imperial Federation" of the British Empire in 1886. Note the prominence of British India, and the racialized imperial subjects along the bottom and sides of the map. Source: Walter Crane / Wikimedia Commons / Public domain.

Formal British rule in India lasted from 1857 to 1947. But this era was preceded by more than a century of commercial and semi-sovereign rule by the famous East India Company. In the western province of Orissa (today's Odisha), British engineers and administrators were intent on instituting a system of capitalist property relations, centered around private land ownership. This move would form a legible system of responsibility and generate tax revenues for the state. But to institute a capitalist property regime, the existing ecological relations in the Orissa delta had to be transformed as well. Delta landscapes, as we saw in the Bengal delta, are shifty regimes. They are neither completely solid land nor fluid ocean – deltas move between and disrupt these categorizations.

Rohan D'Souza (2006) argues that a "flood-dependent" agro-ecological regime persisted in the Orissa delta long before the introduction of privatized land ownership. Delta peasants did not view inundation as catastrophic and confusing. Instead, they relied on regular flooding for the deposition of rich silt-laden soils onto their fields. Moreover, peasant communities developed communal strategies to minimize the risk of flooding through the adoption of risk-distributing cropping practices.

Nonetheless, British imperialists viewed floods as a calamity to be guarded against and a threat to the orderly demarcation of individual land plots. They fought to maintain large embankments (or *bunds*) along the deltaic rivers to contain and redirect flood waters to their channels and to keep water and land separate. In 1863, the colonial government passed the Orissa Canal Scheme to introduce orderly canal-fed irrigation into an agro-ecological landscape reliant on regular flood irrigation. Forced channelization led to greater vulnerability to large flooding events, as streams were not permitted to routinely overflow their banks and deposit silt across a large floodplain. The "colonial dispensation" in Orissa implemented "hydraulic strategies that forced the transformation of the delta from being a flood-dependent agrarian regime to instead dominantly becoming a flood-vulnerable landscape" (D'Souza 2006, p. 45). Importantly, this hydraulic transformation did not occur for exclusively economic reasons. Rather, the British understood their interventions into this landscape as a project of moral uplift. Through the introduction of private property and legible landscapes, the colonizers thought they would transform an idle land (and people) into a prosperous region.

A similar story of hydraulic intervention, infused with colonial moralistic undertones, unfolded on the other side of British India, in the Punjab province in the northeast. Punjab was the last major region of India to be annexed into British rule, in 1849. The British attempted to remake the Punjab into a solid, loyal, and unshakable military and agrarian foundation for their rule in India. Much money and labor were sunk into the Punjabi landscape to construct new infrastructures: roads, rails, and crucially, irrigation canals. British engineers saw in the Indus River System – which included five tributaries that flowed through the province of Punjab – a natural blackboard on which to draw their vision of a conservative, dependable, agrarian paradise (Figure 2.3). Here, on the dusty plains between the rivers of the Indus system in western Punjab, the British meticulously planned and constructed the famous "canal colonies."

The canal colonies were one of the most ambitious socio-ecological planning initiatives in history. The plan of building these towns in the middle of fields meant: (i) constructing thousands of miles of canals, and (ii) settling over a million Punjabi farmers in the newly watered fields. Canal colonies also invoked violence and dispossession. The drylands between the canals were not devoid of people but were inhabited by a rich culture of pastoralists. The British disrupted an existing culture of reciprocal rights and obligations between landlords, cultivators, and rural service classes, and instituted a new property system based on private land rights. The British hoped that this new class of landlords would remain loyal to their empire. Much of the canal colony land was devoted to military purposes (such as military farms and horse breeding facilities) and to demobilized soldiers. Thus, Punjab was constructed, through infrastructural intervention and institutional action, as an agro-militarized bastion of support for the empire (Ali 2014). In both Orissa and Punjab, changes in property relations – creating private property on land and linking those rights to water use – were seen as a civilizing influence on India. Without private property, the thinking went, there would be no propriety.

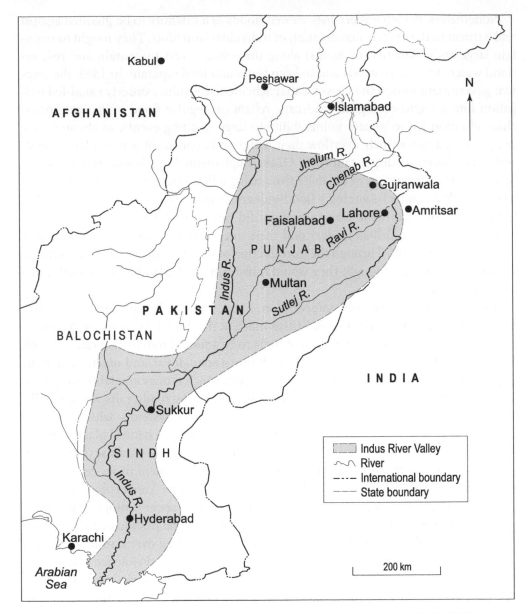

Figure 2.3 Map of the Indus River basin. Source: The Authors, cartography by Philip Stickler.

The hydraulic transformation of Punjab during British imperialism was underpinned by racial and ethnic differences and hierarchies. As discussed, the interfluvial zones in central Punjab were populated before the British arrived. But British imperialists assigned property and cultivation rights only to those populations they deemed "worthy" from a racial perspective – their racial perspective was rooted in a white Christian nation.

Teams of British ethnologists and surveyors descended onto the diverse social landscape of Punjab to divide and categorize the many communities into distinct castes, tribes, and groups. Only some of these were seen to be worthy. Jat Sikhs, for example, were described by British administrators as "the most desirable of colonists" and farmers from the Arain group as "the prince of market gardeners . . . as thrifty as he is prolific" (Malcolm Darling, quoted in Rizvi 2020, p. 76). In contrast, the "nomadic petty traders, pastoralists, and plains- and jungle-dwelling tribal groups who did not conform to the model of settled agricultural and wage labor were either criminalized under the Criminal Tribes Act of 1871 or forced into settlements where they were disciplined into agrarian life" (Rizvi 2020, p. 77). Mapping a racial and ethnic hierarchy onto property relations of colonial Punjab was a major axis of British intervention. This racial logic, by which some communities are deemed by the state to be naturally or biologically better suited to agriculture, was codified into the Land Alienation Act of 1901. This law continues to structure who can buy agrarian land in the interfluvial Indus region today, based on that person's community identity at birth. But did this racial logic that tied property to water and empire start with the British?

The archaeologist Kathleen Morrison argues that hydraulic interventions follow cultural logics of political legitimacy older than either European colonialism or global capitalism. She posits that European imperialism latched onto patterns of political relations already present in India. Morrison's research on irrigation in southern India revolved around artificial reservoirs, described as **tanks** (Figure 2.4). Evidence shows artificial reservoirs have been present in South Asia since 1000 BCE, but they flourished in the Early Middle period (500–1300 CE). Ruins from the Vijayanagara Empire (1336–1646) suggest that tank irrigation was a crucial part of the irrigation-led colonization on these arid lands. Several of these tanks were tall even by today's dam standards (15m or more) and would have required massive amounts of labor.

Morrison argues at least a percentage of this labor was coerced and was recruited according to a caste-based logic; specifically, the Vodda tank digger community. Moreover, there are records of sacrifices being made during times of drought – sacrifices whose "costs were disproportionately borne by women, and occasionally low-caste men" (Morrison 2010, p. 187). Hydraulic infrastructures from the period preceding European imperialism were also constructed at a massive scale, caused ecological degradation, and could worsen social conditions for large swathes of the population. While European empire in India was accompanied by an extended period of racialized interventions into the property regime (D'Souza 2006), the long archaeological record suggests continuity with past modes of making empire through water.

Boundaries and Imperial Space

Turning the crystal yet again, we continue our dialectical journey. Empire is intimately tied to changes in religious identity and racialized property relations. But what other tools are used? The creation of **borders** and **partitions** – the segmentation of political

Figure 2.4 A water tank near Hemakuta Hill, Vijayanagar, India (1856). Tanks served a variety of crucial religious, economic, and water supply functions in South Asia society. Photograph by A.J. Greenlaw. Source: Greenlaw, Alexander John Colonel / Wikimedia Commons / Public domain.

space (including water) into distinct units – is also a crucial part of establishing an empire. A key part of creating new political spaces is the control of the population within that geography. Often, this occurs by encouraging the migration of specific groups to the new political space. We saw this in the newly annexed British Punjab, when settlers from the east were encouraged to inhabit the canal colonies, to the detriment of the pastoral groups already dwelling in the region.

Contrary to what we might assume today, desert spaces were not always spaces of water scarcity. Scarcity is *made*, not ready-made – as Chapter 1 points out – and discourses of scarcity and abundance also work in the service of expanding empire. Samer Alatout (2009) unpacks narratives of resource abundance and scarcity in a different colonial context: early twentieth-century Palestine, a British Mandate between 1920 and 1948. To encourage greater Jewish migration to and settlement in Palestine, a "Zionist network of water abundance, immigration, and colonization" emerged during this period. Alatout explains how the technical category of "absorptive capacity" emerged as the prime determinant of migration flows to Palestine, rather than the more political issues of historical belonging and presence. To promote Jewish migration to Palestine, Palestine was depicted as a potentially water-abundant land (including prospective groundwater sources), joined with Biblical imagery of flowering deserts (Alatout 2009). Moreover, representing a "biblical conception of Palestine" as "vacant, in ruins, and in need of rehabilitation" by Jewish farmer-settlers had the effect of "negat[ing] nearly 2000 years of non-Jewish history of Palestine" (Alatout 2009, p. 384).

The two distinct territories of Palestine, the West Bank (including East Jerusalem) and the Gaza Strip, have been occupied by Israel since the 1967 War, the third Arab–Israeli conflict since the British left in 1947. Around 3 million Palestinians live in the West Bank, along with over half a million Israeli settlers. According to international law encoded in Article 49 (6) of the 1949 Geneva Convention IV, Israeli settlement in Palestine is illegal. Nonetheless, the Israeli state maintains a policy of increasing Israeli settlement in these occupied territories. This policy is supported (or at least not actively opposed) on the international stage by influential actors like the United States. One of the most controversial topics in contemporary geopolitics, this situation is marked by both Israelis and Palestinians arguing that the other side is impinging on their sovereign rights and fundamental security (Rudolph and Kurian 2022).

Water played a key role in settlement. At the outset of the occupation, Israel passed a series of military orders (Nos. 92, 158, and 29) placing water in the West Bank under the direct control of the Israeli military. These orders introduced a permit system for any new water infrastructure and ruled that all prior water agreements were invalid. Echoing settler colonial practices in Australia and North America, the existing uses and customary rules and practices around water were ignored or not formally acknowledged.[1] Palestinians were denied access to the waters of the Jordan River. Before 1967, Palestinian farmers had access to approximately 150 pumps (Rudolph and Kurian 2022, p. 80). New infrastructures and water development were directed to meet the needs of Israeli settlers, not the Indigenous Palestinian population. This inequality was reinforced by the Oslo Accords of the early 1990s.

The Oslo Accords resulted from a mediation process between Israel and the Palestine Liberation Organization (PLO) to pave the way for the eventual self-determination of Palestine. Oslo I was signed in the United States in 1993, and Oslo II was signed in Egypt in 1995. Oslo II gave special attention to water governance and development issues. Article 40 of Oslo II allocated the waters of the Mountain Aquifer (which include the sub-aquifers called the Western Aquifer, the North-Eastern Aquifer, and the Eastern Aquifer) between Israel, Israeli settlers in the occupied territories, and Palestine. Approximately 80% of the aquifers' waters were allocated to Israel and Israeli settlers, and 20% of the water was allocated for Palestine. This allocation was made even though most of the recharge of the aquifer occurs from rainfall

---

[1] There is a long durée legal argument that pre-exists British Mandate in Ottoman water law in Palestine and cases have used that precedent. Like Indigenous Nations in North America, the concept of time in settler discourse of land and resources is very much based on racist, religious cultural myths embodied in law (for more, see Chapter 3).

and snowmelt in the West Bank, and despite a long tradition of Palestinian agriculture in the region. The Jordan River, the only perennial river in the region, was not mentioned in the Accords. Consequently, the major source of water for Palestinian agriculture has been the Eastern Aquifer. The inequality of access to water is established not only by the 1967 the military occupation, but also by the legal-administrative framework of the 1990s Oslo Accords (a theme we return to in Chapter 3).

This structure of spatial occupation has resulted in massive inequalities in water access between Israeli settlers and the Palestinian population (Figure 2.5). For example, settlers are estimated to have 18 times more water than Palestinians for agricultural and residential use. While Palestinians in the West Bank consume about 73 lpcd (liters per person per day), settlers in the Jordan Valley consume 487 lpcd, and settlers in the Northern Dead Sea area consume 727 lpcd (Rudolph and Kurian 2022, p. 85). The World Health Organization estimates that humans need a minimum of 100 lpcd for their daily needs. This means that on average Palestinians do not have enough water to meet the daily minimum, whereas Israelis enjoy above-average consumption in the illegally settled areas. The Israeli occupation of Palestine reinforces settlement politics by controlling people and space through *water*.

Let us consider a parallel example. Occupied and contested by the states of Pakistan and India since 1947, the territory of Kashmir presents a similar case of boundaries, water, and space. When British India was partitioned in 1947, the status of some regions like Kashmir were disputed by the successor states of Pakistan and India. The Indian and Pakistani militaries moved to occupy as much of the territory as they could, and in the process fought their first war in 1948. Pakistan ended up occupying

Figure 2.5   Problems of water supply for Palestinians. Here, an elderly man collects water at a public sink at Khan Yunis Water Authority's wastewater treatment plant in Gaza, Palestine. Source: Muhammad Sabah, B'Tselem / Wikimedia Commons / CC BY 4.0.

one-third of the territory, and India occupied about two-thirds. Since the late 1980s, there has been a movement by Kashmiris for independence – or *azadi* (freedom). This independence movement was heavily repressed and there have been many violent clashes between occupied forces and the Indian government. Since 1990, Kashmir has been occupied under the Armed Forces (Special Powers) Act of India, which allows the military extraordinary powers to kill, invade homes, and escape legal accountability for their actions with respect to the occupied population.

Water enters the partitioning of Kashmir in two major ways. Primarily, a geographic fact is the key streams of the Indus River System originate in the territory of Kashmir (Figure 2.3). The Indus main stem, along with the largest tributaries of the Jhelum and the Chenab rivers, flow through Kashmir before entering Pakistan. This means that the Pakistani government views control of these headwaters as a sensitive and strategic issue. Although the Pakistani government officially supports the quest of Kashmir for independence, the preferred outcome is for Kashmir to become a part of Pakistan. India also considers Kashmir to be an "integral part" of the Indian nation and is invested in developing the river resources for hydroelectric power and irrigation. Although the landmark 1960 Indus Waters Treaty was signed between Pakistan and India and allocated the waters of the international Indus between these two rival states, the treaty studiously ignores the thorny question of Kashmir. Thus, the occupation by Pakistan and India of Kashmiri territory allows them privileged access to the Indus, while the rights of Kashmir to these waters were barely mentioned in the treaty (Haines 2017).

Today, farms in downstream Pakistan and India have flourished, yet human development indicators in Kashmir continue to be some of the lowest in the region. Since the early 2000s, India has shown greater interest in building hydroelectric dams on the major tributaries of the Indus flowing through Kashmir. However, there is skepticism about the benefit to the local population from these dams (Bhan 2014). Moreover, Pakistan continues to invoke the Indus Waters Treaty to try to block these dams from being built in the first place (Akhter 2019). The occupation of Kashmir by Pakistan and India is accompanied by a seizure of the rights to the major stream in the region – to the detriment of the Kashmir people.

Legacies of Empire

Water is a mechanism and key lubricant for empire. At the same time, water is also a site of struggle, sustenance, resurgence, and resistance. To illustrate this duality, think back to the Fremen: fictional inhabitants of planet Arrakis in Herbert's novel *Dune*. For the Fremen, water is life. Their knowledge and practices centered on ingenious use of water, energy, and other resources. Fremeni relations with water and their Arrakis environment represent a site-based political ecology of power and resource use, a relationality used to fight against invasion. Despite attack by imperial off-worlders, the Fremen used their cultural and environmental knowledge as tactics of struggle and resurgence against the Imperium.

What are the legacies of empire? Why do imperial relations persist? What might the future of water hold in a world marked by colonial relations and domination? Turning our crystal one last time, this section considers legacies of water and empire as exercised by contemporary nation-states and as experienced by peoples who are Indigenous to places. Indigenous communities are incredibly diverse, and it is impossible to generalize insights to entire populations and places. Our discussion is necessarily framed by a critical take on the ongoing structural violence and dispossession at the hands of imperialist nation-states, but here we also wish to spotlight Indigenous resurgence, resilience, and struggle. To understand the imperial present in relation to water, we begin by troubling the common-sense notion of the "nation."

The **nation** is typically defined as a homogeneous bounded territory that is ruled over by an undisputed unitary sovereign. Building an empire involves violence, exploitation, enforced cultural and economic change, and physical and symbolic domination. As critical thinkers, we do not take the "nation" for granted as a neutral category (Smith 2020). Rather, we recognize the spatial, cultural, racial, and political differences within a given territory, and we pay close attention to the history of political negotiation and struggle that has resulted in what we might designate as national territory. We continue to live with the practices and legacies of empire, including racialized labor regimes, uneven patterns of wealth accumulation, and restrictions on mobility across borders.

In many parts of the world, nation-building has been informed and enabled by logics and practices of **white supremacy**. A definition and focus on white supremacy as an institutionalized practice that enshrines white superiority in everyday life – as opposed to a set of accidental prejudices or overt discriminatory acts infused with racial intent or bias – helps to explain the ongoing *reproduction* of social and racial inequality, hegemonic structures, and environmental injustices in the world (Bonds and Inwood 2016; Bruno and Jepson 2018; Mascarenhas 2012; Pulido 2015). A critical engagement with white supremacy productively begins with Ruth Wilson Gilmore's (2006, p. 28) definition of racism as "the state-sanctioned or extralegal production and exploitation of group-differentiated vulnerability to premature death." This theorization helps to shift our understanding of empire's legacies from discrete "acts" to hard-baked "logics" and reproducible "practices" backed by the power of the state (Pulido 2017). "If privilege and racism are the symptoms," explain Bonds and Inwood (2016, p. 720), "white supremacy is the disease."

Settler colonialism is a concept that has been developed primarily in Australia, Canada, and South Africa to help explain ongoing "colonial moments" that sustain and strengthen white racial domination in settler societies. Settler colonialism is not the same as white supremacy, but the linkages are fundamental and deep. Anne Bonds and Joshua Inwood (2016, p. 716) helpfully explain key distinctions and lines of connection:

> As a project of empire enabled by white supremacy, settler colonialism is theoretically, politically, and geographically distinct from colonialism. Rather than emphasizing imperial expansion driven primarily by militaristic or economic purposes, which involves the departure of the colonizer, settler colonialism focuses on the permanent

occupation of a territory and removal of indigenous peoples with the express purpose of building an ethnically distinct national community [. . .] Because of the permanence of settler societies, settler colonization is theorized not as an event or moment in history, but as an enduring structure requiring constant maintenance in an effort to disappear indigenous populations (Wolfe 2006). Settler colonialism is therefore premised on "logics of extermination" (Wolfe 2006) as the building of new settlements necessitates the eradication of indigenous populations, the seizure and privatization of their lands, and the exploitation of marginalized peoples in a system of capitalism established by and reinforced through racism. Key examples of settler societies include the United States, Canada, Israel, Australia, New Zealand, South Africa, Argentina, and Brazil.

How are these structures formed and enacted with respect to the hydrosocial cycle? What does settler colonialism have to do with water? In the case of Australia, Indigenous legal scholar and practicing lawyer Virginia Marshall (2017) critiques the socio-legal doctrine of **aqua nullius** as a powerful example of how white European settlers constructed both Australian land *and* water as "empty" to erase Aboriginal communities, claims, livelihoods, and knowledges. Aqua nullius served to divide the "new" Australian territory – including its rivers and water – into private property rights. For example, in the Murray–Darling River basin, long viewed as an "exemplary" model of water policy and tradeable water rights, white settlement and the establishment of irrigation districts and agricultural production "rendered Aboriginal communities politically invisible; they became disenfranchised from economic water entitlements and framed by western cultural constructs" (Marshall 2017, quoted in Quentin Grafton et al. 2020, p. 12). Settler colonialism is thus rooted in the seizure and control of territory and natural resources like water, forests, and minerals – an ongoing, not antiquated, process of hydrosocial domination (Curley 2021a, b). "Whatever settlers may say – and they generally have a lot to say – the primary motive for elimination is not race (or religion, ethnicity, grade of civilization, etc.) but access to territory" (Wolfe 2006, p. 388).

At the same time, Indigenous resurgence in Australia around water raises fundamental questions of **sovereignty**, unsustainable settler legacies, and the future of livable worlds. "First People's water policy challenges the Australian water management status quo," argue Katherine Taylor, Bradley Moggridge, and Anne Poelina (2016, p. 141), "If Australian institutions are to make space for self-determining First Peoples, and all that entails, significant changes will be needed to water governance frameworks and water property rights. This is no small task." Aboriginal and First Peoples in Australia offer compelling and creative paths forward for institutionalizing alternative visions of environmental management that restructure hydrosocial relations in transformative ways. For example, Nyikina scholar Anne Poelina and colleagues describe the Fitzroy (Martuwarra) River Declaration and Council, which is grounded in ancient First Law (Customary or Aboriginal Law), promotes an Indigenous-led co-management model of governance, and successfully established a new water governance council in 2018 (Poelina et al. 2019; see Chapter 3 for a more extensive discussion of First Law).

Can we rectify and transform the structural legacies of empire? A critical lens and dialectical approach show us that any substantive changes must run deep – far beyond technical fixes. Water is relational, any "fix" will be no easy fix. Indeed, as Poelina et al. (2019, p. 238) remind us, "Undoing colonization requires substantive changes to the relationship between Indigenous Australians and other Australians particularly in regard to law and policy, social and economic structures that generate material outcomes." This thinking far outpaces Wittfogel's mechanical approach to empire. Moreover, this mindset challenges us to find ways to make change – to make camp. As Quentin Grafton et al. (2020, p. 12) suggest:

> As a way forward, Recommendation 4 (of nine recommendations) in Marshall (2014) calls for Aboriginal water rights to be enshrined in law, and that the Australian national, state and territorial governments, with leadership from Aboriginal communities and Aboriginal organisations, provide for: 1) the recognition of Aboriginal peoples' special association to water as a First Right before other water rights, 2) the increase of Aboriginal participation in the water market, 3) an increase in the ownership of water property assets by Aboriginal communities, and 4) self-determination.

This final turn of our crystal takes us to the present moment. Nation-states actively perpetuate legacies of empire, especially with respect to Indigenous communities living in settler societies. Systems of violence, dispossession, and control persist, shaping water's material relations with society. At the same time, camp is everywhere – as Indigenous leaders in Australia and elsewhere continue to remind us. Settler nations are not "empty" of Indigenous people, knowledges, practices, or relations – despite violent efforts to do otherwise. A critical perspective of the hydrosocial cycle reveals ongoing structures of social and environmental domination and exploitation, and it also makes room for and elevates the ongoing practices of Indigenous communities, who are leading the way for new models of co-management and environmental stewardship in water.

Summary and What's Next

Empire is not a closed chapter of world history, but an ongoing fact and force within the global hydrosocial cycle. This chapter examined water in the context of geographically expanding states, called empires. Water is central to the operations of empire, especially those of nation-states that carry on legacies of imperial territorial domination. Water also shapes empire in nuanced and complex ways – and, as we argued, in dialectical relation with ecological, cultural, and ecological processes. A mechanistic approach offers a simple but reductive analysis of water and empire, as embodied by Karl Wittfogel's famous work.

Empire operates through techniques that separate population, space, and territory; impose hierarchies; and control life, including water. Water governance and development in imperial contexts are closely tied with racialized logics of property, citizenship, and dispossession. In settler societies, such as Australia, Canada, Israel, and the United States, logics and practices of white supremacy and settler colonialism serve to dispossess and/or marginalize Indigenous communities from water (and land), including through legal agreements and treaties (a topic we explore in more detail in the next chapter). At the same time, water has become a means for resistance, resurgence, and transformation. In efforts to resist dispossession and occupation, from Kashmir to Palestine to Australia to North America, struggles for Indigenous sovereignty and water are front and center. In our hydrosocial understanding, any other understanding of water and empire is insufficient.

It is no accident that many of the examples we gave in this chapter are rooted in law and legal institutions. Indeed, the "law" can be seen as the handmaiden of empire. How did this come about? Why is law so important to understanding the hydrosocial cycle? How is property staked and realized on rivers and waterways? Who benefits from our legal systems of water allocation? What can a critical perspective of legal waters offer to our understanding of our watery past, present, and future? That is the subject of our next chapter.

Further Reading

On Karl Wittfogel

Akhter, M. and Ormerod, K.J. (2015). The Irrigation Technozone: State Power, Expertise, and Agrarian Development in the US West and British Punjab, 1880–1920. *Geoforum* 60 (1): 123–132.

Banister, J.M. (2014). Are You Wittfogel or Against Him? Geophilosophy, Hydro-Sociality, and the State. *Geoforum* 57 (1): 205–214.

Harrower, M. (2009). Is the Hydraulic Hypothesis Dead Yet? Irrigation and Social Change in Ancient Yemen. *World Archaeology* 41 (1): 58–72.

Peet, R. (1985). Introduction to the Life and Thought of Karl Wittfogel. *Antipode* 17 (1): 3–21.

Worster, D. (1992). *Rivers of Empire: Water, Aridity, and the Growth of the American West*. Oxford, UK: Oxford University Press.

Water, race, and property relations

Bosworth, K. (2021). "They're treating us like Indians!": Political Ecologies of Property and Race in North American Pipeline Populism. *Antipode* 53 (3): 665–685.

Gaber, N. (2021). Blue Lines and Blues Infrastructures: Notes on Water, Race, and Space. *Environment and Planning D: Society and Space* 39 (6): 1073–1091.

Ranganathan, M. (2016). Thinking with Flint: Racial Liberalism and the Roots of an American Water Tragedy. *Capitalism Nature Socialism* 27 (3): 17–33.

Ranganathan, M. and Bonds, A. (2022). Racial Regimes of Property: Introduction to the Special Issue. *Environment and Planning D: Society and Space* 40 (2): 197–207.

Caste and water

Binoy, P. (2021). Pollution Governance in the Time of Disasters: Testimonials of Caste/d Women and the Politics of Knowledge in Kathikudam, Kerala. *Geoforum* 124: 175–184.

Joshi, D. (2011). Caste, Gender and the Rhetoric of Reform in India's Drinking Water Sector. *Economic and Political Weekly* 46 (18): 56–63.

Naz, F. (2015). Water, Water Lords, and Caste: A Village Study from Gujarat, *India. Capitalism Nature Socialism* 26 (3): 89–101.

Water, colonialism, and the state

Akhter, M. (2022). Dams, Development, and Racialised Internal Peripheries: Hydraulic Imaginaries as Hegemonic Strategy in Pakistan. *Antipode* 54 (5): 1429–1450.

Doshi, S. (2014). Imperial Water, Urban Crisis: A Political Ecology of Colonial State Formation in Bombay, 1850–1890. *Review (Fernand Braudel Center)* 37 (3–4): 173–218.

Ranganathan, M. (2018). Rule by Difference: Empire, Liberalism, and the Legacies of Urban "Improvement." *Environment and Planning A: Economy and Space* 50 (7): 1386–1406.

Usher, M. (2022). Territory, Hydraulics, Biopolitics: Internal Colonization through Urban Catchment Management in Singapore. *Territory, Politics, Governance* https://doi.org/10.1080/21622671.2022.2056503.

Williamson, F. (2020). Responding to Extremes: Managing Urban Water Scarcity in the Late Nineteenth-Century Straits Settlements. *Water History* 12 (3): 251–263.

Water in Palestine/Israel

Alatout, S. (2009). Bringing Abundance into Environmental Politics: Constructing a Zionist Network of Water Abundance, Immigration, and Colonization. *Social Studies of Science* 39 (3): 363–394.

Braverman, I. (2020). Silent Springs: The Nature of Water and Israel's Military Occupation. *Environment and Planning E: Nature and Space* 3 (2): 527–551.

Gasteyer, S., Isaac, J., Hillal, J., and Hodali, K. (2012). Water Grabbing in Colonial Perspective: Land and Water in Israel/Palestine. *Water Alternatives* 5 (2): 450–468.

Mason, M., Zeitoun, M., and El Sheikh, R. (2011). Conflict and Social Vulnerability to Climate Change: Lessons from Gaza. *Climate and Development* 3 (4): 285–297.

Settler colonialism and white supremacy

Bonds, A. and Inwood, J. (2016). Beyond White Privilege: Geographies of White Supremacy and Settler Colonialism. *Progress in Human Geography* 40 (6): 715–733.

Pulido, L. (2015). Geographies of Race and Ethnicity I: White Supremacy vs White Privilege in Environmental Racism Research. *Progress in Human Geography* 39 (6): 809–817.

Pulido, L. (2018). Geographies of Race and Ethnicity III: Settler Colonialism and Nonnative People of Color. *Progress in Human Geography* 42 (2): 309–318.

Tuck, E. and Yang, K.W. (2012). Decolonization Is Not a Metaphor. *Decolonization: Indigeneity, Education & Society* 1 (1): 1–40.

Veracini, L. (2010). *Settler Colonialism: A Theoretical Overview*. New York: Palgrave Macmillan.

Chapter 3

Legal Waters

The Chilean Water Code

Late in 1970, US President Richard Nixon had a meltdown in the Oval Office of the White House. News wires announced that Chileans had elected Salvador Allende, an established, left-leaning politician, as president. Starting in the 1950s, the Southern Cone countries of Chile, Argentina, and Brazil had adopted a program of developmentalist (state-led) economic policies, and US leaders were not happy. Despite US efforts to undermine Allende by secretly funding his opponents' electoral campaigns, Allende won and became one of the first openly Marxist presidents in Latin America. Nixon was caught off guard, and in front of his shocked ambassador, screamed profanities into the phone.

> Nixon then commenced a seven-minute monologue on how he was going to "smash Allende." He instructed the CIA to "make the economy scream" and over the next three years, Washington spent millions of dollars to destabilize Chile and prod its military to act. It finally did on September 11, 1973, in a coup that brought Augusto Pinochet's seventeen-year long regime to power. (Grandin 2006, p. 59)

Augusto Pinochet was a severe man. Under his long reign, the military government built a neoliberal economic project that eliminated socialist doctrine in law and government institutions – reversing Allende's policies to nationalize major industries and create public services such as health care and pensions. Pinochet's project enforced neoliberal law and order with violence. "The military oppressed leftists and political critics," explain Manuel Prieto, María Christina Fragkou, and Matías Calderón (2020, p. 2589), "resulting in thousands of executions, forced disappearances, torture, internment, and forced exile."

Water was a key part of Pinochet's agenda to transform Chile. As Jessica Budds (2013) explains, water presented the perfect opportunity to consolidate elite power, territory, and alliances of the military regime, corporate interests, large landowners, and government technocrats. Pinochet could not operationalize the new agenda alone and needed a network of experts to manifest neoliberal ideals. Under Pinochet's rule, a group of expert technocrats nicknamed "Los Chicago Boys" devised and institutionalized a free-market model of private, tradeable water rights in Chile, culminating in the 1981 Water Code, a constitutional-level reform.

The Chicago Boys were a group of predominantly male university students who studied neoclassical economics at elite US universities and brought this knowledge back to South America. Knowledge is power – a core plank of this book (Chapter 1) – and the choice of the University of Chicago was no accident. Led by Milton Friedman, the UC Economics Department was widely known as a free-market incubator for young minds, considered far more "out there" in terms of neoclassical economic thinking and policy application than the Keynesians at Harvard and Yale (Klein 2007; see also Grandin 2006).[1] The development of Chicago's "laissez-faire laboratory" started in the mid-twentieth century, when Washington political elites grew alarmed at the leftward movement of Latin American countries. Indeed, the ideology of the Chicago Boys "combined utilitarian economic arguments of state failure and free-market efficiency with a libertarian morality drawn from Hayek" and "conceive[d] the state's role as largely limited to that of establishing the conditions for 'free' markets" (Tecklin et al. 2011, p. 881).

Between 1957 and 1970, roughly a hundred Chilean students received fellowships to study at the University of Chicago. Student tuition was paid for by the US government and later the Ford Foundation. Academic staff traveled to Santiago to conduct research and train students and professors in Chicago School fundamentals. Chilean graduates returned home, purged their universities of state-led economic thinking, and established free-market institutes and think tanks. When Pinochet took power in 1973, the Chicago Boys embarked on a major set of neoliberal reforms at the constitutional level. Trade unions were banned, regulations were cut, public pensions and universal health care were dismantled, and water became a tradable commodity (Bauer 1997, 1998; Budds 2013; Prieto et al. 2020). Chicago was not simply an academic program, as Naomi Klein (2007) argues, but the US government's attempt to transform the ideological landscape of Latin America.

---

[1] As Klein (2007, p. 59) reports, "Friedman was always complaining about how marginal he was, how the Chicago School saw itself as a band of rebels, working at a cult of extreme capitalism." Friedman later advised Pinochet directly and inspired a host of influential policy think tanks in the United States.

Why does this story matter to hydrosocial thinking? The story of the Chilean Water Code conveys an important lesson: the law came first, *before* the "free" market. The state, including the Chilean government, military, and influential foreign governments such as the United States, was fundamental to the creation of so-called "laissez-faire" market trade in water rights. The 1981 Water Code quickly became a textbook example of free-market models of governing water, a model Karen Bakker (2014) calls **market environmentalism**.

The Chicago Boys rewrote the constitution to separate water rights from land so that rights could be priced and traded independently on a market (Bauer 1997; Budds 2013). In theory, the Chicago Boys argued, the commodification of water rights would increase security, efficiency, and productive value (Bauer 1997, 1998). In practice, the water flowed to elites. Other notable features of the Water Code included:

- The Code granted a property-owner the exclusive right to use, trade, and sell their water on a market (Bauer 1997; Prieto et al. 2020).
- Water rights were freely tradable among users and market transactions did not require government approval, thus providing an incentive for hoarding and speculation by powerful actors (Budds 2004).
- The Code created a uniform, nationwide formula for market allocation of legal rights. Climatic variances, customary rights, or the local specificity of water conditions did not matter nor apply (Prieto 2015).
- The Code established a new category of "nonconsumptive" water rights to encourage export-oriented industry and hydroelectric development, especially in the mountainous and relatively water-rich region of southern Chile (Bauer 1998).
- The role and authority of the government water agency (Dirección General de Aguas or DGA) was severely restricted and designed to be as "hands-off" as possible. For example, the DGA had no flexibility or power to adopt discretionary decisions in water management (Prieto et al. 2020).

Even after Chile transitioned from a military regime to a democracy in the late 1980s, the new government retained the Water Code. Attempts to reform the Code in the 1990s were thwarted by large industry and subsequently diluted into a series of minor reforms in 2005 (Budds 2013).

A critical perspective on Chile's legal waters exposes a second key lesson: law is a battlefield, a site of struggle. While Chile is considered extreme in its free-market approach (Bauer 1997, 2004; Budds 2004), the Code became a symbol of international policy debate regarding the role of the market in water, a topic we explore in Chapter 4. Around the world, the "triumphalist analyses of the Chilean water model have led many experts from the World Bank and Inter-American Development Bank to present it as a successful model for international water reforms" (Prieto et al. 2020, p. 3). Chile set the high-water mark for market-oriented legal reform.

Finally, a relational approach to Chile's Water Code reveals a third important lesson: the violent conditions and extractive nature of market-oriented legal reforms. Manuel

Prieto, a Chilean geographer and water scholar, argues that we cannot understand the imposition of the free-market model and its enabling legal architecture *without* understanding the violence invoked by the primitive accumulation of water (Prieto 2015, 2021). Under the Code, water is divorced from the land, a practice that runs counter to local and Indigenous practices of farming and environmental management (Prieto 2015, 2021). Customary use rights to water – some of which had been in exercise for centuries – were silenced or erased overnight. Exploitation was not limited to people. Since 2004, the allocation of groundwater rights has increased and far exceeded the physical volume of available groundwater, leaving dry rivers and parched ecologies (Budds 2004). In his interviews with Indigenous people of the Chiu-Chiu valley (Figure 3.1), Prieto (2015, p. 223) reflects on the violent nature of property regimes:

> The fact that privatization occurred during a military dictatorship made resistance impossible. Several informants told me that they were afraid of the consequences they would suffer if they did not follow official instructions. Simón described the situation as follows:
>
> *We were afraid. We were forced to privatize. The mayor came here and told us that if we did not privatize, [the military] would come and they would beat us with sticks. That is how they measured and privatized the way they wanted.*
>
> *My dad fought hard to maintain his right to irrigate the wetland with his water. We had sheep and [other] animals. But no, it was impossible for him to keep the water. [The DGA] said that the wetland had its own water; but that was not true, they lied, lied, lied! They told us that the wetland was a kind of natural thing, so [my father] had no right to water.*

The story of the Chilean Water Code embodies a core argument of this chapter: the law is *relational*, the outcome of social, spatial, and ecological relations. Law is a process, a struggle bracketed by power, a relation that mediates (and is shaped by) water and society (Angel and Loftus 2019; Cantor 2016; Jepson 2012; Meehan 2014, 2019). Law is also one of the most important entry points to comprehend the biophysical and social flows of water. A critical perspective of the law helps us explain why, how, where, when, and to whom water flows through its worldly movement and circulations. All waters are thus *legal waters*.

Environmental change is a power-laden process. Chile debunks the idea that the so-called "free" market is ready-made, self-maintaining, or molded by an invisible hand. Rules, norms, and institutions build the legal infrastructure in which a market functions or fails in the first place (Bauer 1997, 2004). As such, critical scholars view the law (and legal institutions) as relational: social and historically situated, products of social power, and productive of space and nature, including environmental injustices (Blomley 1994; Braverman et al. 2014; Cantor 2016; Cantor and Emel 2018). A critical perspective explores how law changes and is changed through space due to situated social, technical, and natural relations (Cantor and Emel 2018; Cantor et al. 2020; Jepson 2012).

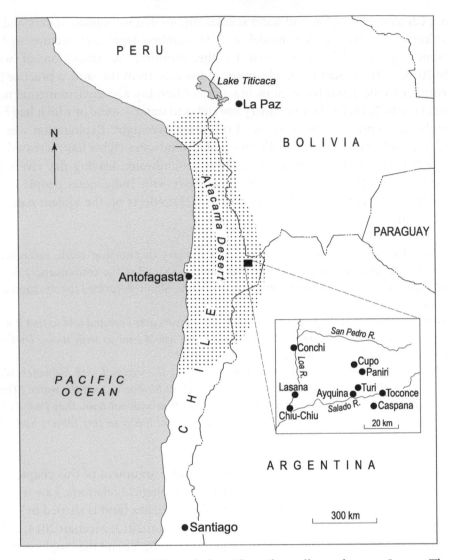

Figure 3.1 Map of northern Chile, including Chiu-Chiu valley and towns. Source: The Authors, cartography by Philip Stickler.

In this chapter, we unpack three common "myths" about legal waters. First, the idea that the law is fixed, static, immutable, and universal (and thus not easy to change). Second, the idea that the law is a construct that just "sits there" on paper, passively waiting to be implemented. And third, the notion that the law is hardwired to deliver justice. You will not be a legal expert by the end of this chapter – that process takes years of training! Rather, our goal is to equip readers with a set of critical thinking tools to probe the production of legal waters. To begin, we set the table by explaining key ideas and debates in natural resource management theory and practice – in short, why institutions matter.

Why Institutions Matter

At our central London campus office, in the old historic Bush House, the kitchen sink on the fifth floor is often a mess: full of dirty dishes, used coffee grounds, limp tea bags, and mysterious leftovers. The fridge is equally plagued with abandoned food containers and cartons of curdling milk, well past their expiration date. On the worst days, this kitchen mess embodies what biologist Garrett Hardin (1968) calls a **tragedy of the commons**. A situation in which a common pool resource – such as a fishery, the climate system, the ocean, or the Bush House kitchen sink – is difficult to enclose as private property and therefore invites overuse and "free riding" from selfish individuals.

The tragedy of the commons is popular in mainstream economics, often used to explain resource scarcity and to justify enclosure and the creation of private property. While the concept has roots in William Forster Lloyd's critiques of peasant access to common land in Great Britain and Ireland, Hardin (1968) applied the model to resource economics and social systems. His theory pivoted on the example of a pasture "open to all." In a commons, when advantages for any self-interested individual or unit are immediate, the costs and implications of over-use – in our kitchen case, dirty dishes! – are diffused across shared space. For Hardin, if all members in a group use common resources for their own gain with no regard for others, all resources would eventually be depleted. His view of the human condition was grim:

> Therein is the tragedy. Each man is locked into a system that compels him to increase his herd without limit – in a world that is limited. Ruin is the destination toward which all men rush, each pursing his own best interest in a society that believes in the freedom of the commons. Freedom in a commons brings ruin to all (Hardin 1968, p. 1244).

Are commons destined to ruin? Elinor Ostrom, a Nobel Prize-winning scholar, disagreed with Hardin and developed major criticisms of his model and explanation. "Instead of presuming that the individuals sharing a commons are inevitably caught in a trap from which they cannot escape," she argued (1990, p. 14), "I argue that the capacity of individuals to extricate themselves from various types of dilemma situations *varies* from situation to situation." Through a lifetime of research, Ostrom debunked many of the core assumptions of Hardin's model. Was his metaphorical pasture truly a common pool resource, she asked, or was it an open access resource? Did all participants have equal knowledge and a stake in the resource? Was privatization or centralization the "only way" to solve commons dilemmas, as Hardin had suggested? Why can't resource problems be solved through institutional cooperation and not conflict?

Ostrom observed a world full of diverse institutional arrangements. For Ostrom, **institutions** are the rules, norms, and organizations that govern collective action, especially in common property contexts like rivers, oceans, groundwater, and the atmosphere. Institutions "can be defined as the sets of working rules that are used to determine who is eligible to make decisions in some arena, what actions are allowed or constrained, what aggregation rules will be used, what procedures must be followed, what information must or must not be provided, and what payoffs will be assigned to individuals dependent on their actions" (1990, p. 51). Legal institutions refer to the "formal" (*de jure*) sphere of governance, though resources are also governed through "informal" (*de facto*) institutions.

Water is referred to as a collective action problem. **Collective action** describes the situation in which individuals benefit more from joint decisions and actions as a community or a government than alone. John R. Commons (1862–1945), a labor historian and institutional economist, argued that "institutions are made up of collective actions that define the economy" – a radical statement at the time, given the ascendancy of neoclassical economics. Rather than pluck rules from the air, Commons developed the idea of "working rules" which are used in practice, monitored, and enforced when individuals make choices about the actions they will take (Commons 1957). "In other words, working rules are common knowledge, and are monitored and enforced."

Building on Commons and others, Elinor Ostrom argued that common pool resources are characterized by extensive working rules (Box 3.1). While Ostrom was not necessarily an avowed anti-capitalist, she provided empirical insights that deflated the mythologies of mainstream resource economics. "Institutions are rarely either private or public – 'the market' or 'the state'," she noted (1990, p. 14). "Many successful CPR institutions are rich mixtures of 'private-like' and 'public-like' institutions defying classification in a sterile dichotomy." Most importantly, she reminded economists to pay attention to the "rules of the game" and institutional architecture. "No market can exist for long without underlying public institutions to support it" (p. 15).

Box 3.1   Rules of the Game

Elinor Ostrom (1993–2012), a Distinguished Professor at Indiana University, was the first woman to win the Nobel Prize in Economics (Figure 3.2). Ostrom spent a lifetime studying the interaction between people and ecosystems, working against the "tragedy of the commons" thesis and Malthusian notions of overpopulation and resource scarcity. She showed how the use of natural resources in fisheries, forests, aquifers, and oceans is underpinned by a set of norms, and both *de jure* and *de facto* institutions. A champion of fieldwork and empirical research, Ostrom was impressed by the "sheer perseverance manifested in these resource systems and institutions" (1990, p. 89).

What makes a commons? Ostrom and her collaborators, principally based at Indiana University, studied dozens of examples from around the world to elicit common trends and patterns, a **theory of common property** (CPT). They generated a set of design principles to identify and characterize a common property

Figure 3.2 Elinor Ostrom, Nobel Prize winner. Source: Courtesy of Indiana University / Wikimedia Commons / CC BY 2.5.

institution that endures over time and space (Ostrom 1990; Schlager and Ostrom 1992). This work has expanded to a variety of institutional contexts, including: a comparative study of water laws and institutions, legal regimes that promote conjunctive groundwater management, and the role of various organizations (districts, cooperatives, companies) in mediating water distribution (Blomquist 2020; Sugg et al. 2016; see Further Reading section).

Examples of these principles include:

1. Clearly defined boundaries
2. Congruence between appropriation and local conditions
3. Collective-choice arrangements
4. A system of monitoring
5. Graduated sanctions
6. Mechanisms for conflict resolution
7. Rights to organize, recognized by external (governmental) authorities
8. Nested enterprises (for common property resources that are parts of larger systems).

Rules can dramatically shape a waterscape. At the same time, institutions are dynamic and adaptable, not static. For example, in arid northwestern India, tubewells are used to pump groundwater for irrigation and domestic needs. As geographer Trevor Birkenholtz (2009a, p. 126) explains, farmers use tubewells "to grow HYVs [high yield variety crops] in support of a capitalist production regime, but through their use the conditions under which those crops could be produced (adequate groundwater quality and quantity, and soil quality) were undermined." Rather than incite a tragedy of the commons, Birkenholtz (2009a) shows how tubewell adoption has led to the emergence of new social institutions of groundwater management, including negotiation, cropping constraints, pooling resources, and land-use change. Water can bring people together and not necessarily tear us apart (Chapter 7).

What happens if institutional management is required across a larger territory – such as a state, province, or nation? In this case, we see a different kind of spatial enclosure and management through water rights.

Water Rights (and Wrongs)

In August 2008, Utah officials informed Mark Miller, a Salt Lake City car dealer, that his cistern (an underground tank for storing rain) violated Utah law. Miller collected rooftop rain at his dealership, using the harvested rain to wash cars instead of piped water. The effort saved money and made his business "green." Boyd Clayton, the deputy state engineer at the time, explained that Miller's collection was a diversion of existing water rights and therefore violated the system of **prior appropriation**, the backbone of water law in much of the American West. "Obviously if you use the water upstream," Clayton said, echoing the "first in time, first in right" principle, "it won't be there for the person to use it downstream."

The story appeared on the evening television news. In response to public outcry, Utah's Division of Water Rights clarified its interpretation in a press release. If rain was merely controlled or channeled – for example, with gutters on a house – a "paper" water right [registered with the state] was unnecessary. However, a water right was needed if rain was stored by people and then used for a purpose other than its release back into the basin (Meehan and Moore 2014). In Utah's strict interpretation of western water law, most forms of rainwater harvesting in the state would be considered illegal.[2] Local journalists lampooned the decision, calling a household rain cistern the not-so-secret "bong" of the backyard garden. "Who owns the rain?" quipped a local news anchor, "Turns out, not you."

Who *does* own the rain? Why? What is a water right? And when might it be considered a "wrong" or an injustice? A recurring theme in water scholarship is that while environmental and social conditions are considered malleable and dynamic, the "law" is often perceived as static, immutable, and universal – and not easy to change (Cantor 2016). In this section, we introduce the fundamentals of property in water, known as **water rights**. We demonstrate how water rights and law are mutable and diverse, a relational outcome, including relations that are uneven, extractive, and even violent.

---

[2] Since then, Utah state law has changed to encourage rainwater harvesting. Utah state law (Code 73-3-1.5) now allows rain catchment up to 2500 gallons per person. Barrels are limited to two (2) per person and must be registered with the state.

The Utah example highlights the power of the state in water rights. Geographer and legal scholar Alida Cantor (2016, p. 51) explains, "[L]egal processes are important for water resources management because law carries state-backed legitimacy." Law is backed by "police" in the broad Foucauldian sense. For Foucault, "police" extends beyond uniformed officers; **police** include the courts, bureaucrats, and officials who determine water rights allocation, adjudicate in rights disputes, or enforce rules (Cantor 2017; Curley 2019b; Meehan 2013). The law enrolls legal infrastructures that extend beyond people to include the settlements, treaties, court opinions, scientific reports, socio-legal practices, and authoritative knowledge built into governance systems (Akhter 2019; Gupta 2012; Meehan 2014). Power is a key ingredient of police. Indeed, "the legal process demarcates the boundaries of water politics because the law determines who holds legitimate power to organize, distribute, and manage a region's physical water resources" (Jepson 2012, p. 615).

But first, the basics. Two major legal traditions influence the geography, institutional structure, and governance of the world's water: civil law and common law (Figure 3.3). A legal tradition is "a set of deeply rooted, historically conditioned attitudes about the nature of law, about the role of law in the society and the polity, about the proper organization and operation of a legal system, and about the way law is or should be made, applied, studied, perfected, and taught" (Merryman and Pérez-Perdomo 2018, p. 2). The **civil law** tradition is based on principles and frameworks originally rooted in Roman law, dating to 450 BCE. Civil law traditions traveled with the colonial reach of continental European powers such as Spain, Portugal, and France. "The civil law tradition is today the dominant legal tradition in Europe, all of Latin

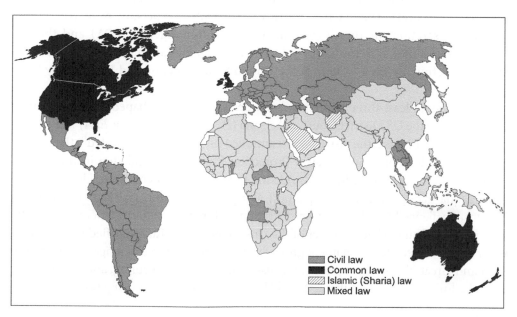

Figure 3.3 Major legal traditions in the world. Source: The Authors, cartography by Philip Stickler.

America, many parts of Asia and Africa, and even a few enclaves in the common law world (Louisiana, Québec, and Puerto Rico)," explain John Merryman and Rogelio Pérez-Perdomo (2018, p. 3). "It bears repeating that the civil law tradition is older, more widely distributed, and more influential than the common law tradition."

Civil law is characterized by an extensive system of codification. In this tradition, core governance principles are codified into a system that serves as the primary source of law. Judges dominate trials and act as "applicators" of codes. A key principle in the civil law tradition is that land and water are linked (Merryman and Pérez-Perdomo 2018).

A second major legal tradition is known as **common law**. "The date commonly used to mark the beginning of the common law tradition is AD 1066, when the Normans conquered England," explain Merryman and Pérez-Perdomo (2018). "It is today the legal tradition in force in Great Britain, Ireland, the United States, Canada, Australia, and New Zealand, and it has substantial influence on the law of many nations in Asia and Africa."

The common law tradition introduces a more "active" role for courts in adjudicating legal norms and property rights. In this system, legal authority is derived from legislation and case law. Judges act as "impartial referees," and lawyers are responsible for presenting and arguing the case. (We suspect this inherent drama is one reason why legal television shows are often based on common law systems!). The law and legal principles can therefore be "created" [technically, interpreted] by sitting judges through legal reasoning and established cases.

In the United States, case law and judicial review have considerably influenced the nation's hydrosocial cycle, even what counts as water. For example, the 1972 amendments to the US Clean Water Act established federal jurisdiction over "navigable waters" – a key litmus test of federal territory and control, especially over rivers – and provided discretion for the Environmental Protection Agency (EPA) and the US Army (home of the US Army Corps of Engineers, a major dam manager in the United States) to define the very "waters of the United States," nicknamed WOTUS by legal scholars.

Since then, at least three Supreme Court decisions have reshaped WOTUS,[3] but a 2021 ruling by the US District Court of Arizona (*Pasqua Yaqui Tribe v. EPA*) is the most compelling. In this case, the district court annulled (in legal terms, "vacated") the federal government's 2020 Navigable Waters Protection Rule, the second step in the Trump administration's two-part effort to (re)define the scope of waters in the United States. The Pasqua Yaqui Tribe, as lead plaintiff alongside five additional Native American Nations, challenged the Trump administration's effort to transform federal waters. In addition to defining what *is* a water of the United States, the Rule clarified what is *not* a water of the United States – namely, groundwater, a major source of irrigation in the US West and Great Plains (Weaver and Drapalski 2020). The court ruling, and a new presidential administration, sent the EPA and US Army back to the drawing board.

---

[3] These include *United States v. Riverside Bayview Homes* in 1985; *Solid Waste Agency of Northern Cook County v. US Army Corps of Engineers* in 2001; and *Rapanos v. United States* in 2006.

Under civil and common law, the legal right to use a designated unit of water is known as a **water right**. In the United States, as in many nations, these rights are *use* rights: a holder has the right to use the water but does not ultimately "own" the water. Unlike other natural resources, running water is recognized as a public resource, even within systems of private ownership of water rights, known as a **public trust doctrine**. Originally based on a legal principle with Roman origins that sought to preserve public interest in rivers, the sea, and tidal lands, the principle was subsequently incorporated into English common law, granting the King (later the State) sovereign ownership of territorial resources. Alida Cantor (2016, p. 52) gives a compelling history of how the public trust doctrine has evolved in the state of California:

> California's public trust doctrine is one of the most protective in the US. The doctrine made a resurgence in California along with the development of environmental law in the 1970s, when law professor Joseph Sax wrote an influential article arguing that the public trust doctrine could and should be expanded to include recreational and ecological values (Sax 1970). Shortly after, the public trust was applied in the 1971 case *Marks v Whitney*, which expanded California's public trust doctrine to encompass recreation and ecology alongside navigation, commerce, and fishing. The court emphasized that the doctrine is "sufficiently flexible to encompass changing public needs."

Flexibility further explains how the doctrine was used by conservation organizations to wedge environmental conservation principles into water rights systems, exemplified in a suit brought by the National Audubon Society. "The *Audubon* case was seen as setting [an] influential new precedent in the use of the public trust doctrine to challenge existing appropriative water rights and to provide protection for instream flow even in small bodies of water that did not meet the 'navigability' requirement" (Cantor 2016, p. 52).

Under English common law, any rights to water are based on the rights to the "permanent and immovable" land below. Australia and the United States, as former British colonies, initially adopted England's system of **riparian rights** to water. "The riparian right entitles the owner of land bordering a surface water body ('riparian' land) to use the water on this land as long as it does not unduly interfere with the rights of other riparian owners," explain Hanemann and Young (2020, p. 109–110). "No specific quantity attaches to a riparian right; there is no recording of the quantity diverted; non-use does not terminate the right. The right is attached to the riparian land through changes of ownership" (Table 3.1). Riparian rights favored landowners and continue to be in use in the eastern half of the United States and many former British colonies.

Settlers in Australia and the United States quickly found that the riparian rights system was ill-suited to their dry, arid climates. "When water was used for hydraulic mining in California," for example, "this violated riparian requirements – the gold deposits were *not* located on riparian land, and the miners did *not* own that land" (Hanemann and Young 2020, p. 110). A new system of water rights emerged on the settler frontier: **prior appropriation**.

Table 3.1   Key Differences in US Surface Water Allocation Doctrines[4]

Prior appropriation
- Predominant system of water rights allocation used by states in the US West.
- Developed during the Gold Rush era, when users fought to establish mining rights over stream access and water use.
- The right to use water is based on historical order ("senior rights" have precedence in water usage over "junior rights" holders), hence the saying "first in time, first in right." In the case of a shortage, senior rights holders are allowed to use up all the water first before water is allocated to junior rights holders. Therefore, senior rights holders are favored in this system.
- Prohibits the impairment of other water rights, such as upstream water users or older water rights.
- Water must be put to "beneficial use" (defined as agricultural, industrial, or household uses) and kept in continuous use. The reallocation of water rights from less valuable to "more valuable" uses is generally not permitted (i.e. once "beneficial," always beneficial). As of writing this book, preserving the environment does not count as a beneficial use.
- Rights can be transferred (bought and sold) at any time, including to owners who are not adjacent to the water body (a situation described as "off-tract use"). Return flows do not need to return to the original water system.

Riparian rights
- The sole system of water rights allocation used by 29 states in the United States.
- Adapted from England, this doctrine grants the owners of the land that abuts a natural stream, river, lake, or pond the right to use that water. Therefore, riparian landowners are favored in this system.
- Most riparian rights allow riparian landowners unlimited use of water for drinking, washing, and animal/garden needs ("reasonable use").
- In the case of a shortage, the principle of "fair participation" means all users share equally in the shortage.
- Rights cannot be transferred to off-tract use. If the land is sold, water rights transfer to the new owner for use on-tract.

As an ontology and system of legal waters, prior appropriation is rooted in a period of empire-building and settler colonial expansion in the late nineteenth century, based on the US ideology of **manifest destiny** (Curley 2019a). In the United States, the 1860s Gold Rush and subsequent settlement did not simply "happen" on the backs of white pioneers. Rather, the US federal government actively *encouraged* land and water grabbing, exploitation of natural resources, and westward expansion through legal institutions such as the Homestead Act (1862), the Land Grant Act (1862), the General Mining Act (1872), and the General Allotment Act of 1887 (Dunbar-Ortiz 2014).

Key differences exist between the appropriative right and riparian rights (Table 3.1). The doctrine of prior appropriation – the system of water rights that dominates law in

---

[4] These systems described in Table 3.1 apply to US surface waters only. Groundwater is governed differently in the United States, typically on a state-by-state basis, and often with weaker laws, less regulations, and rarely in governance models that encourage "conjunctive" management in tandem with surface waters (for exceptions see Sugg et al. 2016). Because of the uneven nature of groundwater governance around the world, aquifers are under grave threat of over-exploitation (Birkenholtz 2015; Blomquist 2020).

the US West and parts of Australia – separates land from water rights. The appropriative right "is based on time and the quantity of the initial diversion creating the right. Unlike a riparian right, the water can be used at a location different from where it was diverted. The imprecise riparian sharing is replaced by a specific priority ranking based on seniority" (Hanemann and Young 2020, p. 110).

Panning for gold and wealth in western streams, miners established a system of water rights allocation, which was later formalized into legal principles. Principles such as "first in time, first in right" – a method to stake mining claims in time and seniority – later served as the backbone of prior appropriation, the legal doctrine that guides nearly all water rights allocation in the US West. Senior rights holders "win" in this system, even over the river itself. For example, "if there is too little stream flow, the senior appropriators can divert their full quantity until the stream is exhausted, while the remaining (junior) appropriators receive nothing" (Hanemann and Young 2020, p. 110). Under prior appropriation, water rights are subject to the test of "reasonable use" and water must be put to "beneficial use" (Table 3.1).

What counts as "beneficial use"? By design, any economic benefit tends to pass the test. Ecological rights have no business in the appropriative system. When the US state of California shifted from a riparian to an appropriative system in 1851, new meanings of "beneficial" in water use quickly followed. California's state water law explicitly prohibits "waste or unreasonable use" of water. "Because a water use deemed 'wasteful' does not constitute a valid water right," Alida Cantor explains (2017, p. 1211), "the accusation of 'waste and unreasonable use' is one that water users hope to avoid." Under prior appropriation, a water rights holder cannot simply let the water run free – such "undeveloped water" (e.g. water to sustain ecosystems) does not count as "beneficial."

But law is not immutable: defining and interpreting the legal category of "waste" has been context-dependent and variable. In California, the concept of "waste and unreasonable use" evolved through court decisions, with uneven consequences for places like the Salton Sea, a lake in inland Southern California. "In the past, water diversions in California framed *undeveloped* water (e.g. water in rivers, lakes, wetlands) as waste in order to exert more *extensive* control over water," Cantor argues (2017, p. 1217), but "market-based water transfers impacting the Salton Sea framed *already-developed* water (runoff, leakage, seepage) as waste via inefficiency in order to *intensify* control over water." Waste is used, in both ways, for power.

Knowledge is power, a core plank of this book, and Australia provides a telling example of how settlers modified systems of legal waters to suit expansion and economic development. Like the United States, Australia initially adopted a riparian rights system to govern surface waters – a legacy of British colonialism. By the 1870s, the Australian agrarian economy had shifted from grazing to crop farming, which is more water-intensive, and farmers began to pressure the government to increase their access to rivers. In response, the Victoria government established a Royal Commission (1884–1886) to "investigate what other countries were doing to promote the development of irrigation" (Hanemann and Young 2020, p. 114).

The Deakin Commission, as it became known, toured irrigation institutions in India, Italy, Egypt, several western states of the US, and Mexico. The Commissions recommended strongly against the prior appropriation system it observed in California and Colorado. It was dismayed by the notion of water rights as private property, the indulgence in costly litigation over water rights, the speculation in water, and the attempts at monopolization of water triggered by the separate ownership of land and water (Hanemann and Young 2020, p. 114).

Eventually, Australia adopted a volumetric allocation and entitlement system. Despite their early rejection of Californian and Colorado systems, Australian water rights reforms between 1980 and 2008 took a decided turn toward the market: water (along with railways, telecommunications, and electricity) was to be governed "in a business-like manner according to rules designed to achieve a more productive and efficient economy" (Hanemann and Young 2020, p. 117). New rules and organizations were created to encourage water trading; the places such as the Murray–Darling River basin became a major test case for implementing market-led principles in water management (see Garrick et al. 2009; Hanemann and Young 2020).

The fact that prior appropriation and its market-based variants are founded on settler logic and resource extraction is not lost on Indigenous communities (Box 3.2). In the United States, for example, Indigenous communities were overlooked or systematically denied water rights by the US government from the very beginning. "In western states and provinces," Andrew Curley explains (2019a, p. 63), "water laws became expressions of nineteenth-century utilitarian logics of resource exploitation and commodification. These laws 'produced' or constructed nature according to the

---

Box 3.2 First Law in Australia

Sovereignty is a key point of struggle for many Indigenous communities in water governance. In Australia, for example, the sovereignty of First Peoples is reflected in the *Echuca Declaration*, a document that explains responsibilities and obligations under **First Law**. As *Echuca* states: "We understand that the Federal and State Governments of Australia say that they have lawfully acquired sovereignty over our lands but we deny and reject that statement" (quoted in Taylor et al. 2016, p. 136). Grounded in Aboriginal Law and customary norms and governance, First Law is a tradition of legal waters that "promotes the holistic natural laws for managing the balance of life" (Poelina et al. 2019, p. 237). Questions of sovereignty, stewardship, future balance, and rivers' "right to life" are core issues (Poelina et al. 2019; Taylor et al. 2016).

How does First Law work in practice? The Martuwarra (Fitzroy) River Council, located in northwestern Australia, provides vibrant and insightful lessons (Poelina et al. 2019; Poelina and McDuffie 2017). Anne Poelina, a professor, filmmaker, community leader, and Nykina Warrwa Indigenous woman, writes about the formation and practice of Indigenous **co-management** as a model of river governance, led by First Law principles:

> In 2016, Traditional Owners came together to discuss their collective vision for the Martuwarra, expressed in the Fitzroy River Declaration. Traditional Owners established the Martuwarra Fitzroy River Council (MFRC) as a collective governance model to maintain the spiritual, cultural and environmental health of the catchment. Traditional Owners advocate a collaborative approach for an inclusive water governance model and catchment management plan (Poelina et al. 2019, p. 236).

> With Poelina and collaborators, the MFRC has made a set of short films describing their philosophy, process, evolution, and lessons learned (see Further Reading). Camp is indeed everywhere – opening new opportunities and futures for governing water and life as we know it.

ontology of the settler colonialist and exerted claims over Indigenous jurisdictions." Indeed, such jurisdictional and regulatory practices advanced by settler states, coupled with a legal structure and ontology of water at odds with Indigenous knowledge and governance systems, continue to marginalize Indigenous peoples and facilitate conditions of water insecurity (Wilson et al. 2021).

Legal Pluralism

Water rights systems are fundamentally diverse. In her comparative study of water rights reform in Bolivia and Peru, Miriam Seemann (2016) argues that water scarcity and insecurity are not "natural" products of hydrologic lack or absence, but are the results of unequal water distribution, access, quality, and water-related benefits, mediated by power asymmetries and relations that produce inequality in the first place. This claim – that scarcity is made, not born (a tenet of this book) – builds on a long-standing argument brought to prominence by geographer Erik Swyngedouw (2004).

While political ecologists like Swyngedouw tend to focus on the political-economic factors that produce inequality, Seemann turns to the law. She interrogates a key assumption in debates over property rights: the idea that "formal recognition of local customary water rights is essential to provide water security" (2016, p. 6). Drawing on theories of **legal pluralism**, Seemann tracks the different pathways followed by the national governments of Peru and Bolivia as they transformed diverse *usos y costumbres* (customary water use rules and norms) into coherent legal rights, regulations, and institutional structures (see Boelens et al. 2010; Perreault 2008). This process, called **policy formalization**, describes an incremental transformation of social norms and practice into a hegemonic system of "formal" water rules and rights (Meehan and Moore 2014; Meehan 2019).

The comparison is important. Peru was considered a neoliberal shining light of South America, a legacy of several decades of Spanish-influenced water law (1902–1969) that concentrated land and water rights in the hands of a small number of landlords. Despite efforts to recognize and integrate Indigenous rights in the 1970s, the Fujimori government (1990–1999) introduced neoliberal water policies in the mode of neighboring Chile, which threatened collectivized management and Indigenous water practices.

In contrast, Bolivia inspired hope for equitable governance structures through its recognition of collective water rights, especially after the 2005 election of its first Indigenous president, Evo Morales. He designated principles such as "living well" (*Buen Vivir*) and "water for all" as guiding coordinates for government policy. Justice through formalization must work here, right? Bolivia, after all, is home to the Cochabamba "Water Wars" of 2000, a series of anti-privatization protests that ricocheted through media headlines and shifted the tenor of water policy conversations worldwide (Chapter 4).

In Seemann's analysis, Peru and Bolivia shared similar challenges in formal rights recognition because of their liberal and positivist – and, we would add, stubbornly colonial – frameworks of law. Legal technologies, such as the local water rights registry, became the prime battleground where customary rights and intracommunal agreements were either recognized or made invisible (Seemann 2016). In practice, many Indigenous users failed to meet state-defined criteria to formalize their rights. As a result, discourses of participation and inclusion, argued Seeman (2016), became capillary "tools for accumulation" and social exclusion, despite their hopeful start as bulwarks against insecurity.

The plurality of legal traditions, water rights, and management systems extends throughout the Americas. As Rutgerd Boelens and Jeroen Vos (2014, p. 55) point out, "[M]ost irrigation systems are managed by small-holder communities and farmer groups" and "these user-managed systems have developed their own traditional, diverse, and often 'hybrid' water rights and management frameworks." Nonetheless, they argue, many countries' legal frameworks tend to deny legal pluralism and disrupt alternative systems of property (Boelens and Vos 2014).

Even within the United States, diverse legal traditions and water rights systems co-mingle and coexist, sometimes uneasily. For example, like many western US states, New Mexico uses the doctrine of prior appropriation for managing its primary system of water rights. At the same time, New Mexico is home to between 600 and 800 communal water allocation systems, known as *acequias* (Perramond 2012). *Acequias* are gravity-fed canals that divert water from a natural stream, following the contour of the land to bring water to farm fields and residences. As geographer Eric Perramond (2012, 2013, 2016) explains, *acequias* originate from a mix of Indigenous water management and Spanish water law, dating from colonial occupation by Spain.

Acequias are not only physical ditches, argues Perramond (2013), they are the kind of common property institutions characterized by Elinor Ostrom. "Ditch governance rules are overseen by a *mayordomo* (ditch boss), annually elected, and three ditch commissioners," Perramond explains (2016, p. 177). He explains that water sharing – known as the *reparto* or *repartimiento* – shaped the hydrosocial flow of water in river valleys and arid landscapes.

> Water, like land, can be either privately held or publicly held. Publicly held water is not communal water. This is a critical distinction: *acequias* assume a communal norm and separate level of controlled access by users to the *acequia* ditch. *Acequia* use rights are considered separate and distinct from a "private right to water" as the state would consider it. (Perramond 2016, p. 175)

Since the 1960s, New Mexico has been engaged in **adjudication**, a process of quantifying and certifying owners of water rights. Here is the friction point between individual water rights systems (prior appropriation) and communal rights systems (*acequias*), where important struggles over power and control are made visible. As Perramond (2013) contends, the process of formalizing water and "mapping rights" in New Mexico reveals broader struggles over power asymmetries and state–society relations. Relations are at the heart of the hydrosocial cycle. A critical way to understand them, as we suggest in the next section (and Chapter 9), is to pay attention to legal waters in action.

Institutions in Action

Once a set of laws and policies are enacted, how do they take shape in practice? The case of Brazil provides important clues. In the late 1990s, Brazil reformed its national water policy and created a new National Water Agency (the ANA), housed in the Ministry of Environment. Alongside these changes, Brazil devolved major water-related decision-making powers – a process known as **decentralization** – to the level of the watershed, "a territorial unit that had no precedent in constitutional norms or historical practices" (Abers and Keck 2013, p. 5). This move was seen as crucial to rebuild democracy and public participation in Brazil, which was emerging from two decades of repressive military rule. "Brazil was gaining fame for experimentation with new participatory arenas," explain political scientists Rebecca Abers and Margaret Keck (2013, p. xvii), "which has spread like wildfire through its political system."

For over a decade, Abers and Keck tracked reforms on the ground, doing research in 23 river basins. "It took us some time before we understood that our original objects of study – functioning river basin committees – were not necessarily there," they confess in their book (2013, p. xix). (We love a good confession.) Nor did the Brazilian watershed committees fit well with existing theories in their field of comparative politics. Such complexity forced the duo to pose a critical question "that we rarely ask in political science: how do new organizations of this kind become institutions at all?" (2013, p. 27).

They posit the answer is **practical authority**: a set of experimental tactics and strategies to "develop capabilities and win recognition within a particular policy area, enabling them to influence the behavior of other actors" (2013, p. 2). Practical authority involves a gradual accrual of capabilities and recognition over time, expressed in a politics of perseverance (2013, p. 143). Drawing on philosopher John Dewey's theory of institutional action, Abers and Keck viewed Brazilian water institutions as political experiments, forged by "conducting policy experiments, building networks, discussing ideas, and pooling resources" toward the accrual of broader capacities, engagement, and authority (2013, p. 19). This process is not tidy and linear, they warn, and perhaps best characterized by the metaphor of entanglement (2013, p. 21).

A favorite example takes place over drinks in a backyard patio garden in São Paulo. Every Friday afternoon, a group of São Paulo public employees met in a home garden to strategize about improving the water management process. "Most were career civil servants," explain Abers and Keck (2013, p.1), "who had risen to positions of influence in their agencies by honing their technical and managerial skills, in the hope of using their positions to bring about change." Abers and Keck explain how this "informal" assemblage of water experts yielded one of the more successful examples of decentralization policy implementation in Brazil.

What does this approach reveal about legal waters? First, a focus on practical authority reveals the everyday practices and informal spaces that constitute policy action and state power. "The relationship between civil society and the state is not reducible to making demands and providing services," Abers and Keck (2013, p. 169) argue, "nor is the relationship between social movements and the state always adversarial."

Second, legal waters are not always tidy and neat. One of their key findings is **institutional entanglement**: the idea that "overlapping administrative jurisdictions layered upon ambiguous functional divisions of labor may produce competition for, confusion about, or even gaps in political authority" (Abers and Keck 2013, p. 21). Federalist countries, such as Brazil, are especially prone to institutional entanglement. Brazil's water reforms created a set of new institutions and mandates that were layered but did not eliminate previous institutions – such as the drastically reduced Water Resources Secretariat, also housed in the Ministry of Environment (2013, p. 195). This layer-cake approach resulted in tensions between federal and state governments, interagency competition, and a process of "mutual adjustment" in which new and old institutions gradually reconfigure themselves into a coherent field.

At the same time, these points of overlap became sites of intervention for non-state actors. Institutional entanglement in the Brazilian water sector led to a proliferation of new civic groups and organizations, exchanges and networks, and opportunities for state-society interaction – elements that strengthened democratic aims, even as they created entanglement and friction. As political scientist Madeline Baer (2017) argues, a legal-centric focus can risk obscuring the right to water, alongside other legal institutions, as a dynamic and fundamentally social process. This process is subject to a variety of institutional actors and organizations operating at different scales, abilities, and capacities.

Pressure makes water flow and not only through pipes. Institutional actors use pressure tactics to make water available to diverse groups and geographies. "Pressure can be mobilized by using pumps or politicians," writes anthropologist Nikhil Anand (2011, p. 543), "and access to the technologies of pressure is mediated as much by capital as by social connections." In the case of Brazil, the pressure was applied by elite and ordinary actors to push toward more democratic participation through river basin councils.

Does the Law Deliver Justice?

A third prominent narrative is that the "law" is hardwired to deliver justice. In the airbrushed world of legal television shows like *Law & Order* (1990–present) and popular films like *Erin Brockovich* (2000), the path to "justice" in the legal process is portrayed as hampered by "bad guys," big money, and/or weak strategies of litigation.

But what if these screen dramas told only part of the story – the hegemonic half? What if law is a site where water injustices are hardwired and routinely *reproduced*? Within the field of **environmental justice**, for example, activists and scholars criticized the state – specifically, state regulation, enforcement, and legal institutions – as part of the structural and institutional machinery that reproduces injustices, rather than prevents or delivers just outcomes (Pulido et al. 2016; Pulido 2017).

Andrew Curley (2021b) applies this critique to legal waters. Drawing on an analysis of Indian water settlements, the legal agreements between US state governments and Native American tribes to allocate water resources, Curley powerfully demonstrates how US water law functions as an ongoing form of "colonial enclosure." Dating to a 1970 Supreme Court ruling, water settlements have become the standard legal mechanism through which Indigenous nations must navigate to obtain rights to water.

> Indian water settlements serve as forms of legal-political enclosure that transforms Indigenous access and uses of the region's surface waters into practices inherently different from previous kinds of jurisdictions and that are fundamentally more limiting . . . In Indian water settlements, Arizona's Congressional representatives and senators have taken an adversarial role against Navajo water claims, even for communities they are supposed to represent. (Curley 2021b, p. 718)

Any water settlements between the Navajo Nation and the US federal government, for example, must conform to existing systems of water rights allocation (in this case, prior appropriation) and diversion, as defined by the Colorado River Compact of 1922, which excluded tribes and "created a status for Indigenous people that is both less than the authority of state governments and not a part of the state's inherent water interest (Curley 2021b, p. 714). Curley explains how tribes have worked for decades within the legal system to gain formal rights through litigation and settlements, even though the system is "stacked" in favor of the settler colonial state:

> Tribal water "rights", or what the Supreme Court called "reserved rights", work within the legal structure of water governance in the west. Access to water requires forms of legal recognition that are bound to rights established in *Winters*. It was for this reason that tribes and their attorneys petitioned courts for the next 70 years to expand these rights, even though they work within colonial laws and institutions. When conservative justices took control of the Supreme Court in the 1980s, it started to roll back tribal rights and authorities under federal law (McCool 2006). In 1976, the Supreme Court ruled that tribal water claims had to be either adjudicated or settled within state courts, making it even more dangerous for tribes to litigate their water claims (Deloria 1985). This is part of the reason why tribes are pressured to settle their water claims. (Curley 2021b, p. 711)

At the same time, the Navajo Nation has resisted water settlements, defying repeated attempts by the State of Arizona to reach an agreement. The Diné (the people and language of the Navajo Nation) community rejected a water rights settlement with the US federal government in 2012. Their rationale drew on the language of self-determination and sovereignty, and recognition of their community as a "nation," not a tribe of minorities at the mercy of the US government (Curley 2021b).

What might we learn from this case? The story of water rights in the Navajo Nation provides a different set of insights than *Erin Brockovitch*. And this account should disturb you. Water rights and law can be a *generator* of ongoing water injustices, not salvation. Dominant systems of water law, backed by state power, originated in colonialism and development efforts. A critical approach never takes legal waters for granted.

A New Day for Chile

We have come full circle. In Chile, a national reinvention is on its way, starting with its Constitution. In the wake of Pinochet's rule, Chile became one of Latin America's richest nations, yet frustrations mounted over widening levels of inequality, deprivation, and entrenched poverty. Students led massive protests between 2011 and 2013, including Camila Vallejo, a student leader and geographer, and demanded educational reform and an end to for-profit schools and universities (Goldman 2012). In early October 2019, the government announced a metro fare hike in Santiago. The minister of economy at the time, Juan Andrés Fontaine, announced that "those upset with the price rise could wake up earlier and pay a lower rate" (McGowan 2019).

Chileans responded by taking to the streets in mass protest. Among their demands, protesters chanted "It's not drought, it's theft" (Bartlett 2022), channeling the idea that scarcity is produced, not ready-made. They also responded through the ballot box. In December 2021, Chileans elected Gabriel Boric, a left-wing legislator and former student activist, as president. A new constitution was at the top of the administrative agenda.

> After months of protests over social and environmental grievances, 155 Chileans have been elected to write a new constitution amid what they have declared a "climate and ecological emergency" . . . And so, it falls to the Constitutional Convention to decide what kind of country Chile wants to be. Convention members will decide many things, including: How should mining be regulated, and what voice should local communities have over mining? Should Chile retain a presidential system? Should nature have rights? How about future generations? (Sengupta 2022)

Water is central in the reform agenda – just as it was decades earlier, under the rule of Pinochet and the Chicago Boys. "Amid a crippling drought supercharged by climate change," reports Somini Sengupta (2022), "the Convention will decide who owns Chile's water. It will also weigh something more basic: What exactly *is* water?" By the time this book reaches publication, Chile may very well have a new Water Code. In 2021, a proposed constitutional reform reconstructed water as a public good. This reform represents the latest iteration in a history of struggle, and would represent a massive shift away from the free-market model of privatized water rights first introduced in the early 1980s.

What lessons can we learn from Chile? First, the law is not static or immutable. A relational perspective reveals how legal waters shape and are shaped by the world. The law is one of the most important elements of the hydrosocial cycle. In the case of Chiu-Chiu, in northern Chile, the precious wetlands were desiccated over time, drained of their essential substance – water – which was taken from customary users and sold off to the highest bidder as water rights. From 1975 to 2009, as Prieto (2015) shows through satellite imagery, the green vegetation of the wetland died.

But Chile teaches us a second critical lesson: people do not sit back. Relations are made and remade every day. The Atacameño people of the Chiu-Chiu valley now articulate their Indigenous identity with the goal of reclaiming a water commons (Prieto 2021). This marks a massive shift in the arc of the country's history. Camp is everywhere, from the windswept prairies of Standing Rock to the graffitied streets, rugged mountains, and salt flats of Chile. The future of Chile's legal waters now enters a new phase, as does its role as a symbol of international water policy debate. And the struggle continues.

Summary and What's Next

Law is a process, a relational struggle. Water law is also one of the most critical relations that shape the global and local flows of water – how water travels through the hydrosocial cycle, to whom and where it flows, and why. In this chapter, we traveled across different legal systems (in Australia, Chile, Brazil, Peru, Bolivia, and the United States) to probe three persistent myths about water and legal governance.

While the forces that enshrine law are seemingly fixed and permanent, we showed how the institutions and ideas underpinning legal waters are in fact diverse and mutable. A critical take on legal waters views the law (and legal institutions) as relational: as social and historically situated, as products of social power, and as productive of space and nature, including environmental (in)justices. A critical perspective on legal waters refuses to place the law in a vacuum. Rather, we explore how law changes and is changed through space because of situated social, technical, and ecological relations.

The case of Chile bookends this chapter. We used Chile to illustrate how the institutional ideas and legal architecture that unleashed and now restricts "market forces" are powerful enough to change a national constitution – possibly twice! Chile teaches us major lessons about the long reach of market forces and the long durée of resistance and struggle. It also prompts a focus on the role of the market in the hydrosocial cycle: How does market environmentalism translate into water management and practice? Why is privatization so compelling and controversial? Does it work? Who benefits from such systems? The next chapter delves into the origins and legacies of market discourse, lifting the veil on the business of water.

Further Reading

Aboriginal water rights and governance (Australia)

Marshall, V. (2017). *Overturning Aqua Nullius: Securing Aboriginal Water Rights.* Canberra, ACT: Aboriginal Studies Press.

Poelina, A. and McDuffie, M. (2017). *Mardoowarra's Right to Life.* Broome, Australia: Madjulla Association. https://vimeo.com/205996720 (accessed 10 July 2022, password: Kimberley).

Poelina, A. (producer) and McDuffie, M (director). (2021). *A Voice for Martuwarra.* Broome, Australia: Madjulla Inc. https://vimeo.com/424782302 (accessed 20 July 2022).

Poelina, A., Taylor, K.S., and Perdrisat, I. (2019). Martuwarra Fitzroy Council: An Indigenous Cultural Approach to Collaborative Water Governance. *Australasian Journal of Environmental Management* 26 (3): 236–254.

Taylor, K.S., Moggridge, B.J., and Poelina, A. (2016). Australian Indigenous Water Policy and the Impacts of the Ever-Changing Political Cycle. *Australasian Journal of Water Resources* 20 (2): 132–147.

Water rights

Bruns, B.R. and Meinzen-Dick, R. (2005). Framework for Water Rights: An Overview of Institutional Options. In: *Water Rights Reform: Lessons for Institutional Design* (ed. B. Randolph Bruns, C. Ringler, and R. Meinzen-Dick), 3–26. Washington, DC: International Food Policy Research Institute.

Kenney, D.S. (2005). Prior Appropriation and Water Rights Reform in the Western United States. In: *Water Rights Reform: Lessons for Institutional Design* (ed. B. Randolph Bruns, C. Ringler, and R. Meinzen-Dick), 167–182. Washington, DC: International Food Policy Research Institute.

Wilson, N.J., Montoya, T., Arsenault, R., and Curley, A. (2021). Governing Water Insecurity: Navigating Indigenous Water Rights and Regulatory Politics in Settler Colonial States. *Water International* 46 (6): 783–801.

Legal pluralism

Boelens, R. (2011). Luchas y Defensas Escondidas. Pluralismo Legal y Cultural como una Práctica de Resistencia Activa y Creativa en la Gestión Local del Agua en los Andes. *Anuario de Estudios Americanos* 68 (2): 673–703.

Boelens, R. and Doornbos, B. (2001). The Battlefield of Water Rights: Rule Making Amidst Conflicting Normative Frameworks in the Ecuadorian Highlands. *Human Organization* 60: 343–355.

Boelens, R. and Vos, J. (2014). Legal Pluralism, Hydraulic Property Creation and Sustainability: The Materialized Nature of Water Rights in User-Managed Systems. *Current Opinion in Environmental Sustainability* 11: 55–62.

Boelens, R., Getches, D., and Guevara-Gil, A. ed. (2010). *Out of the Mainstream: Water Rights, Politics and Identity*. London: Earthscan.

de Vos, H., Boelens, R., and Bustamante, R. (2006). Formal Law and Local Water Control in the Andean Region: A Fiercely Contested Field. *International Journal of Water Resources Development* 22: 37–48.

Roth, D., Boelens, R., and Zwarteveen, M. ed. (2005). *Liquid Relations: Contested Water Rights and Legal Complexity*. New Brunswick, NJ: Rutgers University Press.

Seemann, M. (2016). *Water Security, Justice and the Politics of Water Rights in Peru and Bolivia*. New York: Palgrave Macmillan.

Common property theory and institutions

Birkenholtz, T. (2009). Irrigated Landscapes, Produced Scarcity, and Adaptive Social Institutions in Rajasthan, India. *Annals of the Association of American Geographers* 99 (1): 118–137.

Blomquist, W., Schlager, E., and Heikkila, T. (2004a). *Common Waters, Diverging Streams: Linking Institutions and Water Management in Arizona, California, and Colorado*. Washington, DC: Resources for the Future.

Blomquist, W., Schlager, E., and Heikkila, T. (2004b). Building the Agenda for Institutional Research in Water Management Research. *Journal of the American Water Resources Association* 40 (4): 925–936.

Ostrom, E. (1990). *Governing the Commons: The Evolution of Institutions for Collective Action*. New York and Cambridge, UK: Cambridge University Press.

Schlager, E. and Ostrom, E. (1992). Common Property and Natural Resources: A Conceptual Analysis. *Land Economics* 68 (3): 249–252.

Critical legal geographies of water

Cantor, A. (2016). The Public Trust Doctrine and Critical Legal Geographies of Water in California. *Geoforum* 72 (1): 49–57.

Cantor, A., Kay, K., and Knudson, C. (2020). Legal Geographies and Political Ecologies of Water Allocation in Maui, *Hawai'i*. *Geoforum* 110 (1): 168–179.

Curley, A. (2019). "Our 'Winters' Rights": Challenging Colonial Water Laws. *Global Environmental Politics* 19 (3): 57–76.

Jepson, W. (2012). Claiming Space, Claiming Water: Contested Legal Geographies of Water in South Texas. *Annals of the Association of American Geographers* 102 (3): 614–631.

Groundwater governance

Birkenholtz, T. (2009). Groundwater Governmentality: Hegemony and Technologies of Resistance in Rajasthan's (India) Groundwater Governance. *The Geographical Journal* 175 (3): 208–220.

Birkenholtz, T. (2015). Recentralizing Groundwater Governmentality: Rendering Groundwater and Its Users Visible and Governable. *WIREs Water* 2 (1): 21–30.

Blomquist, W. (2020). Beneath the Surface: Complexities and Groundwater Policy-Making. *Oxford Review of Economic Policy* 36 (1): 154–170.

Owen, D., Cantor, A., Nylen, N.G., Harter, T., and Kiparsky, M. (2019). California Groundwater Management, Science-Policy Interfaces, and the Legacies of Artificial Legal Distinctions. *Environment Research Letters* 14 (4): 045016.

Sugg, Z., Ziaja, S., and Schlager, E. (2016). Conjunctive Groundwater Management as a Response to Social Ecological Disturbances: A Comparison of Four Western US States. *Texas Water Journal* 7 (1): 1–24.

Chile and water

Bauer, C.J. (1997). Bringing Water Markets Down to Earth: The Political Economy of Water Rights in Chile, 1976–1995. *World Development* 25 (5): 639–656.

Bauer, C.J. (1998). *Against the Current: Privatization, Water Markets, and the State in Chile*. Boston, MA: Kluwer.

Bauer, C.J. (2004). *Siren Song: Chilean Water Law as a Model for International Reform*. Washington, DC: Resources for the Future.

Budds, J. (2004). Power, Nature and Neoliberalism: The Political Ecology of Water in Chile. *Singapore Journal of Tropical Geography* 25 (3): 322–342.

Budds, J. (2009a). Contested H_2O: Science, Policy and Politics in Water Resources Management in Chile. *Geoforum* 40 (3): 418–430.

Budds, J. (2009b). The 1981 Water Code: The Impacts of Private Tradable Water Rights on Peasant and Indigenous Communities in Northern Chile. In: *Lost in the Long Transition: Struggles for Social Justice in Neoliberal Chile* (ed. W.L. Alexander), 41–60. Latham, MD: Lexington Books.

Budds, J. (2013). Water, Power, and the Production of Neoliberalism in Chile, 1973–2005. *Environment and Planning D: Society and Space* 31 (2): 301–318.

Kelley, S.H. and Valdés Negroni, J.M. (2021). Tracing Institutional Surprises in the Water-Energy Nexus: Stalled Projects of Chile's Small Hydropower Boom. *Environment and Planning E: Nature and Space* 4 (3): 1171–1195.

Prieto, M. (2015). Privatizing Water in the Chilean Andes: The Case of Las Vegas de Chiu-Chiu. *Mountain Research and Development* 35 (3): 220–229.

Prieto, M. (2021). Indigenous Resurgence, Identity Politics, and the Anticommodification of Nature: The Chilean Water Market and the Atacameño People. *Annals of the American Association of Geographers* 112 (2): 487–504.

Prieto, M. and Bauer, C. (2012). Hydroelectric Power Generation in Chile: An Institutional Critique of the Neutrality of Market Mechanisms. *Water International* 37 (2): 131–146.

Prieto, M., Calderón, M., and Fragkou, M.C. (2020). Water Policy and Management in Chile. In: *Encyclopedia of Water: Science, Technology, and Society* (ed. P.A. Maurice), 2589–2600. Oxford, UK: Wiley-Blackwell.

Tecklin, D., Bauer, C., and Prieto, M. (2011). Making Environmental Law for the Market: The Emergence, Character, and Implications of Chile's Environmental Regime. *Environmental Politics* 20 (6): 879–898.

Chapter 4

The Business of Water

Paying for the Pipes

"Water is a gift from the gods," the saying goes. "[B]ut the gods forgot to pay for the pipes," added the chief executive officer from a large French multinational water utility in the early 2000s. Many experts who work in the water sector echo his riposte. They view themselves as the careful stewards in utility management, including the work of maintaining aging systems, balancing costs with investment, and keeping customers supplied and shareholders happy, a complex operation we might call the "business of water" (Bakker 2014). Without their expertise, they argue, water would run "scarce" and people around the world would lack this necessity. The CEO's quip is a sharp edge undercutting a normalized insistence that water must be paid for. The "joke" is often used as a back-handed chiding of the global justice movement's demand for "free water for all."

Across the globe, many of us pay for basics like clothing, food, and shelter. Is it a surprise that we are expected to pay for the pipes that "the gods forgot to provide"? Perhaps there is nothing more fundamental about water than, say, food. But there is a difference between paying a large multinational for water and paying a local

Water: A Critical Introduction, First Edition. Katie Meehan, Naho Mirumachi, Alex Loftus, and Majed Akhter.
© 2023 John Wiley & Sons Ltd. Published 2023 by John Wiley & Sons Ltd.

municipality for the same resource. And there are diverse *ways* of paying for basic services. One option enables people to access water even if they don't have cash on a particular day. Another policy can distribute the financial burden more equally. Alternative forms of payment may result in dividends for shareholders. Other approaches result in increased investments for the pipes the gods forgot to install.

Returning to our French CEO, water infrastructure must be paid for in most societies. However, this case presents a series of questions. Who should pay? How should they pay? Who should receive payments and at what level should those payments be set?

In this chapter, we delve into the thorny issues in the marketization of water. Following the previous chapter, we argue that the political economy of water, the "business of water," has changed tenor over time and emerged in different places and contexts – even if those places are related to each other. At some political moments, public providers have been preferred over private. Distinctive forms of the market have developed for water: from pre-payment meters to wetland water banks, from Fiji bottled water to water that is free at the point of delivery.

This chapter adheres to the fundamental axiom, water is relational. The business of water cannot be understood without a critical take on those relations. In this chapter, we contextualize the emergence of market-based approaches in the water sector. Our goal is to develop a vocabulary, framework, and theoretical context to understand the increasing presence of the market in global water – as well as its social response and pushback (see also Chapter 9). In so doing, we ask: should water be managed like a business?

We question the policy paradigm in which a large French multinational corporation is *necessarily* the best provider of those divine albeit forgotten pipes. We challenge the myth that water is already always a common resource (or "a gift from the gods"). Whether (or not) water becomes a common resource, and whether (or not) a multinational company will profit from its provision, relies on a set of contested political relationships. The business of water is an outcome of struggles that shape its biophysical, social, and spatial flows – not only the quantity of water, but its temporality, quality, cycles, and meaning. By adopting a critical approach, we begin to see how a thing like a "market" – like Chile's water rights market, described in Chapter 3 – is contingent, an outcome of ideological and institutional power, an influential narrative, and a source of on-the-ground struggle. Just as in the story of Chile, water could become a private good traded on a market, or a business shielded behind a public utility, or a commons. Political struggles will decide.

Public or Private?

Building a water network is an expensive business. But if the infrastructure is costly to construct, the upside is that major water infrastructure projects tend to have relatively long lifespans. Cycling through the narrow one-way streets of London's Soho neighborhood, one encounters street closures and signs by Thames Water (the local utility) with apologies to fix London's Victorian water network. While the network may need regular repairs, 160 years old and still going is not too bad. Elsewhere in London, there is a need to re-engineer the nineteenth-century combined sewer overflow system with a GBP 4.1 billion separated system, nicknamed the "Super Sewer." This project reminds the public how investments made over a century ago form the backbone of the city's water and sanitation network.

As in many places in the global North, London relied on disparate public wells drawing from groundwater sources, prior to the development of piped infrastructure networks. Population rapidly increased in the eighteenth and nineteenth centuries and the quality of wells deteriorated. Private companies emerged to provide water to the wealthiest parts of the growing city. Writing for the World Bank, Michel Kerf (1998, p. 11) emphasized the importance of these experiments in water provision, "[P]rivate companies developed much of the early water infrastructure in France, Britain and the United States." What the authors neglected to add is that such water infrastructure was in most cases pretty useless, prone to disease outbreaks (such as cholera), and supply was limited to rich neighborhoods (Budds and McGranahan 2003; see also Gandy 2002; Melosi 2011).

In short, these private sector experiments in supplying water to the public were largely failures due to their rudimentary nature and limited coverage. People gradually recognized that the lack of a universal piped water supply posed a grave health risk to the entire city, which prompted the construction of a large-scale network. In the case of London, this new water delivery system would be a public network. Funded by the Metropolitan Board of Works, the system was made possible by the local government's ability to draw on the tax contributions across London, along with a nascent municipal bond market.

Similar initiatives had commenced in cities around the world. Matthew Gandy (2002) chronicles New York City through the development of public wells and their declining quality (1658–1774); ill-fated private interventions that sought to bring the first piped supplies (1774–1830); and the development of a comprehensive public supply from the 1830s onwards. In Mexico City, Jeff Banister and Stacie Widdifield (2014) explain how the development of the modern potable water network in the early twentieth century was seen as an impressive visual and technoscientific achievement of engineering and public health. The construction of the Mexico City network was made possible by an authoritarian centralized regime that replaced individual providers with the state – and its extension was by no means "universal" (see also Castro 2006).

Focusing on two Spanish cities, Hug March (2015) contrasts the public utility in Madrid, widely seen as the country's most successful large-scale water project, with the more limited private initiative in Barcelona. Each city confronted multiple sets of challenges with very different relationships with the state. But, even without such contextual differences, March demonstrates the sense that most large-scale – and most successful – water infrastructure projects throughout Western Europe and

North America were made possible by the financial backing of a wide tax base and, very often, the state.

The most prominent exception to this trend is France, where municipalities contracted out the provision of water to private utilities. Two important corporations emerged in this context: Générale des Eaux and Suez-Lyonnaise des Eaux. By gaining the necessary experience, expertise, and capital in France, these two companies bid for concession contracts in other parts of the world, when the sale of municipal assets to private providers (a process known as **privatization**) became popular in the 1990s. Profits soared. Générale des Eaux and Suez-Lyonnaise des Eaux became two of the largest multinational companies operating in France, with interests vested across an unusually wide range of different sectors, from entertainment to waste management.

With these important French exceptions in mind, during much of the twentieth century, the consensus was that large public institutions – what Philippus Wester and colleagues (2009) call **hydrocracies** – were best placed to provide the knowledge, managerial, and financial resources for supporting the development of a universal piped network. Stephen Graham and Simon Marvin (2001) term this period the growth of the **modern infrastructural ideal**, in which a new social contract was forged between states and citizens. This process took place in lockstep with the emergence of the Keynesian welfare state. Following the writings of the liberal economist John Maynard Keynes, a "Keynesian" state was managed by governments that would frequently intervene in the market economy to mitigate periodic crises.

In putting this ideal into practice, Graham and Marvin (2001) explain, modern water networks were characterized by at least three major elements: (i) networked systems that bundle and deliver resources (*integration*), (ii) a logic of resource provision that privileges the same quality and quantity of resources (*uniformity*); and (iii) the idea that resources ought to be provided to all people (*universality*).[1]

> How can we imagine the massive technical systems that interlace, infuse, and underpin cities and urban life? In the Western World especially, a powerful ideology, built up particularly since World War II, dominates the way we consider such urban infrastructure networks. Here, street, power, water, waste, or communication networks are usually imagined to deliver broadly similar, essential services to (virtually) everyone at similar costs across cities and regions, most often on a monopolistic basis. (Graham and Marvin 2001, p. 8)

Water is essential for national growth and a healthy workforce: a well-functioning capitalist economy could not have relied on the unpredictable private initiatives seen in many cities. Public works projects could stimulate the economy during a downturn in economic activity. And the state was "welfarist" in that the social contract assumed

---

[1] Furlong (2014) discusses important divergences from the modern infrastructural ideal as experienced in the global South.

a level of solidarity in which citizens should be treated equally, e.g. access to water should not, therefore, rest on the ability to pay. Instead, the state should guarantee universal access to this basic resource.

By the late twentieth century, the Keynesian compromise was running into trouble. Spiraling national debts within many countries in the global North, stagnant national economies, and rising inflation suggested that the tools through which states had tended to stimulate their economies were no longer working. Making the situation worse, debt crises in the global South crushed the post-independence hopes of many decolonizing nations. Back in the North, election victories for Margaret Thatcher and Ronald Reagan, who had campaigned explicitly for a break with the post-War compromise, began to precipitate a shift to what some refer to as neoliberalism: a set of policy prescriptions that, as we saw in the previous chapter, had been incubated in Chile under the authoritarian leader General Augusto Pinochet.

In the UK, following the privatization of various sectors such as gas, electricity, and aviation, Margaret Thatcher fully divested the water boards of England and Wales to the private sector in 1989. This fundamental shift would have ripple effects around the world. Indeed, while our focus on England and Wales in this section might seem a bit parochial and Eurocentric, the privatization of water in England and Wales – as a thought experiment put into practice – had significant repercussions in many other parts of the world (Bakker 2010, 2013). Privatization would precipitate a massive boom in the business of water, and English water companies were set up to profit well. French water utilities were even better set.

State Failure?

Two of the main arguments used by donors and international financial institutions for privatization are: (i) water is best treated as an economic good, and (ii) the state failed to fulfill its responsibility to citizens (Budds and McGranahan 2003; Bakker 2010). The latter argument gained traction throughout the 1980s. Major debt crises were partially manufactured by new administrations in the global North, which increased interest rates to control domestic inflation. These rates and a relatively small tax base from which to draw the revenues needed for water networks were compounded by rapidly urbanizing populations and a dramatic increase in the demand for piped water supplies. Whether the public provider was effective or not did not matter under these circumstances. There was simply no money to ensure the upkeep of a network, let alone the resources to expand that network to new groups of people.

Toward the end of the twentieth century, policymakers and donor agencies (based, ironically, in the global North) began to argue that the outright failure of states in the global South to provide adequate water services required a turn to the private sector. As part of the broader "Washington consensus" (Williamson 1993), a remarkable about-face from previous policy paradigms, these experts argued that only the private sector was able to find the resources to roll out rapid infrastructural improvements (Bakker 2010; Goldman 2007). Advisors from the World Bank and the International

Monetary Fund (IMF) worked to propagate a radically different vision of the state's role. Rather than a provider of services, they argued the state should regulate services not under its direct control. Public sector institutions should take a backseat role, fostering competition between providers, while ensuring that standards are met on water quality, the environment, and consumer costs (Conca 2005).

The privatization of water in England and Wales was regarded as a successful model to be emulated by other countries (Bakker 2010). In practice, the English and Welsh model of outright divestiture – the sale of all assets associated with the water network – was rarely considered a viable option elsewhere, akin to Chile's "extreme" experiment in water rights trading (Chapter 3). Nonetheless, the creation of long-term concession contracts, granted to the private sector (legally obliging them to operate a city's water network for a given period), was seen as an effective alternative. When a water utility was privatized, a competitive tendering process would be established. The favored bidder would be contracted to run the service for a period in the region of 20–30 years. In other variants of this model, the private sector could absorb more or less risk, and public sector bodies could take on a more or less limited role.

Fast forward from the 1980s to today and a variety of new financial actors and institutional strategies are part of water governance (Table 4.1). The range of actors and their involvement in water is staggering – and has evolved well beyond privatization. Broadly, this shift is described by Bakker (2010) and Conca (2005) as **market environmentalism** and covers a diverse set of market-based strategies in water. For example, bottled water, a fully commodified form, is one of the main sources of drinking water in countries such as Mexico, Indonesia, and the United States (Kooy and

Table 4.1 Market Environmentalism in Water. This table illustrates the ways in which market relations can be introduced into water management. Adapted from Bakker (2014).

Strategy	Examples	References
Financialization A process – often led by new financial actors such as pension funds – in which profits come to be amassed not from the direct sale of water but from derivative income streams.	The securitization of Thames Water's revenue streams (predominantly household bills) in 2007 under its new owners, the Macquarie Group.	Allen and Pryke (2013); Loftus and March (2016); Loftus et al. (2019)
Privatization The transfer of ownership or management of resources to the private sector.	Divestiture (asset sale)	Bakker (2003)
	Private sector participation, e.g. "public–private partnerships" (PPPs)	Castro (2008)

(Continued)

Table 4.1 (Continued)

Strategy	Examples	References
Commercialization The incorporation of market institutions, principles, and models in resource management.	Land/water grabbing Corporatization	Mehta et al. (2012) McDonald and Ruiters (2005)
	Alternative service delivery	Furlong (2016)
Marketization The introduction of water markets as trading mechanisms, predicated on the existence of secure private property rights.	Full-cost pricing Creation of water markets	O'Toole (2014) Water rights markets in Chile (Budds 2004; Bauer 1997, 2004; see also Chapter 3)
Liberalization The transfer of decision-making and oversight from government to non-governmental actors.	Deregulation	Regulatory reforms removing state's role in overseeing water quality, such as in Ontario, Canada (Prudham 2004)
	Decentralization of decision-making and authority	Decentralization of water management to subnational level, such as watershed councils in Brazil (Abers and Keck 2013)
Commodification	Transformation of water into a commodity	Bottled water, sachet water (see Bottled Water: Explanations of a Global Trend, a 2020 special issue in *WIREs Water*)

Furlong 2020; Prasetiawan et al. 2017; Greene 2018). Techniques such as "full-cost pricing" have been introduced into public utilities in places such as Durban and Dublin, with mixed results. For example, following the "Water Rebellion" of 2014, market pricing in Ireland was subsequently rolled back (O'Toole 2014). As we will explore in this chapter, household bills in London were transformed into financial commodities in 2007, a process some have described as financialization.

Privatization Debates

In each of these forms of market environmentalism, the main arguments used in favor of privatization advance the idea of water's role as an "economic good" and state governance as a source of failure. The main arguments used against privatization are water is a *public* good with two distinctions: (i) as a natural monopoly, water is not easy to introduce competition; and (ii) as a human right, water should be democratically managed within the public realm (Budds and McGranahan 2003). In assessing the business of water and the various myths surrounding this business, we consider each of these arguments – and one or two others – in turn.

Arguments in favor of privatization

Perhaps the most damning argument used against public sector provision is that individual states have failed to provide water for their populations. Population increase and rapid urbanization, coinciding with massive fiscal and debt crises, placed many public services in the global South in an impossible situation. Public institutions were unable to expand networks in response to rising demand. With those same institutions simultaneously unable to maintain existing networks, donors responded to the debt crises of the 1980s not with offers of support but with demands for new loan conditionalities. Foremost was a demand that the private sector be involved in public service provision.

The privatization of water service provision became almost inevitable. Indeed, while periods of crisis can open the space for democratic alternatives to the status quo, they can also legitimize political economic change under the guise of apolitical technical expertise. The seemingly commonsensical assumption that the private sector would be the best provider of water services in the global South contradicted over a century of experience elsewhere in the world. And the fact that technical advice came largely from the global North was a bitter twist for many water activists.

Thus, claims of state failure were compounded by the growing prominence of free-market ideologues within expert networks, donor agencies, and international financial institutions. The dominant view shifted from water as a public good to water as a private resource. This view was effectively enshrined within the **Dublin Principles** (Box 4.1). The Dublin Principles were the product of conversations and consultations among an elite set of global water experts (Conca 2005). Prior to the Rio Earth Summit in June 1992, in which debates around sustainable development were growing, the Dublin Principles crystallized signature ideas and contradictions in governing the world's water.

Of those contradictions, Principle 4 is perhaps the most notable, declaring "[W]ater has an economic value in all its competing uses and should be recognized as an economic good" (Box 4.1). On the surface, this principle advances an economistic approach to water, one in which property rights are privileged. And from this perspective, this declaration seems to contradict the claim in Principle 2 that a "participatory approach" to water management should be adopted. One may well be justified in asking whether those who view water as a non-economic or even spiritual good can still "participate" in its management as per the second principle.

The text that follows the initial outline of Principle 4 confuses things even further: "Within this principle, it is vital to recognize first the basic right of all human beings to have access to clean water and sanitation at an affordable price." The fourth Dublin

Box 4.1   The Dublin Principles

Created in 1992 at the International Conference on Water and the Environment in Dublin, Ireland, the Dublin Principles outlined a policy blueprint for how water should be regarded, managed, protected, and governed around the world. While the production of the Dublin Principles was hotly debated among conference delegates, especially Principle 4, it has since formed the backbone of mainstream policy approaches such as integrated water resources management or IWRM (Conca 2005). The principles include the following:

1. Fresh water is a finite and vulnerable resource, essential to sustain life, development, and the environment.
2. Water development and management should be based on a participatory approach, involving users, planners, and policymakers at all levels.
3. Women play a central part in the provision, management, and safeguarding of water.
4. Water has an economic value in all its competing uses and should be recognized as an economic good.

Principle frames water as an economic good and, simultaneously, as a basic human right. Unsurprisingly, this principle has been both contested and upheld by nongovernmental organizations (NGOs) and water activists seeking fairer access to water (Bakker 2007).

Treating water as an economic good, so the logic goes, will help to preserve what is called a "finite and vulnerable resource." Such logic is captured effectively in Kathleen McAffee's (1999) framing of "selling nature" in order to "save it." If environmental governance refers to the set of regulatory processes, mechanisms, and organizations through which political actors shape environmental actions and outcomes (Bakker 2014, p. 476), then market environmentalism entails a broad set of market-based influences and strategies used to govern nature (Bakker 2014; Conca 2005). Similar principles have driven a range of strategies, from payment for ecosystem services to emissions trading schemes.

Similar arguments were used to advocate for the privatization of water services in England and Wales. A Green Paper (published prior to the Act through which the divestiture of water boards was enshrined in law) claimed that introducing market pricing for water, in conjunction with a process of privatization, would drive prices up in times of short supply (such as a drought) and would lower prices when water resources were more plentiful. Utilities operating in parts of England and Wales with higher rates of precipitation could presumably export water to drought-prone parts of those two countries. Such a system of market pricing never developed. Not only is there a lack of a fully integrated network covering England and Wales but there are safety concerns over multiple providers using the same network.

Others argue that the social and physical characteristics of water necessitate a rethinking of the core principles of neoclassical economics (Garrick et al. 2020). Savenije and van der Zaag (2002) are critical of the implication that water pricing should be developed in a way to manage demand. In contrast, they suggest the principles of financial sustainability and equity should be integrated when designing tariff systems for water provision. They emphasize how the principles of integrated water resources management (IWRM) vary with the notion that water is an economic good. Unsurprisingly, interpretations of the fourth Dublin principle continue to pose significant problems and disagreement in the business of water.

What are other, secondary arguments in favor of privatization? The "profit motive" has often been invested with a strangely fetishistic power. Simply introducing the hidden hand of the market will, to some, enable improved customer relationships and ensure leaking pipes are fixed. With a clear eye on the bottom dollar, private companies are assumed to be more likely to retain paying customers and preserve their main revenue generator. Furthermore, private companies are assumed to be the leading innovators in water supply technologies. Such arguments have been advanced with little or no evidence and have been roundly dismissed by heterodox economists such as Mariana Mazzucato (2011), who point to a largely parasitic relationship between many of the largest tech "innovators" and the public sector that enabled their rise.

For others, embedding the profit motive in the provision of a basic service such as water will magically lead to a more responsive utility. This hypothetical private utility will be more concerned with the needs of its consumers, willing to fix leaks (which are an economic loss for the company), and prone to innovate when it comes to producing and distributing potable water. State-owned companies, these critics would argue, are inefficient and require the revolutionary change unleashed in privatization.

Arguments against privatization

Critics of water privatization argue that water is a public good best managed within the public realm. They recognize assets exceed economic gain and range from healthcare to ecological benefits, and this argument has been used to dismiss narrowly framed economistic arguments. Many of these arguments merged as demands for the human right to water and sanitation, which moved from place-based struggles over water access to a global movement of pressure on individual nation-states to fulfill the social contract. There was a certain irony during the announcement of the UN General Assembly's recognition of the right to water in June 2010, when one of the most prominent voices for private water utilities, *Global Water International*, declared the UN vote a major failure for the water justice movement. We begin Chapter 9 with this irony.

For Karen Bakker (2007), the dismissiveness by *Global Water International* could have been predicted. While those calling for the right to water might position their demands against the private sector, Bakker points out that conceptual slippage within human rights discourse permits the private sector to appropriate the core argument. For example, private utilities can claim they are the ones with the resources, expertise, and incentives to render the change needed for the fulfillment of a right to water, as opposed to "failing states" or "inefficient and bureaucratic public utilities" (Bakker 2007).

For Bakker (2007), a more compelling argument frames water as a **commons**: a common property resource to be shared and democratically managed by all participants. Moreover, Bakker points to further conceptual slippage, as the human right to water conflates with calls to impose property (water) rights on this common resource. We unpack some of these claims in Chapter 9, noting how social movements deftly move between moral and political claims, between tactical gains and more strategic victories. The demand for the human right to water, in short, continues to be a major force in the struggles against water privatization.

Why is the experience of privatization mixed? Water, Bakker (2005) argues, actively resists easy commodification. Water is an "uncooperative commodity," or in the language of resource economists, water is a **natural monopoly**. Let us break this down. If you take a break from reading this chapter and grab a glass of water, many of you will choose water that flows directly from your kitchen tap (or a filtered version taken from a different tap). If you take a shower, you will also turn on the tap. In both cases, you won't have to choose whether to buy your water from a local provider or a multinational utility.

Water is a **flow resource**: a type of resource that is "naturally" replenished in shorter time cycles and is therefore "renewable" if that use remains at or below its capacity to regenerate (Bakker 2005, 2010).[2] Unlike so-called "stock resources" that are fixed in the earth, such as minerals or oil, a flow resource is "fugitive" in character, meaning that it moves freely across property boundaries and is difficult to control across space and time. Water is heavy, bulky, and expensive to transport. As a result of these physical properties, there will likely be one tap and one provider bringing water through a single network to your home or workplace. For other basic services, such as electricity and gas, although there is only one network arriving at a building, you might be able to change your supplier – as you can do in London, where the authors live. However, water prevents such market competition due to its material properties, the absence of competing networks, and concerns over quality and safety if multiple water providers were to use the same infrastructure (Bakker 2005, 2010).

By considering the material characteristics of water and spooling these characteristics into a critique based on neoclassical economics' own references to natural monopolies, Bakker (2005) provides a convincing argument as to why water privatization

---

[2] Groundwater can be an important exception to this rule, as rates of aquifer recharge can take thousands of years. For discussion and scholarship on groundwater governance, see Chapter 3 and Further Reading section.

fails on the very terms its advocates argue for success. The hidden hand of the market – if it even exists – cannot have any regulating influence over water pricing.

Following these arguments, the claim that the private sector will necessarily be more efficient than the public sector is a little hollow. Although privatization in England and Wales is generally considered one of the more successful attempts at water privatization, consumers have seen real terms price rises of 40% in household water bills since 1989 (National Audit Office 2015). Leakage rates, generally referred to as "unaccounted for water," have declined but remain unacceptably high as the investment in fixing leaks declines. At the same time, dividends paid out to shareholders in many of the newly privatized entities have soared.

In the example of England and Wales, the success of privatization is often tied to effective **regulation**. The creation of a separate consumer regulator, the Water Services Regulation Authority (OFWAT), and environmental regulator (the Environment Agency) was seen to be a wise move, and both entities were given powers that enabled them to be champions of both consumer rights and environmental protections. Having granted substantial price increases to the privatized utilities as chief executive of OFWAT, Cathryn Ross (now an executive team member of Thames Water) was able to claim that the regulator had ensured a saving of £120 on household bills since privatization. Nevertheless, the gains are small: as Ford and Plimmer (2018) write in the *Financial Times*, "[T]hat may sound a large number, but it equates to an annual productivity improvement of just 1 per cent. It is well below even the anemic 1.5 per cent average rate for the UK economy over the same period."

Regulation helps to set the prices that utilities charge customers for water. The "productivity improvement" Ford and Plimmer refer to is part of the system of regulation, whereby prices are capped according to "RPI minus x." Designed in a matter of weeks for the newly privatized utilities under the Thatcher administration, RPI minus x permits price increases within a sector based on the Retail Price Index (the rate of inflation) minus the "efficiency gains" or productivity increases within a given sector. For most utilities, this simple formula also reflects the capital outlays of a firm (K) on infrastructure and so on. The simple formula would therefore become RPI – X + K. While tweaked slightly over the years, the basic principles for price cap regulation remain the same. Such a model of regulation has served utilities well and, as we shall explore in the next section, also facilitated the massive increase in profits under financialization.

Elsewhere in the world, profits have been maintained through what Joseph Stiglitz refers to as "briberization." Prominent among early examples of water privatization in the global South, the Buenos Aires water concession faced such criticisms before eventual collapse in 2005 (Loftus and McDonald 2001). In the 1990s, Argentina underwent a rapid process of neoliberal reforms under then President Carlos Menem. This period of rapid political economic change was dubbed a period of "Menemismo economics" and was characterized by the privatization of public services, massive layoffs of former public sector workers, and cuts to the welfare state.

In the case of water, the Buenos Aires account, the largest public utility in the country, was sold to the private sector in 1993 through a concession contract intended to last for 30 years. What was expected to be a competitive tendering process resulted in only three viable bids. The winning bid came from a consortium combining the two largest French water utilities (Suez-Lyonnaise des Eaux and Vivendi, as they were then known). These two utilities were joined by Anglian Water from the UK, Aguas de Barcelona from Spain, the Banco de Galicia, and the International Financial Corporation.

Initial outcomes seemed to be positive. The potable water network expanded; yet the concurrent expansion of the sewerage network failed to keep pace, resulting in complaints that groundwater sources were becoming polluted. The cost of water appeared to be discounted in the new concession contract. In reality, this discount masked a very sharp increase in costs immediately prior to privatization, leading to the claim of a "manufactured reduction" in costs, part of a public relations effort by the government on behalf of the private sector (Loftus and McDonald 2001).

Most notable in the post-privatization era, Aguas Argentinas had a rapid renegotiation of the contract in 1997, and the inclusion of various new charges to cover network expansion and "universal coverage." The Minister for the Environment was responsible for the contract renegotiation and was later tried in court for several illicit financial transactions under the Menem administration, serving 21 months in prison and an additional two years under house arrest. The beneficial terms awarded to Aguas Argentinas under the contract renegotiation are a clear case of how privatization actively fosters corruption. As Joseph Stiglitz, the former chief economist at the World Bank, argues (2022, p. 58–59), this trend goes against "the rhetoric of market fundamentalism" which assumes that privatization will reduce the ability of government officials to skim profits or award contracts. It is little surprise, he notes, "that in many countries today privatization is jokingly referred to as 'briberization.'" Having been heralded as a success story by the international financial institutions (IFIs) around the turn of the century, the Buenos Aires concession collapsed in 2005 after a period of sustained economic crisis in the country. The crises include various wranglings over whether the profits of the multinational companies involved could be protected against the currency devaluations taking place in Argentina.

While the Buenos Aires concession is a good example of water privatization brought down by its own internal contradictions, other examples demonstrate how water privatization has been challenged by grassroots movements. Among the iconic movements is the 2000 Cochabamba Water War in which a massive mobilization of peasants, union groups, and public sector workers overturned the privatization of the water utility in Bolivia's second-largest city (Wutich 2010). Occurring at a time of growing international movements targeting corporate globalization, the Cochabamba Water War became an emblematic moment in the struggle to achieve water justice around the world.

If the challenge to water privatization has developed from grassroots movements – often allied with larger NGOs and networks – critiques from within academia have provided further support. For David Harvey (2003), water privatization is part of a broader process he refers to as **accumulation by dispossession**. This term describes a process of fixing capital during the periodic crises of capitalism. For Harvey, such crises are characterized by the overaccumulation of capital, and at the same time, limited opportunities to put that capital into profitable investments. Identifying what he refers to as spatial, temporal, and spatio-temporal "fixes" to overaccumulation, Harvey argues how to offset crises across space (with new investments in different territories), through time (with new investments with a longer turnover time), and a combination of the two (for example US-based capitalists investing in a 50-year dam project in Argentina).

Capitalism loves a fix. With territorial limits on any spatial fix soon being realized – and with a temporal fix only permitting a brief displacement of any crisis prior to more serious overaccumulation – Harvey argues that new fixes need to be found, including profitable investment in resources formerly considered to be part of a commons. The latter is encapsulated in accumulation by dispossession. Harvey uses the privatization of water to illustrate this process. Other analyses of water privatization have combined scholarly contributions with activist insights. For example, the Municipal Services Project, Transnational Network, Public Services International Research Unit, and various organizations have produced brilliant critiques both of individual privatizations and of the process (the websites of each of these organizations provide the most up-to-date studies).

Perhaps the most damming critique of privatization comes from the industry itself, which seems less and less interested in the ambitious contracts that many assumed would be signed in the 1990s. Rates of privatization have not only slowed but more importantly have begun to reverse. Incidences of re-municipalization now outstrip those of privatization around the world. Extensive research conducted by Public Services International Research Unit and the Transnational Institute (Kishimoto et al. 2015) presents the data in stark terms: privatization is in reverse, and municipalization is now the dominant trend. While this trend is a "victory" for anti-privatization advocates on one level and a reversal of decades of outsourcing to the private sector, the process can take many forms and can be influenced by conflicting economic priorities. As Cumbers and Paul (2022) demonstrate, these trends are crucially important to situate alongside broader political and economic shifts in different contexts.

New Frontiers of Financialization

To add a final wrinkle to our account, we consider recent efforts to transform water into something unprecedented: a viable financial investment capable of generating solid economic returns. As discussed, it isn't easy to make money out of expensive pipes that crumble into the ground. And it isn't easy to make money out of selling water to people who don't have the cash to pay for it. Bottled water producers and tanker truck suppliers might be able to generate high returns, but the process of capturing, containing, purifying, and distributing water through a network has never made for a particularly good business. Treatment and proper disposal of sewage is an even less profitable business. Hence, a comprehensive water network has tended to be the job of states rather than the private sector. Turning these large and bulky forms of investment into instruments that generate returns on a similar level to other sectors is something of a holy grail in the business of water.

To make sense of why water networks tended to be poor investment vehicles in the past and how they might become profitable ones in the future, we lay out a few basic facts. First, water infrastructure cannot generate profits in its own right. Instead, profits are generated when people install pipes, treatment facilities, and dams to make a commodity (potable water) that can be sold profitably to consumers. Those consumers range from industry to individual households. Second, depending on a variety of different factors, such as ownership, the labor being employed, and the society in which it operates, water infrastructure can either be thought of as an example of "fixed capital" (a tool to be employed by labor in the generation of surplus value) or as part of a "consumption fund" (used to facilitate the flow of commodities to consumers). In some instances, water infrastructure serves *both* as fixed capital *and also* as part of the consumption fund. Abstracting from such an understanding, David Harvey (1982) has shown how fixed capital poses serious problems for capital circulation. He terms it a "value imprisoned within a specific use value" (in the case of water, it is money imprisoned within those crumbling pipes) that must command future labor (e.g. the people who do the work in producing clean drinking water) if its value is to be realized.

In the case of Thames Water, investments made in the Victorian era are still being put to work by the corporation's workforce to sell water to individual households. Those Victorian pipes require further investments to keep them functioning and, eventually, they will need to be replaced with better pipes and new technologies. Investment strategies are a long-term gamble, involving decisions around diverse factors including whether technologies will become outdated; where populations and industries will continue to be located; and when climate change might make certain infrastructures defunct. If you make the wrong decision, your money will sit in the ground as a useless monument to your foolishness.

While fixed capital proves to be a barrier to circulation, Harvey (1982) explains how it is the culmination of collective ideas (right and wrong decisions), knowledges, technological advances, and competitive processes when viewed from the point of view of production. Furthermore, the long lifespan of fixed capital investments can prove a significant advantage in moments of crisis, when over-accumulated capital builds up within the circuit of production. In short, as profitable opportunities diminish within the wider economy, individual capitalists may divert capital to fixed capital investments – a dam, a new factory, new machinery – that could realize the embedded value over a longer period of time.

Playing the long game means investing in fixed capital at specific conjunctures. Harvey refers to the processes in which capital circulates through fixed capital and the consumption fund as part of the **secondary circuit** of capital. At key moments, water infrastructure has functioned as part of this secondary circuit. Switching capital between the primary and the secondary circuit relies on a well-developed credit network. This credit system affords a key role to finance in preventing, while also deepening, crises of capital accumulation. The privatization of water in places like England and Wales enabled water infrastructure to be more clearly integrated into a secondary circuit of capital, thereby opening new opportunities for capital switching and the displacement of crises through time and space. However, as the credit system has developed it has also helped to ensure that water is now much more deeply implicated in a shift to what some people refer to as financialization. While Thames Water is often seen as the iconic example of a financialized water utility, the implications of this shift have been felt across the water and sanitation sectors around the world (Box 4.2).

Brett Christophers (2015) claims that if "globalization" was the buzzword of the 1990s and "neoliberalism" that of the 2000s, then "financialization" was the trend for

---

Box 4.2   Thames Water, An Iconic Case

Thames Water is an iconic example of financialization. Thames Water is responsible for providing water to over 10 million consumers in the Thames River basin, including the entire population of London. In 1989, the government of Margaret Thatcher privatized Thames Water, in line with the divestiture of the entire water sector. While the UK government held a golden share in each of the privatized water utilities for the first five years (to prevent monopolistic ownership in the early years), Thames Water was acquired by the German utility firm RWE in 2000. In 2006, the holding company responsible for Thames Water was acquired by the Macquarie Group, an Australian investment bank.

With significant experience in opening infrastructure projects to new investment schemes – having been an early pioneer in Australia's toll motorway schemes – the Macquarie Group transformed the Thames Water business model. They sold debt and borrowed large sums, made possible through a finance arm established in the Cayman Islands. Various accounting practices ensured the utility was able to pay almost no corporation tax in the UK. And while indebtedness went up, shareholder dividends were maintained. Since becoming a private equity firm in the late 2000s, the range of different actors making up Thames Water has changed fundamentally. The sovereign wealth funds of China, Kuwait, and Abu Dhabi all have significant holdings, as does the UK's largest pension fund, held by academics in higher education. The utility has benefited from a complex corporate structure in which OFWAT, the UK's water regulator, has limited oversight of a small part of the entity. In a crucial price review in 2008, OFWAT failed to see that the cost of borrowing would plummet, therefore enabling Thames Water to benefit from the revenue streams made possible through its high level of gearing.

New infrastructure projects – such as the Thames Water Desalination Plant or the Super Sewer – should be situated in relation to this complex corporate structure, the regulatory mechanisms, and dependence of Thames Water on the guaranteed revenue streams from household bills (Allen and Pryke 2013; Loftus and March 2016, 2019). Thames Water is the quintessential actor in the business of water in the early twenty-first century.

the 2010s. As a buzzword, Christophers suggests financialization can hide as much as illuminate. Meaning different things to different people, the analytical fuzziness associated with the expression can lead to several contradictory processes being described at the same time.

For heterodox economists, financialization refers to a shift from profits being made in the "real" economy to profits being made in the financial sector. Placing emphasis on slightly different aspects of this process, others have emphasized the relative strength of financial institutions and actors compared to those within manufacturing industries. Among Marxists, financialization has been referred to as a process of "profiting without producing" (Lapavitsas 2013), a phrase that points to an emerging parasitic class and a new wave of rentierism. Other Marxists have described financialization as the expansion of interest-bearing capital within the economy (Fine 2013), a subtly different explanation that continues to emphasize the role of labor in the production of surplus value, albeit within a quite different global economy.

What has changed in the past decade? Perhaps the most obvious shift since the 1990s has been the type of actor investing in water. During the 1990s, multinational corporations and utilities were the most likely actors to bid for concession contracts to run water services. By the mid-2000s, a new type of investor had emerged: one that represented pension funds or sovereign wealth funds. Ostensibly, this shift is a sensible match. Pension funds, notoriously, seek longer-term investments (see Clark 2000) and are therefore well-suited to fixed capital or consumption fund investments. However, the change goes much deeper and has transformed the ways some utilities now operate.

Above all, the production and sale of potable water play a less important role. Instead, bundling up the guaranteed revenue streams from household bills and turning these into globules of risk to be traded in secondary markets appears to have opened the way for a period of incredibly high dividend payments for investors in water utilities. At the same time, many large utilities have become very heavily indebted as they have leveraged debt to facilitate new speculation within secondary financial markets. These "financialized utilities" are incredibly difficult to regulate, with many of their operations lying outside of the traditional ring-fence imposed by OFWAT (Allen and Pryke 2013).

While this might seem to be the perfect example of "profiting without producing," scholars such as Kate Bayliss (2017) and Tom Purcell and colleagues (2020) ascribe this phenomenon to the rising importance of **rents** being generated within the water sector (see also Loftus et al. 2019). Rent is a social relation and "refers to payment for access to a resource such as housing, land, or patented knowledge" and in the case of the water sector, infrastructural assets (Loftus et al. 2019, p. 5). The degree to which financialization will continue to influence the water sector will largely correspond to the ability of major utilities to continue to generate these rents. Currently promoted by groups such as the United Nations Environmental Programme (UNEP) and others, the financialization of the water sector is likely to expand in the coming years.

Streams of Revenue

Important shifts in the business of water are not confined to pipes. Beyond the provision of water for drinking and irrigation, market environmentalism has come to influence dam construction, desalination, as well as river and wetland restoration. The latter is part of an important shift in environmental policy, one that has come to emphasize the role of markets in restoring ecosystems and achieving environmental benefits through **payment for ecosystem services** (PES). Placing a monetary value on what the environment does for humans enables economists to quantify the costs and the benefits involved in different projects. Framing nature in economic language makes it easier for economists to think about what the environment provides (often free of charge) to business.

In the case of wetland and river restoration, a slightly more complicated market has emerged in which the damage done by development in one location – such as draining a wetland or channelizing a stream – can be offset through investments in restoration at another location. Depending on the policy framework, a developer wishing to build a shopping mall in one catchment can proceed with that construction, with inevitable ecological disruption, on the condition that they compensate for the restoration of a river elsewhere. As a result, new private sector actors have emerged to sell their services in stream restoration, supported by a cottage industry of scientific experts, knowledge production, and certification (Lave et al. 2008; Lave 2012). In effect, stream mitigation banking trades on the broader desire to "restore" damaged riverine environments – negatively impacted by past growth – while introducing an opportunity for market forces to accumulate and "govern" water in strategic new ways.

Stream mitigation banking is closely associated with a similar system of wetland restoration in which land might be purchased by private actors, transformed into a wetland, and credits issued to be bought by those causing environmental harms elsewhere (Robertson 2004, 2006). Markets for wetland ecosystems services were created in the United States in the early 1990s, prompted by the Clean Water Act and its requirement that new wetland function be created as other wetlands are developed and destroyed (Robertson 2004, 2006). In this system, the credit commodity – the thing to be traded – is defined in units of "ecological function." What counts, then, as a healthy wetland? As Morgan Robertson (2006, p. 368) explains, "the definition of such units requires the application of ecological assessment techniques that draw botanists, ecologists, and hydrogeologists into the world of capital circulation as never before – as integral parts of the entrepreneurial production of large acreages of wetland commodities that require certification for their subsequent sale."

What is the future of such market-based experiments? Early experiments in stream mitigation banking tended to measure the success of river restoration through relatively crude assessments of quantity (the length of restoration) and quality (usually associated with the shape of the river), raising debate within the scientific community (Lave 2012). In looking forward, Rebecca Lave and colleagues (2008) raise three important concerns around stream mitigation banking as a technique for protecting habitats: (i) the outcome of existing experiments is unclear; (ii) the construction of new ecosystems is privileged over the preservation of existing ones; and (iii) the measures of "success" place greater emphasis on form (the shape of a river) than function (what rivers do).

Uncooperative Waters?

These attempts we describe to impose market relationships within the hydrosocial cycle represent only the most recent effort to craft a business out of water. Nevertheless, as Karen Bakker and others have shown, water does not always cooperate. The material properties of H_2O mean that it is heavy and therefore very costly to transport from one location to another. Seasonal variations ensure that supplies are unpredictable. And the cost of mitigating or coping with seasonal variations through new infrastructure is often unprofitable. Moreover, the very fact that life depends on access to water has meant – unsurprisingly – resistance to the idea that water should be treated like any other commodity, left to the market. Nevertheless, and perhaps in part because it is such a universal necessity, efforts to try to make money out of water are ongoing, constantly pushing the frontiers of accumulation.

While efforts to make a business of water have been successful in different historical moments, these initiatives have been countered by those seeking to ensure that water is a common resource. Thus, efforts to counter privatization throughout the 1990s resulted in significant victories for the water justice movement and the collapse of concession contracts in Cochabamba, most notably, and other cities across the global South. In Cochabamba, Bolivia, an alliance was formed between urban and rural dwellers, many of whom objected to the undermining of a moral economy of water when private actors prevented their *usos y costumbres* (customary use rules and norms) of water (Perreault 2006; Wutich 2010; see also Chapter 3). Recently, new successes can be seen in the process of remunicipalization, although this trend is a complicated one resulting in different levels of access to water. In short, water is a resource that will be struggled over by competing interests, with each putting forward arguments for why the private sector should or should not be involved.

Writing about resource struggles, Nik Heynen, Maria Kaika, and Erik Sywngedouw (2006, p. 12) state that political ecology "attempts to tease out who gains from and who pays for, who benefits from and who suffers (and in what ways) from particular processes of metabolic circulatory change." This book – and our approach to the hydrosocial cycle more broadly – demonstrates how the flow of water is influenced by numerous relations between companies and states, between scientists and regulators, between financiers and working people.

Summary and What's Next

In this chapter, we explained how the business of water has developed around the world in different ways. We introduced the many faces of market environmentalism in the water sector, plumbed the historical roots of market-based thinking, reviewed arguments used for and against water privatization, and surveyed the current state of play around the world. We placed special emphasis on the role of the capitalist state in setting up market governance, as exemplified by England and Wales in the 1980s. Noting that most water networks have been constructed by large public utilities, we explained why it is difficult for the private sector to return a profit from the provision of potable water. Nevertheless, the discourse of greater private sector involvement in water continues. While privatization schemes have slowed globally and have even been reversed in some parts of the world, efforts to squeeze profit out of water are ongoing.

Financialization is one of the most recent efforts to make money from water. In the case of water, financialization is perhaps most evident when the revenue streams associated with the bills that most of us pay are transformed into financial commodities. Stream and wetland restoration are two additional domains where market discourse and banking instruments have been developed to "sell nature in order to save it." Still, camp is everywhere. Despite the growth and evolution of the business of water, many people actively resist and counter efforts to inject market principles in water. Some actors seek to profit from the sale of water, while others view water as common property. Which of these perspectives will shape policy in a given context is the result of ongoing struggles around the hydrosocial cycle. In the next chapter, we will step into grocery stores, farm fields, and commodity markets to explore how some of these interests have shaped the greatest water consumer in the world: agriculture and food production.

Further Reading

On the history of urban water networks

Agostoni, C. (2003). *Monuments of Progress: Modernization and Public Health in Mexico City, 1876–1910*. Calgary and Mexico City: University of Calgary Press and UNAM Press.

Gandy, M. (2002). *Concrete and Clay: Reworking Nature in New York City*. Cambridge, MA: MIT Press.

Graham, S. and S. Marvin. (2001). *Splintering Urbanism: Networked Infrastructures, Technological Mobilities, and the Urban Condition*. Abingdon, UK: Routledge.

March, H. (2015). Taming, Controlling and Metabolizing Flows: Water and the Urbanization Process of Barcelona and Madrid (1850–2012). *European Urban and Regional Studies* 22 (4): 350–367.

Melosi, M.V. (2011). *Precious Commodity: Providing Water for America's Cities*. Pittsburgh, PA: University of Pittsburgh Press.

On water privatization

Bakker, K. (2003). Archipelagos and Networks: Urbanization and Water Privatization in the South. *The Geographical Journal* 169 (4): 328–341.

Bakker, K. (2010). *Privatizing Water: Governance Failure and the World's Urban Water Crisis*. Ithaca, NY: Cornell University Press.

Bakker, K. (2014). The Business of Water. *Annual Review of Environment and Resources* 39 (1): 469–494.

Budds, J. and McGranahan, G. (2003). Are the Debates on Water Privatization Missing the Point? Experiences from Africa, Asia and Latin America. *Environment and Urbanization* 15 (2): 87–114.

Conca, K. (2005). *Governing Water: Contentious Transnational Politics and Global Institution Building*. Cambridge, MA: MIT Press.

Loftus, A.J. and McDonald, D.A. (2001). Of Liquid Dreams: A Political Ecology of Water Privatization in Buenos Aires. *Environment and Urbanization* 13 (2): 179–199.

On remunicipalization

Cumbers, A. and Paul, F. (2022). Remunicipalisation, Mutating Neoliberalism, and the Conjuncture. *Antipode* 54 (1): 197–217.

Kishimoto, S., Lobina, E., and Petitjean, O. (2015). *Our Public Water Future: The Global Experience with Remunicipalisation*. Amsterdam, London, Paris, Cape Town and Brussels: Transnational Institute (TNI)/Public Services International Research Unit (PSIRU)/Multinationals Observatory/Municipal Services Project (MSP)/European Federation of Public Service Unions (EPSU).

McDonald, D.A. (2018). Remunicipalization: The Future of Water Services? *Geoforum* 91: 47–56.

On financialization

Ahlers, R. and V. Merme. (2016). Financialization, Water Governance, and Uneven Development. *WIREs Water* 3 (6): 766–774.

Allen, J. and Pryke, M. (2013). Financialising Household Water: Thames Water, MEIF, and 'Ring-Fenced' Politics. *Cambridge Journal of Regions, Economy and Society* 6 (3): 419–439.

Christophers, B. (2015). The Limits to Financialization. *Dialogues in Human Geography* 5 (2): 183–200.

Furlong, K. (2021). Full-Cost Recovery = Debt Recovery: How Infrastructure Financing Models Lead to Overcapacity, Debt, and Disconnection. *WIREs Water* 8 (2): e1503.

Loftus, A., March, H., and Purcell, T.F. (2019). The Political Economy of Water Infrastructure: An Introduction to Financialization. *WIREs Water* 6 (1): e1326.

Stream and wetland banking

Lave, R. (2012). *Fields and Streams: Stream Restoration, Neoliberalism, and the Future of Environmental Science*. Athens, GA: University of Georgia Press.

Lave, R., Doyle, M., and Robertson, M. (2010). Privatizing Stream Restoration in the US. *Social Studies of Science* 40 (5): 677–703.

Lave, R., Robertson, M.M., and Doyle, M.W. (2008). Why You Should Pay Attention to Stream Mitigation Banking. *Ecological Restoration* 4: 287–289.

Robertson, M. (2006). The Nature That Capital Can See: Science, State, and Market in the Commodification of Ecosystem Services. *Environment and Planning D: Society and Space* 24 (3): 367–387.

Robertson, M. (2012). Measurement and Alienation: Making a World of Ecosystem Services. *Transactions of the Institute of British Geographers* 37 (3): 386–401.

Part 2
Big Waters

Chapter 5

Eating Water

London in May. A time when asparagus, peppers, spring onions, new potatoes, and strawberries are in season.

Tuesday 5 p.m. at Tesco, the largest supermarket chain in the UK. Their motto, "Every little helps," printed on their entrance, greets customers.

In the aisles, British carrots and potatoes are stacked high. British asparagus is on sale for £2.25. Further down on the shelves, asparagus flown in from Peru sold for £2. An international lineup of vegetables is displayed with runner beans from Guatemala next to fine beans from Egypt. Peppers crowd a shelf, some British, some Dutch. A bunch of spring onions from Senegal goes for 40p. Avocados from Peru, two for £2.50.

Punnets of British strawberries are being picked up by customers, their eyes caught by the vibrant color. Tesco sells irregular-sized fruit to reduce food waste, calling them "perfectly imperfect." But no wonky strawberry packs or bags of misshapen apples are in sight, at least not today.

The London living wage is £11.05 an hour, roughly £1 and some pennies more than the national living wage, reflecting higher costs in the capital city.

Cheap food sells. For every product, there is a supermarket brand. A loaf of bread goes for 65p and a pack of pasta for 75p. Dried chickpeas come from Russia and bulgur from Turkey. A can of trusty baked beans for 35p and six eggs for 95p.

At the back of the store, cool shelves are full of packets of meat. Pork is "Reared in Holland, Slaughtered in Germany" for the sum of £2.80 a packet. Lured by the sticker promoting any 2 for £8, customers pick up beef steaks.

Water: A Critical Introduction, First Edition. Katie Meehan, Naho Mirumachi, Alex Loftus, and Majed Akhter.
© 2023 John Wiley & Sons Ltd. Published 2023 by John Wiley & Sons Ltd.

Back in homes, the fruit, salad, bread, and other bits of food get binned. Forgotten and unopened, food is thrown away daily in half a million British households.

How Much Water Do You Eat?

Do you know how many liters of water you eat every day? Yes, *eat*, not drink. Asking this question is germane because we use most of our available fresh water for food. While we might drink about a liter of water a day, we consume significant volumes of water via our meals of grains, vegetables, dairy, and meat. A cheese sandwich requires approximately 90 liters of water to produce the wheat for the slices of bread and the milk for the cheese. That sandwich is equivalent to the amount of water you would use if you flushed a modern dual toilet 15 times. An apple requires 125 liters of water to be produced. That is approximately three-quarters of a bathtub. A cup of coffee requires 130 liters of water to produce just 7 grams of the coffee beans. An average British person eats and uses over 4500 liters of water a day. Just that steak from the supermarket would be about 3400 liters.

In all these food items, water is necessary to grow the crop itself or the foodstuffs that feed animals, which provide dairy and meat. Just as for humans, the amount of water that livestock requires for drinking is minor compared to the water used to grow the fodder they eat. This water used to develop crops and products, called **virtual water**, gives a sense of how much water is hidden in the agricultural production which puts food on our table. The water footprint of our everyday food is enormous. Throwing away food is wasting water.

The agricultural sector makes up 72% of global water use. Water for drinking, flushing the toilet, washing, and keeping clean is important for health and hygiene. However, the domestic sector uses just 16% of total water use. Industry accounts for 12% (UN-Water 2020). From a global perspective, the water footprint of agriculture, or the total amount of fresh water consumed and polluted during production, is 8362 km^3/year (Mekonnenn and Gerbens-Leenes 2020).

Food has a hidden geography of water (Box 5.1). Most countries have agricultural sectors but not all can produce enough food to be self-sufficient. This means that food is actively traded throughout the globe, creating a pattern of virtual water flows from one part of the world to another. Arid countries, like those in the Middle East, and countries with limited land to develop agriculture at scale, such as Japan and the United Kingdom, "import" virtual water through the trade of food staples and other agricultural products. London supermarket shelves reflect this hidden geography of water trade.

Countries like the United States, Brazil, and India produce goods at scale, resulting in an "export" of virtual water.[1] Virtual water flows of agricultural and industrial

---

[1] We discuss virtual water "trade," including virtual water "export" and "import," denoted with quotation marks. Technically, food and other commodities are traded between countries – not virtual water *per se*. Flows of virtual water exist because of the way crops and products are exchanged spatially.

products traded across the world are estimated at 2320 km³/year. Approximately 90% of this figure relates to the trade of food, products based on crops and animal products (Hoekstra and Mekonnen 2012). Water for agriculture is important in areas where crops are grown. But the allocation and use of this water are part of a globalized process of producing, trading, and distributing food.

The agricultural sector has high **consumptive use** whereby there are volumes of water not possible to reuse or retrieve due to evapotranspiration, evaporation, or drainage to saline water bodies including the sea. One way to account for hidden water consumption is through a categorical "footprint." A water footprint is described as "a measure of humanity's appropriation of fresh water in volumes of water consumed and/or polluted" (Water Footprint Network 2022). Experts categorize the water that makes up a footprint in three ways (Figure 5.1). **Green water**, or rainwater and moisture in the soil profile, is vital to farming and represents 80% of agricultural water consumption. **Blue water**, or water from rivers, lakes, wetlands, and aquifers, makes up 11% of the water footprint of the agricultural sector. There is water required to assimilate any pollution caused during the production and this **gray water** footprint accounts for the remaining 9% (Mekonnen and Gerbens-Leenes 2020). Overall, and despite technological advances in farming and water management, water consumption in the agricultural sector is expected to grow steadily over the next decades and reach nearly 1500 km³ by 2040 (UN-Water 2019).

The water footprint of Figure 5.1 reveals the enormous role of agriculture and farming practices in global water use, especially when compared to domestic or residential water consumption. Clearly, the sustainable management of water and soil is

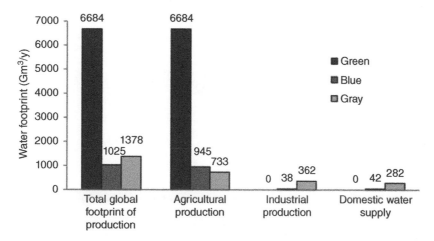

Figure 5.1 The water footprint of agriculture. This graph depicts three categories of water (green, blue, and gray) that make up the world's water consumption. The food and agricultural sector is the world's major consumer of water, despite efficiency gains and advanced technologies. In other words, most of the water we "drink" is hidden in the food and products we eat. Note: 1 Gm³ is equal to 1 km³ or 1 billion m³ (cubic meters). Source: Chenoweth et al. (2014, p. 9433) / Copernicus Publications / CC BY 3.0.

Box 5.1 What is Virtual Water?

The original concept of virtual water was first developed by J.A. (Tony) Allan in the early 1980s (Figure 5.2). Tony was a prominent geographer, specializing in water, natural resources management, and agricultural development. His decades of fieldwork in countries such as Egypt, Libya, Jordan, and Yemen provided essential insights and policy solutions. Most important, Tony never took the idea of water "scarcity" for granted, and the concept of virtual water reflects his thinking about the production and political economy of scarcity. Tony recalled the moment he realized the significance of water embedded in food and the role of food import:

> It came when I was working in a library in Cairo in the early '80s. I discovered that the imports of food and especially of grain and flour in Egypt had suddenly kicked up in 1972. It was a political moment in the region when the Russians were being asked to leave and US influence was coming back in. The Russians could not meet Egypt's growing demand for staple food commodities. The United States could and had a PL480 funding policy [Title II Food Aid of the Agricultural Trade Development and Assistance Act of 1953 (Public Law 480)] which helped countries with food aid. Egypt could import inexpensive, cheap, affordable food produced

Figure 5.2 Portrait of J.A. Allan, or Tony, as he liked to be called. Source: Illustration by Francesca Greco and reproduced with permission.

with US water. [...] At that point I realized that an economy can get into this very serious predicament of being food insecure – which could destabilize society and bring people on to the streets. However, this elemental problem of water and food insecurity could be solved by importing underpriced food, which is what happened in Egypt. (Whittington and Thomas 2020, pp. 1971003–1971004)

At that time, Tony called the concept "embedded water." By the 1990s, it was renamed "virtual water," which had more traction in policy. Since then, virtual water has been a central discourse in the water policy of governments worldwide. In 2008, the Stockholm Water Prize, the most prestigious water prize in the world, recognized the major advances to scientific understanding and policy made by this pioneering concept.

Virtual water debunked the myth about water wars. Tony worked in the Middle East and North Africa, a water-stressed region prone to stereotypes about water and conflict. Despite the severe pressures, nations did not engage in acute military conflict over water. With the import of food, a country's water supply did not have to be devoted in huge volumes to agriculture and, instead, could be used for other purposes. Tony argued that examining the water budgets of a country does not yield appropriate solutions to water management. Rather, he argued we should focus on the global level and on the political economy of the food trade. He extended this idea by terming water used for agriculture as "food water" to emphasize the significance of food and its trade. This myth-busting also advanced studies in transboundary water management and led to further research that focused on the role of power to explain the problem of unequal water allocation in lieu of water wars. Tony inspired and mentored an interdisciplinary group of researchers who focused on the role of power in water governance, now known as the London School (Chapter 7).

important. Yet, the current food system is organized and incentivized to make short-term returns at the expense of environmental sustainability (UNCCD 2017). Land degradation is experienced in 1 billion hectares (ha) across 100 countries, due to soil salinization caused by poor irrigation and drainage practices (IPBES 2018). Over half of the blue water footprint for crop production is deemed "unsustainable," with seven crops – wheat, rice, maize, sugar cane, fodder, and cotton – as the main culprits (Mekonnen and Hoekstra 2020).

From a virtual water perspective, the scale of the linkage between water for food is extensive, spanning many river basins and regions of arable land. This linkage connects, for example, smallholder farmers in India with Tesco consumers in England. From the vantage point of virtual water, two mythologies deserve unpacking.

The first myth is that farming relies on rivers and large bodies of fresh water for irrigation. In fact, farming relies more on green water than blue water. This means that we need to consider how we manage soil and land as well as water. A critical perspective elucidates the tensions and struggles around access to land, land rights, and ownership to address water for food.

A second myth refers to scarcity, a narrative often couched in the language of the "global water crisis." The nature and impact of scarce water, however, is local: in scenarios where farmers don't have enough water in their fields; when irrigation canals run dry; and when land becomes salinized. Consumers in London are often shielded from the price squeeze when supermarket shelves are full of food flown in from various regions. Most agricultural activity in the world does not require irrigation, but local problems of water "scarcity" persist because there is a track record of over-abstracting rivers, lakes, and aquifers. Why is this so? In the following sections, we unpack these myths by asking how water is the product of social, spatial, and ecological relations unfolding over agriculture, food, and land.

The Water of Land Grabbing

The boomerang-shaped Niger River in west Africa flows through nine countries. The Inner Niger Delta in Mali, located in the mid reaches of the river, floods during the short wet season providing water to the soils. In the lands managed by the Office du Niger, a governmental agency tasked with irrigation using these river waters, the number of agricultural investments increased 20-fold between 2002 and 2009. In the span of seven years, land leases covering 870 000 ha of land were applied for. In a place where rainfed (i.e. green water) agriculture is widespread and less than 5% of the land of the Office du Niger is under irrigation, these investments modernized farming. This invested land is primarily designated for the cultivation of rice along with oil crops such as sunflower, soybean, peanut, and jatropha. More than half of these are foreign investments by Libya, Saudi Arabia, France, Canada, the United States, China, and Malaysia (Adamczewki et al. 2013).

One of these investments is the 100 000 ha Malibya project signed in 2008, dubbed as "one of Africa's largest and most secretive foreign agricultural investment deals" (Arsenault 2015; Figure 5.3). The project is financed by Libya's sovereign wealth fund benefiting from oil revenues and set up by the son of Colonel Muammar Gaddafi, the country's leader at that time. The Malibya project aims to extend rice cultivation and holds a 50-year renewable land lease. Granted water access by the Office du Niger, the project built a 40 km canal, the fourth largest in the region. The construction removed lands where small-scale farmers grow crops for subsistence and market. At least 150 households were relocated, with compensation insufficient, if given at all. The project highlighted two food insecurity concerns for those who lost their lands: (i) uncertainty for farmers who held customary land rights when exact project lands were unknown; and (ii) potential water competition between smallholders and the agribusiness. The Gaddafi regime collapsed in 2011, grinding the project to a halt. In the meantime, farmers have not been removed from their

Figure 5.3   Head regulator of the Malibya canal new area, Office du Niger, Mali.
Source: Jean-Yves Jamin.

land, but they continue to live in uncertainty over their future land and water access as the lease is still in place (Larder 2015).

The case of foreign investment in land and water in Mali is not unusual. There has been a global rise in the number of **large-scale land acquisitions** (LSLAs), which have implications for the access and allocation of both green and blue water. LSLAs are referred to as "land grabs" where deals are struck over the rights to land or leases, typically land over 200 ha. In 2009, 56 million ha of land were signed off under such deals (Deininger et al. 2011). In 2021, land deals covered over 81 million ha, suggesting that LSLAs have not waned in the last decade (Land Matrix 2021). These acquisitions are widespread in Sub-Saharan Africa, Latin America, and South and Southeast Asia. The limited public data and opaque nature of these deals make calculations difficult, but in 2009 USD 20–30 billion were exchanged for 15 to 20 million ha of land (Holden and Pagel 2013).

What do land grabs have to do with water? An overwhelming 80% of these deals are for agricultural projects, and the most desirable deals involve easy access to water resources or convenient methods to abstract water (Quick and Woodhouse 2014). By investing in lands, project developers can control green water and secure blue water. Land and water grabs are presumed to provide economic opportunities from relatively high crop yields in inexpensive lands and help manage climate uncertainty (Chiarelli et al. 2021). Green water cannot be stored or transferred because it is an integral part of soil and therefore land. The high green water footprint of agriculture shows that rainfall is important. For land grabbers, seeking "attractive" land is important. But particularly in arid regions, green water is commonly supplemented with blue water through irrigation. Blue water can be stored and pumped out of rivers, lakes, wetlands, and groundwater bodies. Consequently, blue water management involves infrastructure, such as dams, pipes, and canals to pump, store, divert, and transfer the resource. Cases like Malibya are justified, at least by the Office du Niger, on the grounds that the state cannot develop all the required infrastructure and external investments are necessary (Larder 2015). In this regard, land acquisitions change who has access to green and blue water and how and when the water is used for what kind of crops.

A critical lens into green and blue water allows us to see the social, political, and ecological relations that shape agricultural water flows within and between places through land grabs. Differentiating blue and green water adds further texture to the hydrosocial cycle and helps us question long-held assumptions about water and farming. Historically, accounting for water tended to focus on blue water only. This is due in part to rainfed agriculture seemingly having no cost for precipitation compared to the cost required to dam rivers or line canals. Green water was often left unaccounted, despite being the "most valuable variable" to water supply (Allan 2001, p. 14). Examining land brings into sharp focus who is included and excluded in the social production of water – both green and blue water.

Large-scale land acquisitions are not a new phenomenon. For centuries, land acquisition has been the foundational template for colonialism and the geographical expansion of business operations. Land has a long and contested history of being enclosed and commodified. But recent trends in large-scale grabbing are particularly contentious for several reasons.

First, recent large-scale land grabs were spurred by the 2008 food crisis that prompted governments to acquire lands elsewhere to buffer against potential food shortages. In 2013, the scale of acquisitions could feed potentially 300–550 million people if those lands were farmed with improved agricultural practices and closed the yield gap (Rulli and D'Odorico 2014). The total blue, green, and gray water grabbed was 490 km^3/year for an approximate 43 million ha of land acquired. The majority was green water, amounting to 379 km^3/year.

Studies suggest that large-scale acquisitions contribute to significant increases in blue water demand because of the kinds of crops grown to suit investors (Figure 5.4). In Africa, where there a major proportion of acquisitions take place, there is reportedly a higher potential for blue water abstraction compared to other investment regions (Rulli and D'Odorico 2013). For example, jatropha, sugarcane, and eucalyptus grown in acquired lands in Sub-Saharan Africa amount to more than 9000m^3/ha. The increased water demand is met through irrigation, meaning that blue water inputs can be 76–86% higher than for staple crops (Johansson et al. 2016)

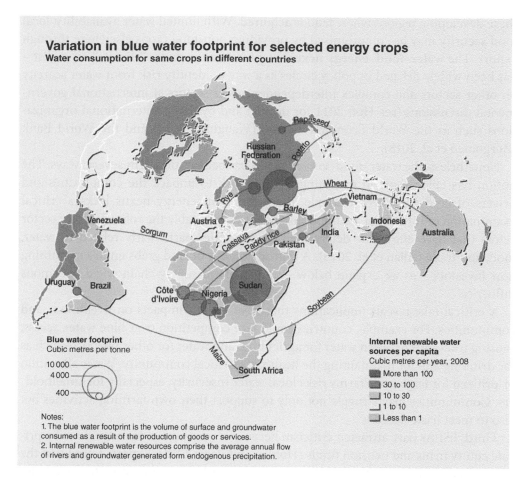

Figure 5.4 Variation in blue water footprints for selected energy crops. This map depicts the uneven spatial distribution of "renewable" water supply and agricultural use for energy crops – a distribution that is mediated by global trade and relations. Source: Riccardo Pravettoni, UNEP/GRID-Arendal. www.grida.no/resources/6198.

Second, the recent land grabs pose vexing questions on the food–energy nexus because a significant proportion of acquisitions are for biofuel production. During 2000–2010, one estimate suggested that over 8 million ha was for jatropha production. Countries such as India, Indonesia, Malaysia, and Philippines have a large number of deals for jatropha as well as oil palm and sugarcane (Holden and Pagel 2013). As of 2021, the Land Matrix reports over 3 million ha of LSLAs for biofuels; these three crops make up close to 80% of all LSLAs (Land Matrix 2021). The interest in these crops is against a backdrop of countries adopting biofuel policies to tackle climate change. The 2009 EU Biofuel Directive and 2007 US Energy Bill are widely regarded as driving demand.

This situation raises questions on how water should be best used – for food or for energy? This question is pertinent, especially where poverty is prevalent in many of these developing regions where land is acquired. With limited water availability, local food security may be compromised by providing non-food crops elsewhere through export. The **water–food–energy nexus** – which often includes "climate" on its list – has been widely debated in policy circles as a way to identify risk from water scarcity on other sectors and complex interdependencies of sectors at international governmental discussions (see Hoff 2011 for details) and by major international organizations such as the World Economic Forum (Waughray 2011) and the World Bank (Borgomeo et al. 2018).

Nonetheless, the water–food–energy nexus has been critiqued in several ways. The notion has fallen short in providing a framework to unpack the complexities and drivers of land grabs. For example, the water–food–energy nexus lacks a critical examination of influential assumptions and drivers, notably the role of private sector actors that have effectively determined the priorities particularly related to water, food, and trade (Allan et al. 2015). A critical account of land grabs entails examining how investors – as we explain below – and the food supply chain wield enormous influence.

A critical take reveals implications that have serious impacts on livelihoods and communities. For example, countries could face competition over blue water access, causing trade-offs between water for agriculture and water for other purposes such as for drinking and for maintaining the health of the local community. Water allocation prioritized for large-scale farms risks local water insecurity, especially for smallholders. Communities may struggle not only to support their own farming activities but also to meet basic needs.

Third, LSLAs have attracted criticism because some of these deals are done by private equity firms and pension funds (Holden and Pagel 2013). Experts found that the 2008 global financial crisis and following recession induced investors to seek "safer ventures," and agriculture has a long-term nature (Hunt 2015, p. 4). Sovereign wealth funds of oil-rich but water-scarce countries, like those investing in the Malibya, secured over 200 million ha (Cotula 2012). Qatar established Hassad Food, which is a sovereign wealth fund specializing in agribusiness with investments in Sudan and operations in places like Australia and Canada. The United Arab Emirates is said to have invested over 15% (USD 627 billion) of all sovereign wealth funds in 2009 that went to the agribusiness sector in Sub-Saharan Africa (Lawrence et al. 2015). However, there is a speculative nature to large-scale land acquisition. In Africa, only 3% of contracted deals are in production out of 22 million ha at the point of analysis in 2016 (Johansson et al. 2016). For example, of the known LSLA deals for agriculture in 2021 – a rough total of 47 million ha – 8 million ha is intended for production (failed LSLAs amounted to 14 million ha of land in total). Out of the deals that have been concluded, 3 million ha are left untouched with projects unbegun or abandoned (Land Matrix 2021).

GRAIN, an international non-governmental organization for sustainable farming and food systems, reported that at least 135 deals totaling 17.5 million ha had fallen through between 2007 and 2017. Reasons include the incompetency of investors and bankruptcy; governments canceling projects or reducing the scope of land permits and concessions; and strong local opposition by communities who were often not consulted on these large-scale projects. Strikingly, while these projects may be canceled or rescaled, they are not "failed" land grabs but simply failed investment projects because investors changed and land was not returned to local communities (GRAIN 2018). Large-scale land acquisitions dispossess communities of their land and water.

Once land acquisitions are in place, even without seeds sown or crops harvested, these deals change the spatiality of green and blue water. There are tensions between customary and formal land rights and water permits, which have shaped and reshaped social relations between farmers and the state authorities. Land grabs create new social relations between local communities and investors, along with their contractors who bring in labor, resources, and expertise. They reconfigure social relations within communities. In the case of Malibya, some farmers want to be included in the irrigation project while others remain skeptical, dividing the local communities over the potential benefits of the project (Larder 2015).

Land-water grabs underscore the idea that water is life. The connections between people and their land and water cannot be reduced to the simplified logics of "safe" investments, low-risk financial gains, or even national food security. This idea resonates when the investments install a discourse of land left to languish instead of being put to good use. Land deals facilitate "unused" or "marginal" resources, which tells the story of water flowing "in vain," underutilized, or "idle" lands (Spagnuolo 2017, p. 295). But for those living on these lands, they are not idle or without value and water is part of social reproduction that weaves through their everyday lives. Lands and water are not simply assets but a way of life and the base for their culture, livelihoods, and heritage.

Cheap Food and Unsustainable Water Use

Who are the world's major water eaters? Five countries contribute an overwhelming 70% of the unsustainable blue water footprint: the United States, India, China, Pakistan, and Iran (Mekonnen and Gerbens-Leenes 2020). The US, India, and China are the top three virtual water "exporters" of agricultural and industrial goods in the world, with a combined water footprint of 582 km^3/year (Hoekstra and Mekonnen 2012). The production of six crops – wheat, rice, cotton, sugar cane, fodder, and maize – contributes to the over-abstraction of blue water sources. The water footprint of agriculture reveals how existing agricultural production is uneven across space and place.

Virtual water flows are controlled by entities beyond states. Four agribusinesses command the lion's share of the global agricultural commodity trade. Archer Daniels Midland, Bunge, Cargill, and Louis Dreyfus – often nicknamed the ABCDs – control 70–90% of the global agro-food commodity trade. They have major influence over how, when, and at what price major commodities such as soy, maize, cotton, wheat, and sugar are traded.

Consumers also shape virtual water flows by eating cheap food at unsustainable rates (Box 5.2). Cheap food is destructive to the environment as the true cost of environmental stewardship is not reflected in the price we pay when we shop (Allan 2019). With industrially produced food, cheap and readily available for consumers, there is no way that water (and more broadly, environmental) stewardship can be undertaken to match the scale of production. The combination of dominant food producers, agribusinesses, and consumers has created and reinforced a food system destructive to the environment.

Cheap food is produced, not accidental. Cheap food is encouraged through government subsidies and agricultural policies at the farm level. The state has provided protection and support to agricultural livelihoods which can be faced with risks and unpredictability. However, this support is not sufficient for farmers to be effective water stewards and practice environmentally sustainable agriculture. Investors in land and agriculture, like those we examined in the land grabs section, may adversely affect the food system as they seek profit while ignoring environmental capital and the welfare of producers. Furthermore, agribusinesses, supermarkets, and other food companies do not account for environmental capital, including water. Companies are subject to accounting rules, but these focus on commercial performance, rather than the costs to the environment (Allan 2019). Consumers have ever-tightening budgets and cheap food continues to be on the shelves of our shops.

Virtual water "trade" addresses growing water demand amidst an increasingly limited blue water supply. The decision to import water-intensive crops and food is "politically silent" (Allan 2001). This is because virtual water is invisible, not just in crops but also in national water budgets. Importing food entails paying for foreign crops, rather than the water embedded in those crops. The national water deficit may not be an urgent problem as long as trade is possible and food is cheap enough to import. Consequently, managing water demand in this way is not a deliberate political strategy to address the impending environmental limits of water use or to consider environmental capital.

Box 5.2   Water on the Menu

How much water do you eat? We posed this question to people visiting cafés in London and Milan through a collaborative art-science project called Wonderwater. The project was originally created by designers Jane Withers and Kari Korkman to raise awareness of water sustainability. Naho Mirumachi worked with Jane, her studio, and university research assistants to transform a typical café menu into a water footprint menu, depicting calculations of virtual water hidden in the dishes and drinks (Figure 5.5). The footprint menus

Figure 5.5 The Wonderwater Project, an art-science collaboration that showed the hidden flows of virtual water in café dishes and drinks. Wonderwater at Triennale Caffè by Jane Withers Studio for XXII Triennale di Milano Broken Nature: Design Takes on Human Survival. Photo: Gianluca Di Ioia.

illustrated quantities of blue, green, and gray water footprints, and provided short commentaries about the origins of ingredients and how water-thirsty they were.

Taking the stigma out of "big" water footprints is vital. Many customers were surprised by the number of liters it took to create a simple-looking fried egg dish or pizza. Wonderwater initiated a conversation with a coffee producer who was unsatisfied with the high water footprint given to his drink. The producer was highly selective with his beans and used minimal water to process the beans, which should have been reflected in the water footprint. Regardless of how the beans are processed, a coffee plant naturally needs a lot of water.

Will eating organic save us? Wonderwater innovated artwork and commentaries to educate consumers about the hidden thirst of organic food (Figure 5.6). Eating organic does not necessarily mean "better," at least from a water footprint perspective. A water-thirsty crop like coffee grown in a water-scarce land is not sustainable, even if it is organically produced. In a creative blend of art and science, the Wonderwater project offers a window into how global relations shape

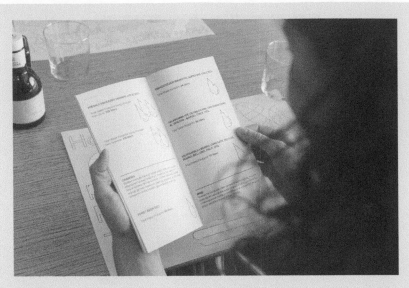

Figure 5.6 Café customers in London and Milan read the menu and learned about virtual water. The Wonderwater Project @ Triennale Caffè for Broken Nature: Design Takes on Human Survival was developed by Jane Withers and Kari Korkman in collaboration with Naho Mirumachi and Arthur Fuest. Photo source: Wonderwater @ Triennale Caffè for Broken Nature: Design Takes on Human Survival, XXII Triennale di Milano, 2019, Gianluca Di Ioia, reproduced with permission.

our daily lives – and vice versa – and suggests that any "solution" to footprint dilemmas must move beyond technical innovations and engage a wholesale transformation of our agricultural industry and relations.

Virtual Water Hegemony

Cheap food and unsustainable blue water use are enabled by the global food market which is heavily influenced by agribusiness powerhouses (e.g. the ABCDs) and a handful of major supermarket chains. Together, they control the food supply chain, a market that takes the shape of an **hourglass** connecting significant numbers of producers and consumers at both ends. This market perpetuates cheap food and systematically undermines efforts to value the environment (Allan 2019). Unsurprisingly, these private companies are Western entities. The handful of companies in the middle of the hourglass determines what consumers eat. They also risk the livelihoods of farmers through their demands and restrictions on crops they buy and trade.

Consider the example of Bunge, one of the ABCDs, and its role in soybean production and trade in Brazil. Bunge is a publicly listed company headquartered in the United States with a total revenue of USD 59 billion in 2021, operating in 40 countries.

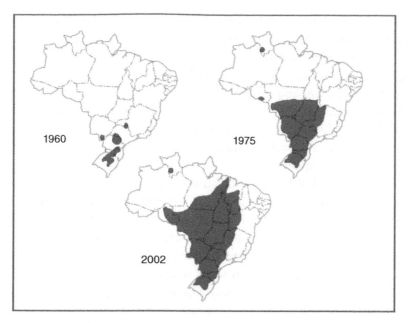

Figure 5.7   The expansion of soybean production in Brazil between 1960 and 2002.
Source: FIAN (2018, p. 26) / FIAN International.

For over a century, Brazil has been a key location for Bunge's grain trade business. In recent years, Brazil has become the top competitor for global soybean production, rivaling the United States. Soybean production has expanded to areas of the *cerrado* and Amazon, two highly important biodiversity hotspots (Figure 5.7). Approximately 38.5 million ha of land is dedicated to soy monoculture, concentrated in the north and northeast region of country (Pitta et al. 2022). Bunge maintains large storage and processing facilities in the region and has an extensive network of ports, transportation, and logistics for global export. Bunge's scale of processing and trade is regarded as a monopoly over soybean production and it is seen as the "dominant buyer" of farmers' soybeans in the region (Drost et al. 2017, p. 2).

Environmental activists have accused Bunge of playing a role in deforestation and silently sustaining land grabs. Through coercion and violence toward local communities, large tracts of land have been deforested illegally to be converted for soybean production. Allegedly, some of the land grabs are speculative and take advantage of strategic locations that will be convenient to sell to farmers who will then sell their soybeans to Bunge for processing (Pitta et al. 2022; Pitta and Mendonça 2022). This land has traditionally been used by peasant, *quilombola*, and Indigenous communities, who face a dual challenge of land rights recognition and worsening food insecurity.[2]

---

[2] *Quilombos* are Afro-Brazilian (and also Indigenous) settlements and communities of freed, escaped, and former enslaved peoples seeking freedom, autonomy, and self-rule within exploitative systems. Within the Americas, *quilombos* are often cited as examples of subaltern agency, Black spatial knowledge, fugitive infrastructure, and abolitionism. For more information, see Bledsoe (2017), Ferretti (2019), and Winston (2021).

Bunge has been named as one of the worst deforestation offenders (along with Cargill, another company of the ABCDs), contributing to deforestation of 60 000 ha of which more than a third is in protected areas, pointing to the illegality of land acquisitions. Major supermarkets in the United States and Europe like Costco, Walmart, Kroger, Tesco, Carrefour, EDEKA, and Albert Heijn buy soy from Bunge (and Cargill) to sell food ready for consumers (Mighty Earth 2022). Far away from these supermarket shelves and dinner tables, local communities in Brazil face intimidation, destruction of their cultural ties to land and livelihood, and loss of biodiversity including changes to water availability and quality.

Consequently, virtual water flows are subject to Western virtual water hegemony and its exercise of power (Sojamo et al. 2012). **Virtual water hegemony** is a structure that conditions and determines the flows of virtual water through an overwhelming exercise of power, in this case, Western agribusinesses (a corollary concept called hydro-hegemony will be examined in Chapter 7). Virtual water hegemony explains how the governance of blue and green water is subject to power and power relationships. These companies influence not only where crops are grown through managing food supply but also how demand is met and managed through supply and value chains. In other words, companies have control over the supply chain by overseeing farming inputs such as seeds and fertilizers, processing, manufacturing, and transportation of products (Sojamo et al. 2012). The financial resources available to agribusinesses are significant and complex. The scale of operations is truly global and, through risk management, they also play a role in the financialization of the food sector.

How agribusinesses, supermarkets, and food companies operate matters because they wield significant power by adding 90% of the value to farm products. They also generate vital economic activity through job creation and employment. These businesses have a major role in determining what gets eaten and enabling cheap food, the costs of which are borne in decreased sustainability of land and water. Nevertheless, corporations are reluctant to change their accounting rules, unless forced by the state with new legal frameworks, which has done little:

> When it comes to the environment they [the accounting industry] do not take into [account] information that is material. Information is material if *omitting it* or *misstating it* could influence decisions that users make on the basis of financial information about a specific reporting entity [emphasis in original]. (Allan 2019, p. 17)

Hegemony over how water is governed is supported by these various mechanisms that favor agribusinesses (Sojamo et al. 2012). This argument demonstrates that power does not have to be extended over physical rivers or groundwater bodies. Virtual water hegemony influences blue, green, and gray water accounts at the global and national levels. Power over decision-making concerning blue and green water access and allocation is distributed unequally between actors vested in managing food.

Virtual water hegemony and the power structures in place are a result of historical processes in which states incentivized and regulated trade. The so-called "first food regime" is represented by the British colonial strategies during the late nineteenth century. They shipped grain, sugar, cotton, and rubber from their territories around the world. The "second food regime" centered in US policy after World War II and enabled agricultural subsidies and the export of surplus grains to a food-hungry market in developing economies. Through these regimes, the United States and European countries secured their role in the food supply chain. Agribusinesses have extended their geographical reach over time. The operations of the private sector are an extension of how Western countries have managed control of the food supply chain (Keulertz and Woertz 2016). The history and the broader structure of the food trade forged a close relationship between Western states and agribusinesses, a partnership enabling virtual water hegemony.

In recent decades, the power hierarchy in the food supply chain has changed over time, and key food exporters have started to shift. Between 1986 and 2011, the volume of food trade and consequently virtual water "trade" grew approximately threefold (D'Odorico et al. 2019). These flows are enhanced partially by new exporters. A notable example is Brazil, competing with the United States as the top exporter of soy since the late 1980s. Since the middle of the 2000s, China's expanding role as a food importer has enabled countries in South America and Southeast Asia to become active exporters (D'Odorico et al. 2019). Further, the food price volatility of 2008 resulted in emerging economies being less willing to rely on the existing food regime and instead make their own agricultural investments abroad. In other words, large-scale land acquisitions, examined in the previous section, are part of a strategy by emerging economies to challenge virtual water hegemony (Sojamo et al. 2012). Major players are no longer limited to those in the established food regimes of the last century.

In addition to China and South Korea, the Gulf countries of Saudi Arabia, Qatar, United Arab Emirates, and Kuwait have been active in investments. Typical of the land grab logic, investments of the Gulf countries have aimed to provide food and reduce pressure on domestic water supply with projects in Sudan, Pakistan, Philippines, Ethiopia, and Egypt as well as regions in East Africa. The agricultural sector in countries like Saudi Arabia has depleted its water supply due to livestock production. The new investment strategy of the Gulf states replaces the need to secure significant volumes of water to grow feedstock domestically. Hence, they import crops such as alfalfa and barley. These investments were made in "geographically and politically close" countries, suggesting these factors expedite projects (Keulertz and Woertz 2016, p. 38). In reality, implementation has been slow and at a much smaller scale than originally envisioned. Nevertheless, this example shows how Western virtual water hegemony is not set in stone, and the combination of financial resources and political leverage can generate alternative ways of securing water and food.

The challenges to virtual water hegemony may shift the control of the food trade, including how and where water is used for agriculture. However, as long as states continue to provide subsidies and investors operate according to existing accounting rules, the issues of over-abstraction of water, pollution, soil degradation, and impacts on ecosystems will remain unaddressed. These problems manifest as a result of technology, infrastructure, institutions, and discourses on abundance and scarcity, which we will turn to in the next section.

Closed Basins, Fragmented Rivers, and Empty Aquifers

Green water resources benefit from being able to resist the greed of society. Farmers cannot consume last year's rainfall. Irrigators on the other hand can consume water stored from previous wet seasons and 'borrow' from future years. And they do. (Allan 2019, p. 6)

The over-abstraction of blue water sources results in rivers and aquifers running dry. Many large rivers in the world are facing water scarcity. Major basins such as those of the Indus, Yellow, and Colorado rivers rank among the most water-stressed in the world (Gassert et al. 2013). Large transboundary rivers which run through multiple countries are particularly notable because the effects of over-abstraction are felt across borders and multiple communities. Rivers such as the Jordan, Nile, and Tigris-Euphrates are well-studied basins that demonstrate the challenges of addressing water use in an arid climate. Drilling into the sub-basin level in these transboundary rivers reveals an even starker picture of unsustainable water use. Calculations show that 1.6 billion people or approximately 64% of people living within transboundary river and lake basins experience severe water scarcity for at least one month during a year (Degefu et al. 2018). In situations of severe water scarcity, the monthly blue water footprint exceeds 40% of natural runoff (Hoekstra et al. 2012). The impact on environmental health is evident in these cases as approximately 20–50% of the mean annual river flow is needed to maintain ecosystem integrity (Smakhtin et al. 2004). Furthermore, there are regions that experience severe water scarcity in all four seasons, such as the sub-basins of the Nile in Egypt and Eritrea, of the Indus in Pakistan, and of the Tigris-Euphrates/Shatt al Arab in Jordan (Degefu et al. 2018).

Groundwater depletion is also a serious cause of blue water over-abstraction in agriculture (Figure 5.8). Groundwater aquifers are recharged in the hydrological cycle, but out of the 37 largest aquifers in the world, 21 face depletion rates faster than recharge. The Arabian Aquifer is likely to experience no recharge due to abstraction for irrigation (Richey et al. 2015). Surface water accounts for 60% of the water used for irrigation with the remaining 40% relying on groundwater. Some argue that these figures grossly underestimate the role of groundwater because the data does not include the baseflow to which groundwater seepage contributes. Consequently, a more accurate figure would be that 70% of irrigation is dependent on groundwater use, as indirect groundwater contributes to approximately half of the surface water used (Wood and Cherry 2021). The combined use of surface and groundwater for irrigation has led to a situation where 68% of the world faces blue water scarcity

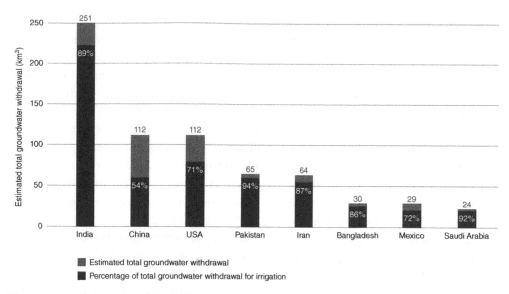

Figure 5.8 Estimation of groundwater withdrawals for irrigation. Source: UN-Water (2022, p. 50) / United Nations / CC BY-SA 3.0 IGO.

for one month of a year. This amounts to 0.10–0.15 billion ha of agricultural land, equivalent to the combined size of Chile and Zambia. Areas that experience blue water scarcity for five months of a year make up 37% of all irrigated cropland (Rosa et al. 2020).

Extensive irrigation has contributed to a human-induced process of **basin closure**. Basin closure occurs when there is insufficient water flow in a river to perform vital functions such as flushing out sediment, diluting pollution, and managing salinity intrusion in coastal and delta areas. Basin closure also indicates poor water quality due to pollution. In these basins, water use is overcommitted such that runoff is too depleted for river and coastal ecosystems to be maintained throughout the year. Many basins, such as those of the Yellow, Colorado, Indus, and Murray-Darling rivers, face basin closure (Molle et al. 2010). The Jordan and the Amu Darya and Syr Darya are closed basins; they no longer replenish the Dead Sea and the Aral Sea respectively (Molle et al. 2007). Basin closure is not restricted to arid regions because basins in water-abundant regions can still have low river discharge, particularly in the dry season, resulting in partial or complete closure during a year. Furthermore, there can be cases where the larger basin is open but sub-basins face closure, rendering inequalities of water access within a basin (Molle et al. 2010).

The material properties of the river and aquifers shape the way people have managed water by storing and pumping in strategic locations within the basin. The materiality of water shaped a waterscape of large and small infrastructure and the technologies, policies, and institutions to manage them. Basin closure has led to a situation of both physical water scarcity and constructed scarcity. Water infrastructure and the management paradigms sustaining them bring about water scarcity. They also create water scarcity through a discourse of controlling and managing nature. Basin closure is the result of the **hydraulic mission**, a management paradigm in which state-led control of water resources is consolidated through the extensive use of infrastructure and institutions for water supply. Through centralized systems of irrigation projects and agricultural expansion initiatives, investments are made for infrastructure that will enable the storage, transfer, and diversion of water.

Dams – a topic we explore in more depth in Chapter 6 – have been core to the hydraulic mission, which mobilizes significant capital and engineering expertise. Dams, as Allan suggests in the opening quote to this section, allow irrigators to control blue water flows. Iconic dams also represent the power and prestige of the state. For example, the Aswan High Dam on the Nile River symbolizes Egypt's control over the flow of the Nile waters to neighboring upstream countries.

Large dams built as part of the hydraulic mission are often hydropower dams with multiple functions, including irrigation. For example, the Akosombo Dam in Ghana was designed to generate power for the aluminum industry. The construction created the largest artificial lake in the world, Lake Volta, and irrigation capacity was extended below the dam to produce water-intensive crops such as rice (Barry et al. 2005). The hydraulic mission does not simply increase irrigation capacity but also secures the state's role in controlling nature (Chapter 6). Hydrosocial relations are born through techno-politics and socio-material impacts.

The discourse of controlling nature and tackling water scarcity is exemplified through wells and pumps for groundwater abstraction. The hydraulic mission institutes a discourse on the productive use of land and water. Unlike megaprojects, pumps and wells are small scale but numerous and dispersed across large tracts of land. In the case of Mexico, the hydraulic mission rapidly advanced in the 1920s, paving the way for 3 million ha of land irrigated with groundwater, largely in the central and northern regions of the country. However, this expansion has been at the expense of aquifer depletion and pollution where 105 aquifers out of 653 are deemed over-abstracted. Groundwater use facilitated by the state promoted pumping for irrigation and provided subsidies. This use was spurred by close ties between the state and the private sector to enable an extension of commercial farming. Led by the state, large commercial producers sprung up to supply domestic and export markets for vegetables, cotton, and fodder crops for dairy production (Hoogesteger 2018). Large land ownership of these commercial producers enabled unfettered groundwater access. Groundwater-fed irrigation was "a means to increase agricultural production and make inhospitable arid lands 'verdant and prosperous'" (Hoogesteger 2018, p. 558). In this discourse, water and land are merely inputs to the economy, and little value is given to environmental capital.

Despite water and land being indispensable resources for agriculture, the environment is the biggest "loser" at the end of the hydraulic mission. Rivers are now fragmented, and flows are altered as a result of dams being built. Dams affect up to 50% of the river volume. Further construction of large dams (estimated to tally 3700) could result in over 90% of the river volume being altered (Barbarossa et al. 2020). Out of the 91 rivers that are longer than 1000 km, only 21 rivers flow directly to the sea (Grill et al. 2019).

The loss of such free-flowing rivers contributes to the decline of the health of rivers, loss of biodiversity, and reduction of ecosystem services, which in turn could affect the livelihoods of communities. The current 40 000 or so existing large dams have impacted the habitat of freshwater fish significantly in the United States, Europe, South Africa, India, and China. If the 3700 large dams being planned or under construction are completed in the future, fish habitats will be fragmented, especially in the tropics. In these cases, medium to small species will be the most affected, which is problematic as they make up more than half of the overall freshwater fish diversity (Barbarossa et al. 2020). While there may be regions that benefit from the production of crops for food, there are trade-offs with alternative food sources and livelihoods that rely on fish. The hydraulic mission cumulates in increased blue and green water use to the extent that supply management is no longer tenable, with trade-offs on other uses and a broken ecosystem.

More Crop Per Drop, More Dollar Per Drop

When basins are closed and water supply is limited, the notion of "more crop per drop" is often borrowed in water management discourses to incentivize the maximization of productivity per unit of water. This entails improving crop water productivity or recycling water so that water is not "lost" downstream. In many regions and countries, that is still not enough and there is a need to realize "more dollar per drop." This motto highlights the economic value of water use and prompts agricultural production to shift to higher value items. There is an implied shift toward giving less priority to the agricultural sector in allocating water. The reasoning is that the agricultural sector provides a low return for the volume of water used. "More dollar per drop" justifies the allocation to industries and urban sectors, for example in the form of **rural–urban water transfer**.

In central Mexico, the El Realito dam was constructed in 2008 to provide the city of San Luis Potosí with a third of its urban water demand as groundwater sources dwindled. The 132 km pipeline running from the dam to the city is a reminder of how the materiality of water shapes water supply solutions. But this materiality shapes and is shaped by discourses that produce water as well as differences between the urban and rural communities. These discourses center around the idea of dams as being a "good" solution to a water "crisis" which could be addressed by "backward" rural communities delivering on their moral duty to support the water transfer. These discourses are constructed by the government's National Water Commission, and the urban inhabitants widely support this discourse based on ideas about the market and efficiency, which allow industries and urban sectors to thrive (Hommes et al. 2020).

Other examples of agricultural-to-urban water transfers are facilitated through the trade of private water rights. Such market mechanisms have been used in places like the United States, Australia, and Chile where major issues of water scarcity exist as well as institutions to allocate rights (Chapter 4). In other cases, the state can reallocate water using permits and quotas, as experienced in places like India and China (Hooper 2015). These interventions are aimed to "optimize" or to provide equity to different users. "More dollar per drop" reprioritizes water allocation for cities, justified on the grounds that shortages to supply in urban areas can affect significant numbers of people compared to the rural sector.

However, such reallocation and transfers are often met with controversy. Transfers are strongly resisted by farming communities, who argue cities are "appropriating" water. Conflicts emerge as historical water rights are ignored, and rural communities face a sense of loss (Hooper 2015). In the case of the El Realito dam, the rural communities reject the dominant discourse and resist the "subjectivity of rural inhabitants as complying citizens that hold back their own necessities for the sake of the greater (urban) good" (Hommes et al. 2020, p. 406). Faced with a situation of insufficient water quantity, the rural communities have attempted to come up with an alternative water supply – low-tech rainwater harvesting – as an act of autonomy and self-determination (Hommes et al. 2020).

These contestations debunk the idea that the agricultural sector is not efficient and that farmers are wasteful. It is a myth that farmers waste a lot of water, because they do not actually dictate the water allocated to their fields and their agricultural practices constantly adapt to conditions of water scarcity through the kinds of crops and cropping calendars they use. This is especially true of smallholders who are at the top end of the "hourglass" market and those who practice subsistence farming. The reliance on external hydrological and climatic conditions inherently makes farmers seek efficiency to fit their context. Deeming the agricultural sector as having low productivity is fraught with challenges as water is not the sole factor that provides value-added to a sector. For example, comparing the number of jobs generated through the same quantity of water use between agricultural and urban sectors provides crude indications (Molle and Berkoff 2009).

Could we reallocate water flows to support the environment – instead of only the agricultural sector? **Environmental flows**, or e-flows, are techniques of river basin planning and management designed to aid the recovery and maintenance of ecosystem health (Poff et al. 1997). For example, in Southern Africa, e-flows have been an important consideration to restore flow and allow priority allocation to the environment. This is particularly important when many of the rivers are transboundary and countries sharing the river are at various stages of river basin development. E-flows have been put in place in legislation and policy, notably in South Africa, where many of the major river basin development projects have taken place. The implementation of e-flows is nevertheless challenging as it requires institutional and policy change, and capacity building of local users (Müller et al. 2022).

"Decoupling," or disassociation, of local water consumption from the existing blue water available requires a paradigm shift: from the hydraulic mission to reflexive modernity in which institutions are set up to consider more efficient and ecological approaches to managing water demand (Gilmont 2014). Reallocation is a very political process and not as invisible as the virtual water "trade." New legislation, regulatory frameworks, and basin development plans need to be debated and enforced. Moreover, reallocation results in only minimal volumes of water being decoupled. In the case of Israel, 500 m³/year of water was decoupled through policies and regulations, which only amounts to 1/16 of what was possible with virtual water "trade" (Gilmont 2014).[3]

"More crop per drop" and "more dollar per drop" approaches only go so far in addressing sustainable water use unless underlying structures related to land, labor, and food are addressed. There are serious challenges and tensions around valuing water as a vital resource in sustaining life – whether as environmental capital for large agribusinesses or as inherent in identity, knowledge, and worldviews. The production of water inhibits or discourages water users from being good water stewards. With pressures to produce cheap food, farmers are hindered from adopting farming and water stewardship practices that would redefine ecological relations. Rural production reveals gendered impacts which prohibit a level playing field for valuing water. Case studies from India indicate how decision-making at the farm level is made by men, leaving a minimal role for women and their autonomy restricted to work considered menial. The difficulties of owning land and acquiring land titles further marginalize them, despite their substantial labor burdens (Pattnaik and Lahiri-Dutt 2020). Decisions on the value of water are not possible with only partial, gendered participation as they risk missing out on the knowledge of women who manage water and land. Water is life and the ontological politics reveal struggles compounded by socio-material aspects of water management and struggles relating to power relations of land.

Summary and What's Next

Food and agriculture consume up to 72% of global water use. This chapter examined the hydrosocial relations that shape – and are shaped by – water and land. The major food systems in place which sustain virtual water hegemony and cheap food produce water scarcity and uneven power relations between states, farmers, consumers, and agribusinesses. The spatiality of water reveals ecological relations that endanger ecosystem health and water bodies. A critical perspective shows that seeking efficiency is underpinned by discourses and reflects how social groups construct water.

Dams, as we pointed out, allow us to "trap" and control blue water flows for irrigation and supply. Dams are highly controversial and influential infrastructures of the global hydrosocial cycle, a topic we turn to in the next chapter. We examine the roots and controversies of large dams and their role in the building of nation-states, exploring the imaginative geography of development in the process of modernization. As perhaps the most iconic symbol of twentieth century techno-modernity, we ask: can dams really save us?

---

[3] 1 m³ is equivalent to 1000 liters.

Further Reading

Green water management

Aldaya, M., Allan, J.A., and Hoekstra, A.Y. (2010). Strategic Importance of Green Water in International Crop Trade. *Environmental Economics* 69 (4): 887–894.

Falkenmark, M. (2020). Water Resilience and Human Life Support – Global Outlook for the Next Half Century. *International Journal of Water Resources Development* 36 (2–3): 377–396. https://doi.org/10.1080/07900627.2019.1693983.

Wani, S.P., Rockström, J., and Oweis, T.Y. (2009). *Rainfed Agriculture: Unlocking the Potential*. Wallingford, UK: CAB International.

Land and water grabs

Adams, E.A., Kuusaana, E.D., Ahmed, A., and Campion, B.B. (2019). Land Dispossessions and Water Appropriations: Political Ecology of Land and Water Grabs in Ghana. *Land Use Policy* 87: 104068.

Birkenholtz, T. (2016). Dispossessing Irrigators: Water Grabbing, Supply-Side Growth and Farmer Resistance in India. *Geoforum* 69: 94–105.

Borras Jr., S.M. and Franco, J.C. (2013). Global Land Grabbing and Political Reactions 'From Below'. *Third World Quarterly* 34 (9): 1723–1747. https://doi.org/10.1080/01436597.2013.843845.

Mehta, L., Veldwisch, G.J., and Franco, J. (2012). Introduction to the Special Issue: Water Grabbing? Focus on the (Re)appropriation of Finite Water Resources. *Water Alternatives* 5 (2): 193–207.

Water–food–energy nexus

Allouche, J., Middleton, C., and Gyawali, D. (2019). *The Water–Food–Energy Nexus: Power, Politics and Justice*. London and New York: Routledge.

Bruns, A., Meisch, S., Ahmed, A., Meissner, R., and Romero-Lankao, P. (2022). Nexus Disrupted: Lived Realities and the Water-Energy-Food Nexus from an Infrastructure Perspective. *Geoforum* 133 (1): 79–88. https://doi.org/10.1016/j.geoforum.2022.05.007.

D'Odorico, P., Davis, K.F., Rosa, L., Carr, J.A., Chiarelli, D., Dell'Angelo, J., Gephart, J., MacDonald, G.K., Seekell, D.A., Suweis, S., et al. (2018). The Global Food-Energy-Water Nexus. *Reviews of Geophysics* 56 (3): 456–531. https://doi.org/10.1029/2017RG000591.

Lebel, L. and Lebel, B. (2018). Nexus Narratives and Resource Insecurities in the Mekong Region. *Environmental Science & Policy* 90: 164–172. https://doi.org/10.1016/j.envsci.2017.08.015.

Critical perspectives on irrigation efficiency and productivity

Boelens, R. and Vos, J. (2012). The Danger of Naturalizing Water Policy Concepts: Water Productivity and Efficiency Discourses from Field Irrigation to Virtual Water Trade. *Agricultural Water Management* 108: 16–26.

Grafton, R.Q., Williams, J., Perry, C.J., Molle, F., Ringler, C., Steduto, P., Udall, B., Wheeler, S.A., Wang, Y., Garrick, D., et al. (2018). The Paradox of Irrigation Efficiency. *Science* 80 (361): 748–750.

Mdee, A. and Harrison, E. (2019). Critical Governance Problems for Farmer-Led Irrigation: Isomorphic Mimicry and Capability Traps. *Water Alternatives* 12 (1): 30–45.

Rural-urban water transfers

Celio, M., Scott, C.A., and Giordano, M. (2010). Urban–Agricultural Water Appropriation: The Hyderabad, India Case. *Geographical Journal* 176: 39–57.

Garrick, D., De Stefano, L., Yu, W., Jorgensen, I., O'Donnell, E., Turley, L., Aguilar-Barajas, I., Dai, X., de Souza Leão, R., Punjabi, B., et al. (2019). Rural Water for Thirsty Cities: A Systematic Review of Water Reallocation from Rural to Urban Regions. *Environmental Research. Letters* 14 (4): 043003.

Komakech, H.C., van der Zaag, P., and van Koppen, B. (2012). The Last Will Be First: Water Transfers from Agriculture to Cities in the Pangani River Basin, *Tanzania*. *Water Alternatives* 5 (3): 700–720.

Liptrot, T. and Hussein, H. (2020). Between Regulation and Targeted Expropriation: Rural-to-urban Groundwater Reallocation in Jordan. *Water Alternatives* 13 (3): 864–885.

Environmental flows

Acerman, M. (2016). Environmental flows – Basics for Novices. *WIREs Water* 3 (5): 622–628. https://doi.org/10.1002/wat2.1160.

Allan, C. and Watts, R.J. (2018). Revealing Adaptive Management of Environmental Flows. *Environmental Management* 61: 520–533.

Arthington, A. (2012). *Environmental Flows: Saving Rivers in the Third Millennium*. Berkeley and Los Angeles, CA: University of California Press.

Poff, N.L. and Matthews, J.H. (2013). Environmental Flows in the Anthropocence: Past Progress and Future Prospects. *Current Opinion in Environmental Sustainability* 5 (6): 667–675.

Chapter 6

Dam Fever

A Fever Spreads

On a crisp autumn day in October 2011, an engineer prepares his dynamite. "Fire in the hole!" he yells, and a blast rips through the base of the Condit Dam on the White Salmon River, a tributary of the Columbia River in Washington state, USA.[1] A plume shoots out, spewing gray debris and black sediment, long trapped behind a 125-foot-tall wall of concrete (Figure 6.1). Water pulses down the river. In the weeks leading up to the explosion, fish biologists had carefully removed as many steelhead and salmon as possible, returning them to the river once the sediment pulse had flushed and settled. Whitewater kayakers cheered and onlookers gaped at the sight of the free-flowing river. Since 1913, the Condit Dam had stood in the way of these glacier-fed flows, feeding electricity to regional customers while blocking the passage of anadromous fish.

If dams are so important, why are we blowing them up? The US Pacific Northwest is a region renowned for its big dams – the Grand Coulee, the Bonneville, and the Dalles on the Columbia River, to name a few (White 1996). Even more dams punctuate the upper reaches of the basin, into Canada, and along the Snake River in Idaho. Some are managed by the US Army Corps of Engineers, others by the US Bureau of

[1] For excellent video footage of the event, refer to "Condit Dam Removal Explained" (a short film by Andy Maser) on Vimeo: https://vimeo.com/33584271 (accessed 13 July 2022).

Figure 6.1 Removal of the Condit Dam on the White Salmon River, WA (USA). Note the plume of debris and dark-colored sediment. Source: OPB.

Reclamation, while others – like the Condit – are privately owned and operated, in this case by PacifiCorps, a regional electricity company serving over 1.6 million customers. In 1991, the aging Condit Dam failed its federal license review. Costs to repair the dam far outstripped its hydroelectricity production capacity. At the same time, stricter protocols to enhance fish passage and protect endangered salmon drove maintenance costs up. Meanwhile, the Confederated Tribes and Bands of the Yakima Nation, together with local environmental groups, negotiated a settlement with PacifiCorps, and the company started the removal process (Fox et al. 2022). Together, these factors combined to produce a landscape of **strategic opportunism** that paved the way for dam removal (Magilligan et al. 2016) – a notable reversal of the dambuilding fever that had gripped the Pacific Northwest for nearly a century. Twenty years later, the Condit came down.

This chapter uses dams – and the rivers, habitats, people, and landscapes they shape – to explore some of the central arguments of this book. Dams are more than just concrete and sand. Dams, we argue, are the material manifestation of social, political, and ecological relations. They are some of the most influential and significant features in the global hydrosocial cycle. Dams reflect and shape relations, and thus the politics of dams – whether building them up or tearing them down – are thick with controversy and complexity. For their admirers, dams represent the power to transform a river into a stream of benefits for economic and social development. These benefits include cheap, "clean" renewable electricity, flood control, and the expansion of irrigation. For detractors, dams are a perfect example of engineering hubris run amok. They represent the impossible and ultimately dangerous fantasy that humans can control nature without adverse effects. Critics of dams point to destroyed ecosystems, depleted fish stocks and aquatic species decline, increased vulnerability to large floods, displaced communities, expensive and inefficient use of resources in the construction phase, and once-fertile floodplains depleted of their regular replenishment.

Dams incite a kind of "fever" in people. Dams are symbols of modernity, progress, and development in a way that few other infrastructures can be. This is because dams are central to issues of water control and distribution. They are our most spectacular technology for interacting with rivers (Doyle 2012). Water is life, as we argue throughout this book, and this means the water that dams block, channel, and control is really about matters of politics and power – about controlling life itself (Sneddon et al. 2021). In this chapter, we will not "cover" all the types and technical aspects of dam infrastructure. Instead, our goal is to cultivate a critical point of view on the "fever" that grips dams and river development in the first place.

Dam construction is often driven more by symbolic and spectacle value than any projected economic benefits (Box 6.1). The power of this symbolic value can be understood as **dam fever**, a strongly held faith in the power of dams to tame nature, catalyze economic development, and move one step closer to a fully modern future. Despite the removal of relatively small dams like the Condit, we remain in the throes of a worldwide fever for building *large* dams, a fever that has been running wild for at least a century. By realizing dams are neither an objective fact nor a necessity of economic development or political progress – and by acknowledging the hydrosocial thesis that dams are relational and contextual – perhaps we can devise a cure.

Box 6.1   What Is a Dam?

The simple answer is that a dam is any structure built across a body of water to raise its level for the purpose of controlling the water flow. Building a large modern dam is far from simple: dams require huge amounts of finance, expertise, and labor (Figure 6.2). Topographical surveys, geological and geotechnical investigations, hydrologic and hydraulic studies, environmental surveys, and complex financial instruments are all required before initiating work on the actual structure. The design of the dam needs to account for multiple factors: the types of materials used, the height and positioning of spillways, the shape and features of the artificial reservoir created by the dam, and how the river will be diverted to enable the construction process. Dams can be categorized by their primary function (storage, diversion, or detention), by their construction material (stone, concrete, earth-fill, or rock-fill), or by their structural design (arched or buttressed). Indeed, there is some controversy around whether a particular water control structure can even be called a dam: for example, in the case "run-of-the-river" dams that allow water to flow for the purpose of generating hydroelectricity (Akhter 2019). The choices of how to design a dam, what material to use, how to design its arch, and where to locate it depend on factors such as topography, geological conditions, seismicity, and the availability of construction materials. We now have a better idea of what a dam is, but pardon our language, why should we give a damn about dams?

Figure 6.2 Dam construction in Nepal. This photograph depicts workers in the powerhouse tunnel of the Upper Tamakoshi Project in Gonggar, Dolakha, Nepal in 2013. The Eastern Himalaya region, including India and Nepal, has witnessed a massive growth in dam construction and hydropower development (Gergan 2020; Lord et al. 2020). Source: Austin Lord, reproduced with permission.

Dams as Development

Dams are one of the most spectacular and controversial technologies of modern development. They also serve as a focal point for a host of modern debates around nature, development, and knowledge. For most of the twentieth century, the global dam-building industry was rooted firmly in the United States and Europe (Doyle 2012). Over the twentieth century, a staggering 45 000 large dams were built. Today, more than 3700 major large dams – with a capacity greater than 1 megawatt (MW) of electricity – are either planned or under construction, mostly in Asia, Africa, and Latin America. Together, these dams are expected to augment global hydroelectricity capacity by 73% to about 1700 gigawatts (GW) (Zarfl et al. 2015) (Table 6.1). As of 2021, the International Commission on Large Dams counts more than 58 000 dams worldwide.

Why do we build so many dams? To answer this question, we need to return to one of the central tenets of this book: scarcity is produced. For their advocates, dams and associated river infrastructures are a response to perceived or projected water scarcity. Temporally, dams store water from periods of excess so that they might be used in times of scarcity. Spatially, canals divert water areas of excess to supply areas of need. Immediately, these dynamics prompt deeper questions about the social and political

Table 6.1 World's Largest Dams by Electric Power Generation Capacity. Source: Compiled from data from the International Commission on Large Dams, World Register of Dams.

Dam	Hydroelectric capacity (GW)	Country
Three Gorges (Yangtze River)	22.5	China
Itaipu (Paraná River)	14	Paraguay/Brazil
Xiluodo (Jinsha River)	13.9	China
Belo Monte (Xingu River)	11.2	Brazil
Guri (Caroní River)	10.2	Venezuela
Tucuri (Tocantins River)	8.4	Brazil
Grand Coulee (Columbia River)	6.8	USA
Longtan Dam (Honshui River)	6.4	China
Xiangjiaba (Jinsha River)	6.4	China
Sayano-Shushenskaya (Sayanogorsk River)	6.4	Russia

conditions that produce scarcity. What is the impounded water used for, and would we consider those uses just or conducive for long-term sustainability? Why are we growing water-needy crops in arid environments? As we explore in Chapter 5, what are the trade-offs of "eating" dammed water? Is it rational to speak of water scarcity in desert cities, such as Phoenix or Lahore, when much of the impounded and imported water maintains amenity landscapes like private lawns, gardens, and golf courses?

Scarcity is produced – as we explain in Chapter 1 – and water scarcity is often produced and represented such that technological solutions like dams seem "natural" and logical. But there is also a more powerful idea animating dam construction than just responding to situations of water scarcity. This powerful idea has moved millions of people to attempt to take control of their destiny by controlling nature and super-charging the economy – the idea of **development**.

To understand the role of dams in social and economic development, we must go back almost a century. In the 1930s, the global Depression forced a reimagining of the role of the state in relation to the economy. Nowhere was this clearer than in the United States, the world's most powerful economy. Franklin Delano Roosevelt (FDR) served as President of the United States from 1933 to 1945, a period coinciding with the Depression and World War II. To mitigate the impact of the greatest collapse the capitalist market system had ever seen, FDR pushed the US federal government into direct forms of economic intervention. He ordered the government to (i) build roads, highways, and public buildings; (ii) sponsor theater, art, and photography projects; and (iii) expand public service provision in parks, rural water infrastructure, and public lands (Melosi 2011). One of the most spectacular ways that FDR increased the role of the federal government was through the finance and construction of large dams.

Run-of-the-river dams were built in the United States in the nineteenth century, for example, to power textile mills in New England (Magilligan et al. 2016). But it wasn't until the 1930s that large dams began to transform the landscape. This was especially true in the US West, where the control of rivers played a huge part in the white settlement, colonization, and seizure of land from Indigenous peoples already living there (Curley 2021a; Worster 1992). These dams were bigger and more ambitious than anything that had been attempted up to this point: a mixture of water storage, electricity generation, and urban growth machine. While the British Empire had been diverting and draining rivers in India for decades (Akhter and Ormerod 2015), large dam construction in the United States in the mid-twentieth century marked a qualitative shift in the way dams were imagined by planners and politicians across the world. Dams shaped and mirrored crucial aspects of the way the state engaged with the economy, especially in terms of mobilizing and regulating labor, popular culture, and a tradition of infrastructure planning for regions considered to be underdeveloped.

As the US empire expanded westward in the nineteenth and early twentieth centuries, settlers encountered a radically different ecology. Instead of the lush forests and gurgling rivers of the US South and Northeast, the vast area west of the 100th meridian was arid, receiving on average less than 20 inches of rainfall in a year. Several rivers came to be the subject of intense intervention efforts in this mostly arid landscape: the Columbia River in the US Pacific Northwest, the San Joaquin River in Central California, and the most prominent of them all, the mighty **Colorado River** (Figure 6.3).

The Colorado River runs through seven US states and two states of Mexico, draining a considerable portion of the US Southwest and ending in the Gulf of California. The river is central to the imaginative geographies of the West, vividly captured by the reports of scientist and settler-explorer John Wesley Powell (1834–1902): deep and dramatic canyons, ferocious rapids, soaring peaks, rich natural resources, and breathtaking vistas of wilderness (Worster 1992). While the Colorado River inspires these imaginaries, it also reflects dreams of westward settler expansion and development – and with it, the realities of exploitation, hegemony, and power. Dams became the technological marvels that promised to translate the raw ethereal beauty of this desert landscape into a hydroelectric-powered economic and social transformation.

The first major dam of this era was the **Hoover Dam** on the Colorado River (originally called the Boulder Canyon Project), located on the border between the US states of Nevada and Arizona. When construction started in 1931, Hoover Dam was an audacious undertaking; never had so much concrete been used in a single project. Concrete required human labor. Constructed during the Depression, desperate labor was not in short supply. Indeed, over 20 000 laborers worked on the dam during the six years of its rapid construction.

Dam workers faced harsh labor conditions. Summers were broiling; winters were freezing. Workers toiled on the job seven days a week, and faced dangers such as dehydration, carbon monoxide poisoning, heat exhaustion, and electrocution. When management demoted some workers to lower-paying jobs in the summer of August 1931, the dam workers decided to strike. Management reacted poorly, refusing to meet worker demands regarding pay and safe working conditions. They vowed to blacklist striking workers. Eventually the harsh economic climate of the Depression – and the need to feed themselves by any means necessary – drove the strikers back to work, with no concessions to show for their struggle (Reisner 1993; Worster 1992).

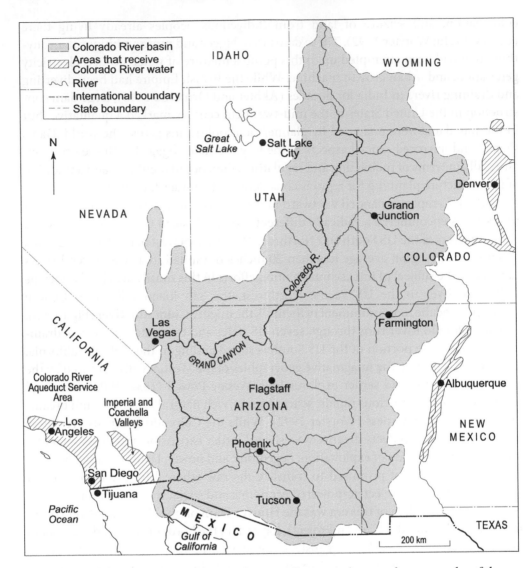

Figure 6.3 The Colorado River basin. This map illustrates the complex geography of the Colorado River watershed and its beneficiaries, a geography shaped by hydrogeology, politics, and national and international law. Source: The Authors, cartography by Philip Stickler.

Not everyone had the "privilege" of working in the dangerous and hostile environment at Hoover Dam. The company building the dam was forbidden by law to hire Asian workers. Only a handful of Black workers were hired to work the dam, and these positions were marginal roles, with little chance of acquiring advanced skills or earning promotions. Most scandalous of all was the near-complete lack of opportunity for the Indigenous or Native American populations of the region, from whom the land had been seized less than a century earlier. Although some Native Americans were hired as high scalers – on the racist assumption that they were naturally suited to high altitudes – most of the employment, electrification, and flood control benefits of the project went to white settlers and the growing cities of the US West.

The Hoover Dam helped change the political and physical geography of the Colorado River basin and with it, the American West (Worster 1992). As Figure 6.3 illustrates, the map of Colorado River system far exceeds its natural basin boundaries. Impounded water is now pumped outside basin boundaries to feed several major urban areas (the Albuquerque, Denver, Los Angeles, San Diego, and Tijuana metropolitan regions) and agricultural zones (the Imperial and Coachella districts). Within the basin, major cities such as Las Vegas, Phoenix, and Tucson – some of the fastest-growing cities in the United States – rely centrally on Colorado water. The desert may be a place of "natural" aridity, but dam-related water infrastructure made it a site of abundance, expansion, and growth.

The Hoover Dam was a key mechanism for expanding US empire in the western states. But the dam project with the most widespread global influence was the **Tennessee Valley Authority** (TVA) project, located on the Tennessee River in the southeastern United States (Figure 6.4). Passed by the federal government in 1931, the TVA Act created and tasked a new agency (the TVA) with maintaining and expanding the navigability of the Tennessee River, managing flood control, and operating the hydroelectric plant at Wilson Dam. Perhaps most important, the Act also empowered the TVA to develop the agricultural, commercial, and industrial prospects of the valley (Klingensmith 2003). As such, the Act explicitly tied regional

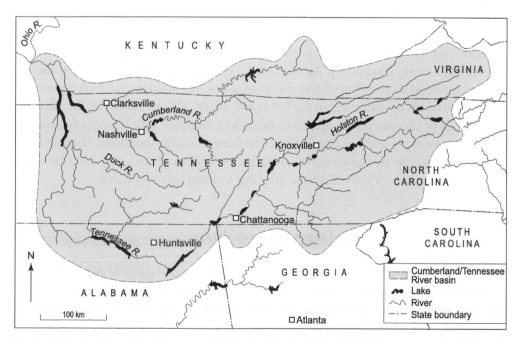

Figure 6.4　Region of the Tennessee Valley Authority, USA. The TVA project was a federal effort to "modernize" the US South, using dams as tools of development and rivers as their lubricant. Source: The Authors, cartography by Philip Stickler.

development and planning to the rational and deliberate development of a river through a networked cascade of dams.

The TVA project offered the stuff of planners' dreams: a scientific, rational, and integrated way to exploit the natural bounty of a region and solve problems of chronic underdevelopment in the US South. Unlike the US West, which the federal government effectively viewed as a *tabula rasa* for development, the government had conceptualized the US South in terms of economic and cultural "backwardness" (Domosh 2015; Klingensmith 2003). The TVA was intended not only to develop economic resources but as a "democracy-building" project to integrate the South into the national fabric. David Lilienthal – the charismatic and powerful administrator who was the face of the TVA from 1941 to 1946 – wrote a book to describe the powerful political impacts of the large dam. The title was *TVA: Democracy on the March* (Lilienthal 1944). This excerpt from the preface of the book gives a sense of how Lilienthal saw river technology, development, and democracy as intimately related:

> I believe men may learn to work in harmony with the forces of nature, neither despoiling what God has given nor helpless to put them to use. I believe in the great potentialities for wellbeing of the machine and technology and science; and though they do hold a real threat of enslavement and frustration for the human spirit, I believe those dangers can be averted. I believe that through the practice of democracy the world of technology holds out the greatest opportunity in all history for the development of the individual, according to his own talents, aspirations, and willingness to carry the responsibilities of a free man. (Lilienthal 1944, p. xii)

Dams are not just concrete artifacts, they are a part of a complex mediation between technology, nature, and society. Reinforcing a key message of this text, dams are the material manifestations of social and spatial relations. In this vein, the TVA embodied a key lesson of US development in the 1930s and 1940s. This message was that dams – in the hands of visionary and competent men like Lilienthal – could not only catalyze the development process but would also contribute to nation-building. This message was received eagerly by the scores of countries in Africa and Asia that declared their independence in the 1950s and 1960s, sparking a global fever for dams.

Dams as Geopolitics

Even before World War II ended, the leaders of colonized regions of the world were planning their independence. By the middle of the twentieth century, most of Africa and Asia (with the notable exception of China) had been colonized by European powers. More than 4000 dams were built per year between the 1950s and 1970s – many of them in Asia (Ahlers et al. 2017, p. 537). Although political independence was achieved by many Asian and African countries during this period, the question of economic independence and autonomy was another matter. As the anti-colonial theorist Frantz

Fanon (1961) predicted in his book *The Wretched of the Earth*, many African and Asian states were unable to achieve true democratic development after independence due to misgovernance by elites. For several decades, large dams became the symbol of development for these postcolonial regimes to leverage hard-fought political independence into tangible economic development.

Rather than political independence ushering in a radical departure from the past, economic planning manifested in terms of continuities. David Ludden (2005) has referred to the way a state articulates and implements infrastructure and improvement projects as a **development regime**. Development regimes are not confined to specific governments or even states. They are ways of thinking and acting across political changes. Ludden observed that the postcolonial state in India continued, rather than interrupted, the process of infrastructure-led development implemented by the British Indian state in the nineteenth and early twentieth centuries. Agrarian development, especially in the arid regions of the subcontinent, sought to control and divert rivers through top-down technical reordering of the physical landscape.

No intervention exemplifies the ambitions and follies of a capitalist development regime better than the **TVA model**. Across not only Asia and Africa but also Latin America, dams symbolized the burning desires of politicians, economists, and even some ordinary people. For example, oral historian Alia Mosallam (2014) records how Egyptian construction workers felt nationalistic pride in building the Aswan Dam. The TVA model and legendary leader David Lilienthal were global celebrities. The TVA model, as it came to be known, entailed the application of a valley-wide technical plan that would exploit the river for hydroelectric power, control the river's pulses to offer flood protection, and store and divert river water for irrigation. In the TVA model, regional planning was carried out by an agency with significant autonomy from both local and federal governments (Klingensmith 2003).

By the early 1950s, more than 30 million people from around the world had visited the TVA, and Lilienthal's book had been translated into 14 languages (Baghel and Nüsser 2010, p. 236). Thus, a global circuit of engineers and representatives traveled to the US South to see how a dam could transform a periphery. In turn, TVA representatives toured many of the newly independent countries of Africa and Asia. But it would be wrong to think of the TVA model as a blueprint picked up from the valleys of Tennessee and applied to places like Ghana and India. Planners, engineers, and administrators adapted and reworked the TVA template in response to the local context (Box 6.2).

Box 6.2   Fighting for Uncle Sam

In the early twentieth century, dams entered mass culture and shaped how millions of people felt and thought about technology, nature, and national development. The Grand Coulee Dam on the Columbia River in the US Pacific Northwest inspired the legendary folk singer Woody Guthrie (Figure 6.5) to sing its praises. Construction on the Grand Coulee Dam began in 1933 and opened almost a decade later. The lyrics of Guthrie's song (recorded in 1941) demonstrate just how strongly the drive to industrialize and modernize had seeped into the popular culture. The lyrics testify to the power people ascribed to large dams for taming rivers and harnessing energy to transform human living conditions.

> *Now from Washington and Oregon*
> *You can hear the factories hum*
> *Making chrome and making manganese*
> *And white aluminum*
> *Now the roar of the Flying Fortress*
> *For to fight for Uncle Sam*
> *On the howling King Columbia*
> *By the big Grand Coulee Dam*

Figure 6.5 Folk singer Woody Guthrie (1912–1967), pictured in 1943. Source: Al Aumuller / Wikimedia Commons / Public domain.

For example, Klingensmith (2003) discusses the influence of the TVA model on a river development intervention in India from the 1950s: the Damodar Valley Corporation (DVC). Faced with opposition from state or devolved legislative assemblies, the DVC was never able to secure the regional autonomy of their operations, essential to the TVA model. In fact, the DVC's ambitious dam building program was scrapped when it became clear the important stakeholders in India's state legislative assemblies would not benefit directly. Instead of understanding the impact of development models in terms of their "successful" replication abroad, Klingensmith argues, we should understand development models as symbols of how technology could master nature if there was a state ready to wield that power. The TVA was not a blueprint, but rather "a highly effective signifier of development." The power of the TVA as a signifier lay in its ability to embody contradictory things to different people at the same time. Klingensmith (2003, p. 138) notes of the TVA model in India:

> It denoted work on behalf of the world's poor, and at the same time "American-style" affluence; the restoration of nature or alternatively, the human triumph over nature; popular control over public planning or, on the other hand, the triumph of the scientific expert and the manager over the politician; "creeping socialism" to some and the expansion of opportunities for capitalist growth to others; attention to the rural population which, in both India and American (US) nationalist discourses, were occasionally taken as the very core of the nation.

Chris Sneddon (2015) highlights the role of another US agency in global dam construction: the US Bureau of Reclamation (the Bureau). Sneddon traces how dam building expertise, embodied in the Bureau, was "exported" as Cold War geopolitical knowledge to foreign countries. Founded in 1902, the Bureau was tasked with water development, primarily in the arid region of the US West. After the successful construction of the Hoover Dam in 1935, the Bureau came into the national and global spotlight, much like the TVA. Indeed "while the TVA . . . became the global symbol of what river basin development might accomplish, it was the US Bureau of Reclamation that spearheaded the global dissemination of the ideal of multi-purpose development projects and basin-oriented planning through its international technical assistance programs" (Sneddon and Fox 2011, p. 454).

Building a large dam, and the associated program of river basin development, requires a vast amount of expertise – a central plank of this book. The Bureau, inspired by President Harry Truman's Point Four program of international aid assistance – and by their failure to overcome domestic politics to implement further river valley developments in the Colorado, Columbia, and Missouri river basins – looked to export this expertise abroad. The Bureau received a flurry of requests from national water agencies in the Middle East, Africa, and Asia throughout the Cold War period. The Bureau's warm response to these requests, and the visible presence of Bureau experts in river basins around the world, further cemented the role of the United States in global efforts at dam-led development (Sneddon 2015; Sneddon and Fox 2011).

As exemplified in the TVA, the purpose of large multipurpose dams was to speed up development in the world's formerly colonized countries. Modernization theory, as formalized by Walter Rostow (1960), was the belief that economic development occurred in stages and the newly independent countries would eventually "grow" to look like the societies of their former colonizers. Modernization theory and the stage-based understanding of history supplied the framework for justifying the argument of large dam supporters. Like other large infrastructures – power plants, railroads, ports, factories – large dams were meant to be strategic interventions by the state to enable a "leap" from one stage of development to the next. If development was a garden, dams and other large infrastructures were the "hothouses" to expedite the supposedly linear process.

The "Narmada Effect"

Large dams gripped the development imagination of postcolonial states into the 1970s. Around this time, several structural changes occurred in the global dam-led development regime. While there had always been critiques of big dams and the scale of the intervention in natural landscapes, these critiques altered dam building in the last decades of the twentieth century. The first major change was the growing chorus of critiques of large dams, from ecological and social justice perspectives. Dam critique – or the defense of river systems and local livelihoods – coincided with more visible Indigenous struggles that challenged dominant understandings of development and environmentalism on a global stage.

Perhaps the most prominent river defense campaign mounted by Indigenous peoples, known in India as *adivasis*, was spearheaded by an organization called the **Narmada Bachao Andolan** (Save Narmada Organization). The Indian case features prominently in the global politics of anti-dam protests. Around 50 million people had been displaced and dispossessed in the name of development during the first 50 years of Indian independence since 1947. Although *adivasis* make up only 8% of the population, they constitute a staggering 50% of the population displaced by dam development (Baviskar 2019, p. 28). This is not only because the *adivasis* are a traditionally marginalized and dominated section of Indian society, but also because many *adivasi* communities reside in forested hills where dams and mining projects are located.

The Narmada Bachao Andolan (the Andolan) was founded by Mumbai-based activist and social worker Medha Patkar (Figure 6.6) in 1985 to protest the planned construction of the Sardar Sarovar Dam on the Narmada River in the western state of Gujarat. The Andolan was part of a growing environmental movement in India, which tended to be more closely tied to issues of livelihood, caste, and economic exploitation than the environmentalist movement in the heavily industrialized and urbanized countries (Baviskar 1999). From People's Science movements in the southern state of Kerala to the anti-deforestation protesters in the Himalayan foothills in Uttarakhand, environmental issues in India were increasingly uniting urban middle-class activists and rural people whose livelihoods depended directly on the land. Throughout the 1980s and 1990s, the Andolan protested the Indian government's attempts to build the Sardar Sarovar Dam in a fragile ecosystem and appealed to the World Bank to repeal its support of the project.

The Andolan captured global media attention by disrupting a seemingly undisputable consensus around the desirability of infrastructure-led development. The movement

Figure 6.6   Social activist Medha Patkar (1954–), pictured in 2002. Source: Kannan Shanmugham / Wikimedia Commons / Public domain.

highlighted how different groups of people experience the state's implementation of development in divergent ways: the benefits (like irrigated water or hydroelectricity) to one class or region could leave another group of people devastated. For this reason, Andolan activists demanded a "class-benefit analysis" of the dam instead of the bureaucratic, depoliticized, and traditional cost-benefit analysis (or its successor, the environmental impact assessment).

In a major victory for the movement, the World Bank removed its support of the project in 1993. Within internal Bank reports and briefings, experts recommended avoiding negative publicity – what they called the "**Narmada effect**" – by requiring a series of public consultations and impact assessments prior to any dam construction, a process that arguably tied large dams up in more red tape, rather than eliminated them from landscape (Goldman 2007). And, despite decades of protest and global prominence, the Indian government pressed ahead with the project. A 2000 Supreme Court ruling held that the dam was lawful and those dams were "good" for the nation. In September 2017, the reservoir was filled for the first time – the people who lived there had been displaced and the valley was flooded. Ultimately, although the Andolan did not meet their objective of halting the construction of the Sardar Sarovar Dam, the activists changed the debate on dams in India and the broader world.

The Andolan also captures one of the main themes of this book: camp is everywhere. A critical approach to the hydrosocial cycle reminds us how political resistance and struggle is waged not only in elections or in theoretical treatises, but in, and for, everyday life. When everyday life was threatened by a large socio-technological intervention, as on the Narmada River, the Andolan organized an eloquent and politically sophisticated response of collective resistance. Even though the Andolan failed to stop the dam construction, many Indigenous and marginalized peoples around the world continue to be inspired by the decades-long fight Andolan mounted against some of the most well-funded state and international agencies. But ground-up resistance is not the only site for critically interrogating dams, that most powerful symbol of modernity and development. A very different type of organized resistance to the hegemony around dams as development arose at the end of the twentieth century and culminated in a most bureaucratic form of resistance: a report.

The World Commission on Dams

The culminating event of ecological and social justice struggles from the 1980s and 1990s was the convening of the World Commission on Dams (WCD) and the publication of the 2000 landmark report. The WCD report came out of a confluence of several ongoing processes (Baghel and Nüsser 2010; Conca 2005). One factor was an internal review process at the World Bank about the efficacy of the large dams the Bank had financed around the world. A second factor was the negative publicity generated around large dams and their impact on marginalized communities, most famously the "Narmada Effect" and Andolan struggles.

Civil society organizations also exerted an influence. A third decisive factor was the influence of the International Union for the Conservation of Nature (IUCN), an international organization that advocates for species protection and park conversation. The IUCN attempted to engage the World Bank to review the evidence and attempt a paradigm shift in global environmental governance and economic development. The WCD consisted of 12 commissioners from government, industry, social movements, academia, and local and international non-governmental organizations (NGOs). The WCD conducted a cross-check survey involving quantitative data on 125 dams around the world and accepted almost a thousand submissions from individuals and institutions (Baghel and Nüsser 2010).

In 2000, the WCD produced its final report after more than two years of the consultative process. The report, titled *Dams and Development: A New Framework for Decision-Making,* (WCD 2000) aimed to evaluate the performance of existing dam-led development and to articulate the best principles for new dam development. The report is filled with detailed analysis of the global impact of dams and in-depth case studies and descriptions of selected dams. The report laid out five key principles for future decision-making: equity, efficiency, participatory decision-making, sustainability, and accountability. With regards to its evaluation of the existing dams, the report stated that while "dams have made an important and significant contribution to human development" in "too many cases an unacceptable and often unnecessary price has been paid to secure those benefits, especially in social and environmental terms, by people displaced, by communities downstream, by taxpayers and by the natural environment" (WCD 2000, p. xxviii; Box 6.3).

Box 6.3 Dams and Displaced Peoples

In 2000, the World Commission on Dams estimated that up to 80 million people had been displaced due to dam construction. This number increased in subsequent decades. In the past 20 years, an estimated 15 million people have been displaced annually due to development projects – many of them large dams. The issue of dam-induced displacement and the closely tied issue of resettlement are a major source of controversy.

Recent examples abound. The construction of the Ilisu dam in Turkey forcibly displaced 55 000 people, and the mismanagement of their resettlement caused several European banks to withdraw their export credit guarantees in 2008. A year later, an international panel of experts criticized the resettlement process around the construction of the Belo Monte dam in Brazil – during which Indigenous people were "bribed" with compensation programs instead of offered meaningful participation (McDonald-Wilmsen and Webber 2010, p.146).

Dams are very likely to shape how climate, conflict, and future development-related resettlement will occur in the future. People from flooded valleys experience a painful sequence of events: the brutality of dispossession, the chaos of eviction, the loss of community, and the slow pace of resettlement. A 2012 review of the resettlement process associated with the construction of 50 large dams found that 82% of displaced people experienced a deterioration in their socioeconomic conditions (Scudder 2012).

What is the future of dam-induced displacement? More recent cases are in projects led by *national* development agencies, instead of international agencies such as the World Bank. Examples include: the Ertan, Shuikou, Xiaolangdi, and Three Gorges dams in China; the Cirata and Saguling dams in Indonesia; and the Diamer-Bhasha Dam in Pakistan. Dam-induced displacement is often catalyzed by a state government – working with a network of development funders, private investors, and construction companies – intent on building a dam. A state infected, perhaps, by dam fever.

While the 12 commissions on the WCD achieved consensus on these conclusions, not all stakeholders received the report enthusiastically. The governments of China and India, the dam-building industry, and the World Bank expressed reservations about elements of the report, especially the guidelines for further development. In March 2001, the World Bank officially stated that it would not "comprehensively adopt" the guidelines laid out by the WCD report. Although there was a notable pause in global dam construction after the release of the report, dam construction picked up not long after publication (Baghel and Nüsser 2010).

Reflecting on the report and the remarkable consensus work of the WCD, the legacy of the report is discursive and institutional. The discursive impact cannot be underestimated. Generations of activists, scientists, and politicians who want complex and nuanced conservations around dam-led development and consequences recognize the WCD as a key moment. As an assembly, the WCD was one of the first moments that actors from non-governmental sectors were systematically involved in a process of global environmental policy-making alongside members of industry and state. This represents an acknowledgment – however small and limited to consultation – that the state and multinational corporations were not the only actors on the international stage. The rights of marginalized peoples from national peripheries and the degraded environments left in the wake of large dams mattered (Baghel and Nüsser 2010).

The Future of Dams

Without a doubt, dam fever has become a global pandemic, spreading to many different countries and regions of the world. The lure of infrastructural-led development and the power of dams to shape hydrosocial relations – in terms of geopolitical influence, "green" energy production, and water control – is undeniable. At the same time, dams remain highly debated and visibly contested elements in the hydrosocial cycle, as Andolan anti-dam activists, the WCD report, and other struggles remind us (Box 6.4). This fever raises the question: what is the future of dams?

If only we could forecast the future! In the absence of a crystal ball, we identify at least three prominent and emerging trends: (i) the continued construction of large dams in relatively "untapped" regions (such as the Eastern Himalayas, Box 6.4) and ongoing reliance on hydropower development; (ii) new trends in the financing, expertise, and development models of dams; and (iii) a concerted push toward dam removal and river restoration in select regions and contexts.

First, the large dam boom continues, but in different locations, under different models of finance and management, and through new networks of expertise – notably, through China. The traditional drivers of development thinking – Europe, Japan, and the United States – are increasingly wary of building large new infrastructures on rivers. Chinese firms have filled this vacuum with their infrastructural finance and expertise. The economic and geopolitical rise of China since the 1980s, and especially since 2000, has introduced more complexity into models of river management, international water development, and geopolitics (Box 6.4).

China has a long history of dams. Along with much of the rest of Asia and Africa, Chinese planners and politicians caught dam fever in the third quarter of the twentieth century. Yet the scientific and administrative tradition of Chinese water control extends back thousands of years. The Grand Canal, which connects the Yellow and Yangtze rivers, has been under construction since 500 BCE (Chapter 2). The authority of the Chinese state entered a long period of decline in the eighteenth and nineteenth

Box 6.4 Himalayan Hydropower, Hunger Strikes, and Geologic Surprises

The Himalayan region of South Asia is home to a staggering number of hydropower plans and large dam projects (Figure 6.7). As of 2020, more than 400 large dams with a total generation capacity exceeding 200 GW are planned in the Himalayan countries of India, Nepal, Bhutan, and Pakistan (Lord et al. 2020). Dam fever took hold of India particularly after independence in 1947, when "Prime Minister Jawaharlal Nehru embarked upon a campaign of modernization and industrialization that aimed to eliminate perceived infrastructural weaknesses and build a bright new postcolonial future – famously referring to dams as temples of modern India" (Lord et al. 2020, p. 3). Today in Sikkim, a small Eastern Himalayan state in India, 27 hydropower projects are proposed under the banner of "the country's future powerhouse" (Gergan 2020, p. 1).

Yet camp is everywhere, as we know, and hydropower schemes have met strong opposition and organized resistance from affected communities. As Amelie Huber and Deepa Joshi (2015, p. 16) explain, "[T]he first anti-dam movement, Concerned Citizens of Sikkim (CCS), emerged in the 1990s in response to the construction of the 30 MW Rathong Chu project in West Sikkim." Since then, opposition to state-led hydropower has included a group of youth

Figure 6.7 Himalayan hydropower and dam development at work. Source: Austin Lord, reproduced with permission.

activists, known as the Affected Citizens of Teesta (ACT), from the Lepcha ethnic community. "In an unprecedented act of civil disobedience, ACT held several rounds of hunger strikes (up to 915 days), street protests, petitions, and
litigation against six projects planned in the tribal 'reserve' Dzongu" (Huber and
Joshi 2015, p. 16; see also Gergan and McCreary 2022).

Resistance to Himalayan hydropower has taken surprising twists and turns.
Mabel Gergan (2020, p. 1) nimbly shows how hydropower projects must encounter and navigate the realities of "unruly" Himalayan geographies: "seismic instability, crumbling hillsides and flashfloods, lay bare the limits of capital
accumulation at geographic and resource frontiers." Conditions are tough, as zinc
chews up the turbines and reservoir design must cope with heavy sediment loads.

Can geology be a site of dam resistance? The answer is tricky. The Indian
government blames construction problems and slow hydropower development
on the region's "geologic surprises" – an ambiguous term used to identify "problematic" geological conditions that delay or impede dam projects. In an important move, Saumya Vaishnava and Jen Baka (2021, p. 4) "draw attention away
from the dam site and towards the institutional configuration that enable these
[hydropower] projects" to push forward, despite considerable risk. Rather than
act as impediments to future development, such "geological surprises" have
drawn together a set of diverse institutional agencies – all fixated on harnessing
Himalayan rivers for power – to *enable* future hydropower development
(Vaishnava and Baka 2021). The fever continues.

centuries. At the time of the Communist Revolution in 1949, there were only 22 dams
in the country.

After the death of Chairman Mao Zedong in 1976, Deng Xiaoping became the
leader of China. He began a process of capital-intensive industrialization geared
toward making China competitive in the global export market. After more than a
decade of turmoil and dislocation, China roared into the capitalist world market.
Dams helped fuel this transition. After China's embrace of capitalist investment and
export markets in 1979, the government proposed 10 areas for the development of
"hydropower bases." By 1989, this number was raised to 12 (Han and Webber 2020b).

Two decades later, all these sites had been developed. Chinese engineering and
construction industry acquired an impressive level of expertise in the construction of
large dams. In part, this expertise was gained through working on dams financed by
development aid from the World Bank and Japan. But technological acumen also
reflected a rapid rise in China's infrastructural capacity. Between 2005 and 2008,
China laid more high-speed rail tracks than Europe had in the previous two decades
(Flyvbjerg 2014, p. 8).

China has also extended its dam expertise to regions outside its borders. By the end
of the twentieth century, China had relinquished its membership in the World Bank's

International Development Association and sought its own agenda to promote infrastructure in the developing world. In 2008, Chinese financiers and builders were involved in no fewer than 93 large dam projects in 38 countries around the world, most in developing countries in Asia, Africa, and Latin America. In 2005, the Chinese **state-owned enterprise** (SOE) Sinohydro, the most prominent international face of Chinse dam construction, was in various stages of completion on more than 57 hydropower projects in 32 countries.

The Chinese drive into hydropower – especially in Southeast Asia and Africa (Chapter 7) – has continued into the twenty-first century. The global reach of Chinese hydropower firms has attracted criticism and concern from politicians, planners, and environmentalists around the world, especially in the US and Europe. Some beneficiaries of China's foreign dam investment, such as Nepal, see this influx of money and expertise as a pathway to energy independence, development dreams, and a bulwark against dominance by other neighboring countries (Box 6.4). One might argue that there are tangible benefits to China's hydropower footprint in the developing world. Between 2010 and 2020, Chinese companies built 8134 MW of hydro-capacity in Sub-Saharan Africa: this represents over half of Africa's added hydro-capacity during that decade (Han and Webber 2020b).

This brings us to our second point: evidence of new trends in the financing, expertise, and management of dams. For example, in contrast to water projects spearheaded by multilateral organizations like the World Bank, which receives its financing (and oversight) from a variety of participating nations, we now see the rise of **bilateral** cooperation between nations. In the case of China, much of their activity in the developing world is spearheaded by bilateral agreements managed by organizations like Sinohydro. In 2003, the major Chinese hydropower companies, including Sinohodryo, Three Gorges Corporation, and Gezhouba, were listed as SOEs by the State Council (Han and Webber 2020a). While these corporations are explicitly state-owned and operate with generous financial and marketing support from China, they can compete for market share and political access.

These SOEs leverage a powerful weapon: knowledge and expert networks. For example, Ching Kwan Lee compares the various strategies pursued by Chinese SOEs and other multinational corporations in the African extraction sector and concludes that the regulatory context is the decisive factor shaping corporate behavior (Lee 2017). Together with a network of financiers, advisors, administrators, engineers, builders, and consultants that make up the Chinese hydropower industry, these actors are fueling a new dam boom across the world that hearkens back to the burst of dam building in the middle of the twentieth century.

Worldwide, major changes have also occurred in how dams are financed. For most of the twentieth century, as this chapter has explained, dams were funded by national governments and managed by large state bureaucracies – think of the TVA, the US Bureau of Reclamation, or the US Army Corps of Engineers. By the late twentieth century, as Rhodante Ahlers and colleagues (2017) have shown, a shift toward more complex and fragmented financing for dam infrastructure has emerged. Private capital has shown great interest in large infrastructure projects, paving the way for the resurgence of large dams (Merme et al. 2014). New forms of public–private partnerships and complex financial instruments have made it increasingly difficult to pin down the financiers and "owner" of the dam – muddying the waters of "public" infrastructure (March and Purcell 2014). For example, the dam project spearheaded by the Nam Theun 2 Power Company in the Mekong River basin is supposedly one of the most complex public–private partnerships in the history of development (Ahlers et al. 2017, p. 11).

Transboundary rivers are already complicated by dams – a topic we turn to in the next chapter – but streams of financialized support are extremely difficult to follow. The Nam Theun 2 project was completed in 2010, and the power purchase agreement stipulates that 95% of the hydropower is destined for Thailand and just 5% for Laos. Since 2011, three companies from three different countries currently share ownership of Nam Theun 2 Power Company. Twenty-seven parties came together to raise USD 1.45 billion to fund the project. This financing was supported by the Laotian government's relaxed labor standards, tax concessions, and mobilized public sector support and financial guarantees. Since the inception of the project, there have been multiple violations reported concerning the social and environmental impacts of the dam. However, the confidential status of the many agreements between the multiple parties makes it difficult to find the party liable for violations. Building on our critical analysis of financialization in Chapter 4, the increasing complexity of the financial and ownership aspects of large dams complicates democratic control and the accountability of operations (Ahlers et al. 2017).

Third, while the number of large dam projects has surged upward in Africa, southern Asia, and Latin America, a new trend is emerging across parts of North America and Europe: **dam removal**, illustrated by the Condit Dam scene at the opening of this chapter. Dam removal, a process known as "decommissioning" to experts, happens for several reasons: economic obsolescence, ecological damage, successful litigation, and/or decreasing social and economic returns (Doyle et al. 2008; Fox et al. 2016; Magilligan et al. 2016). In the United States, more than 1700 dams have been removed to date, with over 70% of these occurring since 1999 (Sneddon et al. 2017; Fox et al. 2022).

What kinds of dams are removed? Most removed dams are relatively small in terms of size and spatial impact (Magilligan et al. 2016). For example, in the New England region of the US Northeast, over 14 000 dams fragment the region's rivers and streams (Magilligan et al. 2016). Since 1990, an estimated 127 dams have been intentionally removed across New England, with more than 50 planned for future removal (Fox et al. 2016). In New England, removed dams are relatively small: 30% of removed dams ranged between 2 and 4 meters in height, and 22% were between 4 and 6 meters (Magilligan et al. 2016).

Just like dam construction, dam removal embodies and triggers a new set of social and ecological relations. An iconic example of ecological response to removal is captured by the 2007 removal of the Marmot Dam on the Sandy River in Oregon, in the northwestern United States. Removing this 15 m-tall dam meant that for the first time in almost a century, the Sandy River would flow freely. The benefits of removing the dam were rapid. Within 48 hours, about 20% of the sediment previously trapped in the reservoir had flushed downstream. Fisheries rebounded quickly, as anadromous fish (namely, endangered salmon) were able to swim upstream to spawn (Parks and Grant 2009).

What catalyzes dam removal? Within a broad landscape of "strategic opportunism" to push for dam removal (Magilligan et al. 2016), Indigenous communities and leaders have been especially vocal, organized, and active (Fox et al. 2022; Gergan and McCreary 2022). For example, Native American Nations in the United States have been leaders in facilitating dam removal and restoring ecosystems on the Elwha River (Washington), the Klamath River (Oregon/California), the Penobscot River (Maine), the Ottaway River (Michigan), and the White Salmon River (Washington), the former home of the Condit Dam (Gosnell and Kelly 2010; Fox et al. 2022). Tribal involvement creates space for new actors, alliances, important knowledges, and practices of ecological restoration, and facilitates opportunities for restorative environmental justice (Fox et al. 2022).

Not everyone embraces the removal of dams. In New England, for example, projects of dam removal and river restoration have been hotly contested and politicized. This region features hundreds of small dams, built during the British colonial and US industrializing eras to generate power for industry (Magilligan et al. 2016). Most small dams in New England are located on private land (or are privately owned) and drain a relatively small watershed – key differences that facilitate removal in comparison to larger dams located in the western United States (Magilligan et al. 2016). State agencies and environmental organizations in New England often appeal to removal projects "on the grounds of dam safety (the structure is 'falling apart') and economics (repairing dams is 'too costly'), particularly when communicating with local residents at public forums" (Fox et al. 2016, p. 96).

> Conversely, the community residents seeking to preserve a particular dam actively contest the notion of ecological restoration, arguing instead that dam removals are quintessentially about cultural and ecological disruption of cherished and historically significant landscapes. (Fox et al. 2016, p. 96)

Dam fever – the power of dams' symbolic value in the world – is a fundamental aspect of dam removal, not just its construction. This enduring struggle for control, knowledge, and the power to define a landscape is captured in some of the citizen responses to dam removal projects in New England: "The swift river (sic) has been part of me since I was a little kid," a respondent wrote, "You kill the dam, you are killing part of me" (Fox et al. 2016, p. 101). Clearly, the search for a cure to the fever continues.

Summary and What's Next

Dams are prominent and controversial features in the hydrosocial cycle. Dams shape riverine flows, aquatic ecologies, electricity production, economic systems, and human livelihoods. As such, dams also reflect broader trends in development, decolonization, nationalism, and social and spatial inequality. For well over a century, dams have been sites of struggles for social justice, the source of vast ecological devastation, and the focus of a seemingly unquenchable thirst for energy and power around the world. As we have reiterated throughout this chapter, dams are shapers and mediators of hydrosocial relations, but the meaning of dams is still situated and contextual.

Dams are so much more than technological marvels. From a critical point of view, the debate over dams cannot be reduced to technical dimensions such as their location or height, electricity production capacity, or the size of their reservoirs, although these features are undeniably important. A hydrosocial perspective allows us to understand dams – and crucially, the rivers, watersheds, communities, and economies they interact with – in all their complexity and nuance, as instruments of social power. Ultimately, only a hydrosocial lens on dams moves beyond sand and concrete to comprehend a phenomenon like dam fever – from its origins to international expansion – and to ask the important question of *why* this fever remains contagious decade after decade.

A hydrosocial perspective helps us understand the very relational nature of "fixed" infrastructures in the world. New trends in dam governance – including the ongoing boom in large dam construction, the shifting regimes of finance and management, and the efforts to remove dams and restore rivers – are not so much major paradigm shifts in hydrosocial relations as they are continuities, adaptations, and accommodations of development regimes, as well as responses to contestation and anti-dam struggle. In the next chapter, we step deeper into the proverbial river to explore a major topic only hinted at here – the politics of sharing waters across national and subnational borders.

Further Reading

Environmental impacts of dams

Allan, J.D., Castillo, M.M., and Capps, K.A. (2021). *Stream Ecology: Structure and Function of Running Waters*. Berlin: Springer Nature.

McCully, P. (1996). *Silenced Rivers: The Ecology and Politics of Large Dams*. London: Zed Books.

Postel, S. and Richter, B. (2012). *Rivers for Life: Managing Water for People and Nature*. Washington, DC: Island Press.

Dams and uneven development

Ahlers, R. and Merme, V. (2016). Financialization, Water Governance, and Uneven Development. *Wiley Interdisciplinary Reviews: Water* 3 (6): 766–774.

Akhter, M. (2022). Dams, Development, and Racialised Internal Peripheries: Hydraulic Imaginaries as Hegemonic Strategy in Pakistan. *Antipode* 54 (5):1429–1450.

Harris, L.M. (2014). Imaginative Geographies of Green: Difference, Postcoloniality, and Affect in Environmental Narratives in Contemporary Turkey. *Annals of the Association of American Geographers* 104 (4): 801–815.

Displacement and resettlement

Del Bene, D., Scheidel, A., and Temper, L. (2018). More Dams, More Violence? A Global Analysis on Resistances and Repression Around Conflictive Dams Through Co-produced Knowledge. *Sustainability Science* 13 (3): 617–633.

Huang, Y., Lin, W., Li, S., and Ning, Y. (2018). Social Impacts of Dam-Induced Displacement and Resettlement: A Comparative Case Study in China. *Sustainability* 10 (11): 4018.

Wilmsen, B. (2018). Is Land-Based Resettlement Still Appropriate for Rural People in China? A Longitudinal Study of Displacement at the Three Gorges Dam. *Development and Change* 49 (1): 170–198.

Nationalism and dams

Allouche, J. (2019). State Building, Nation Making and Post-colonial Hydropolitics in India and Israel: Visible and Hidden Forms of Violence at Multiple Scales. *Political Geography* 75: 102051.

Menga, F. (2015). Building a Nation Through a Dam: The Case of Rogun in Tajikistan. *Nationalities Papers* 43 (3): 479–494.

Menga, F. and Swyngedouw, E. ed. (2018). *Water, Technology, and the Nation-State*. Oxford, UK: Routledge.

Sneddon, C. (2015). *Concrete Revolution: Large Dams, Cold War Geopolitics, and the US Bureau of Reclamation*. Chicago, IL: The University of Chicago Press.

River restoration and dam removal

Doyle, M.W., Stanley, E.H., and Harbor, J.M. (2003). Channel Adjustments Following Two Dam Removals in Wisconsin. *Water Resources Research* 39 (1) https://doi.org/10.1029/2002WR001714

Fox, C.A., Magilligan, F.J., and Sneddon, C.S. (2016). "You Kill the Dam, You are Killing a Part of Me": Dam Removal and the Environmental Politics of River Restoration. *Geoforum* 70 (1): 93–104.

Fox, C.A., Reo, N.J., Fessell, B., and Dituri, F. (2022). Native American Tribes and Dam Removal: Restoring the Ottaway, *Penobscot and Elwha Rivers*. *Water Alternatives* 15 (1): 31–55.

Gosnell, H. and Kelly, E.G. (2010). Peace on the River? Socio-ecological Restoration and Large Dam Removal in the Klamath Basin, USA. *Water Alternatives* 3 (2): 361–383.

Magilligan, F., Graber, B.E., Nislow, K.H., Chipman, J.W., Sneddon, C.S., and Fox, C.A. (2016). River Restoration by Dam Removal: Enhancing Connectivity at Watershed Scales. *Elementa: Science of the Anthropocene* 4: 1–14. https://doi.org/10.12952/journal.elementa.000108.

Sneddon, C., Magilligan, F.J., and Fox, C.A. (2017). Science of the Dammed: Expertise and Knowledge Claims in Contested Dam Removals. *Water Alternatives* 10 (3): 677.

Sneddon, C., Magilligan, F.J., and Fox, C.A. (2021). Peopling the Environmental State: River Restoration and State Power. *Annals of the American Association of Geographers* 112 (1): 1–18.

Himalayan hydropower

Gergan, M.D. (2020). Disastrous Hydropower, Uneven Regional Development, and Decolonization in India's Eastern Himalayan Borderlands. *Political Geography* 80 (1): 102175.

Huber, A. and Joshi, D. (2015). Hydropower, Anti-Politics, and the Opening of New Political Spaces in the Eastern Himalayas. *World Development* 76 (1): 13–25.

Lord, A. (2016). Citizens of a Hydropower Nation: Territory and Agency at the Frontiers of Hydropower Development in Nepal. *Economic Anthropology* 3 (1): 145–160.

Lord, A., Drew, G., and Gergan, M.D. (2020). Timescapes of Himalayan Hydropower: Promises, Project Life Cycles, and Precarities. *Wiley Interdisciplinary Reviews: Water* 7 (6): e1469.

Vaishnava, S. and Baka. J. (2021). Unruly Mountains: Hydropower Assemblages and Geological Surprises in the Indian Himalayas. *Environment and Planning E: Nature and Space* 5 (3) https://doi.org /10.1177/25148486211050780.

Chapter 7

Shared Waters

A River in "Crisis"

During the months between May and September, the Mekong River undergoes a major transformation. Intense rains during these wet season months trigger a rise in the river's water levels. With a dramatic influx of water in the lower basin, the Mekong River – which originates in China and runs through Myanmar, Laos, Thailand, Cambodia, and Vietnam – reverses direction and flows upstream into Tonlé Sap Lake in Cambodia, the largest flood pulse lake in Southeast Asia (Figure 7.1). Water flows unevenly across the 4900-km-long Mekong River. In its headwaters, the upstream basin (covering China and Myanmar) contributes relatively little to the Mekong River in terms of annual mean water flow. Most river water comes from the downstream basin, which contributes over 80% of total flow from seasonal rainfall (MRC 2010). The flood pulse in the lower reaches of this river shapes a vital seasonal rhythm of culture, livelihood, economy, and ecology – the hydrosocial relations between people and the river.

Tonlé Sap grows five times in surface area, making it a unique biodiversity hotspot with at least 149 known species of fish (Campbell et al. 2006). Fish migrate between Cambodia and Vietnam, and fisheries thrive in this rich ecosystem exchange. *Dai* fishery, a traditional Cambodian stationary trawl using rows of nets, has been in practice

Water: A Critical Introduction, First Edition. Katie Meehan, Naho Mirumachi, Alex Loftus, and Majed Akhter.
© 2023 John Wiley & Sons Ltd. Published 2023 by John Wiley & Sons Ltd.

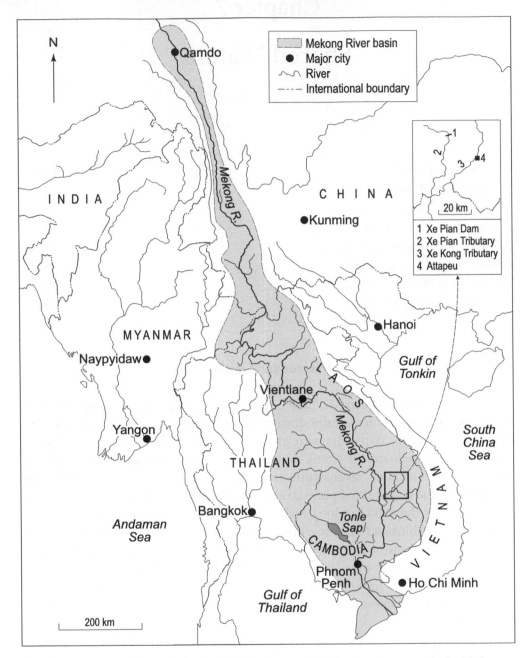

Figure 7.1 Map of the Mekong River basin. Source: The Authors, cartography by Philip Stickler.

for over a hundred years. Approximately USD 150–250 million worth of fish are caught in the lake, a quantity of between 200 000 and 235 000 tons (t), in addition to what is caught in the delta (Halls et al. 2013). During the dry season, Tonlé Sap sustains people in the delta in Vietnam as water stored in the lake is discharged (Kummu et al. 2014).

The Tonlé Sap, Mekong River delta, and other parts of the lower basin stretching across Laos, Thailand, Cambodia, and Vietnam have been governed by the "Agreement on the Cooperation for the Sustainable Development of the Mekong River Basin" (hereinafter, Mekong Agreement) since 1995. This **transboundary institution** establishes a cooperative framework for these four nations to participate in collective management and sustainable development of the river, following international water law principles such as equitable and reasonable utilization of water resources. A transboundary institution includes written law and international agreement to govern shared waters, and – following our expansive and critical approach to legal waters (Chapter 3) – also includes participating organizations, decision-makers, policy ideas, and formal and informal norms and practices, influenced by situated histories, economies, and geometries of power.

The Mekong Agreement is also the governance basis for irrigation projects in various parts of the river and, more recently, hydropower dams. In 2015, 365 dams were completed, in construction, or planned, including three projects which were later canceled (WLE-Mekong 2015). Large dams on the main stem of the Mekong are now in operation. With more dams planned and underway in upstream China, there are concerns that the dams will impede seasonal flooding – the lifeblood of Tonlé Sap – and cause irreversible damage to aquatic life and regional livelihoods (Hoang et al. 2019; Kallio and Kummu 2021). Experts calculate Cambodia will lose 267 429 t of fish and a further 366 570 t in Vietnam, totaling a loss of USD 500 million (Yoshida et al. 2020). Calls have been made for the political union of the Association of Southeast Asian Nations (ASEAN) to address the "crisis" of the Mekong ecosystem (Hoang Thi Ha and Seth 2021).

What produces a "crisis" of shared waters? And who experiences its implications? The Mekong River illustrates the complexity and scale of hydrosocial relations that shape the lives of *dai* fishery workers, decisions of ministries involved in basin planning, operations of dam construction companies, and migrating fish and aquatic organisms. Water is a delicate balance in Southeast Asian regional politics. A major change to the flows of the Mekong holds serious ramifications for local livelihoods and ecologies. A reduction in fish catch will result in a loss of fish protein, making people turn to farmed meat as an alternative source. Livestock are water-intensive food sources, requiring 13–63% more land and 4–17% additional water needs (Orr et al. 2012). Dietary shifts are already reflected in the increase of food imports, highlighting dependence on other regions (Burbano et al. 2020). Even with its abundant seasonal rains, in the lower Mekong basin water scarcity will become increasingly fraught as more water will be needed to make food.

A relational point of view reminds us that a "crisis" of transboundary waters is manufactured, not pre-given. A relational view helps us see how the Mekong River shapes – and is shaped by – the power, decisions, institutions, and livelihoods of people residing near and far beyond its riverbanks. The Mekong River basin is not unique in facing competing demands for water that place national interests above river development plans and expose livelihood and ecosystem challenges to those living with the river. While the geography and hydrology of the world's rivers are distinct, many of the tensions over water allocation are imbued with power relations. Worldwide, a relational lens helps us see how rivers that no longer flow freely from source to delta are generated by ideologies of development that utilize water resources as things to be stored, diverted, and pumped out of the river for economic development. At the scale of nation-states, a relational view asks us to question how crisis is produced, experienced, and negotiated among differently situated actors and governments who share waters.

In this chapter, we explore the cooperative geographies and conflicts among shared waters. We draw on the Mekong River as an illustrative example that usefully brings together critical insights concerning the scale of hydrosocial relations in transboundary river basins. The Mekong River is one of the world's 286 international transboundary river basins – a remarkable number. These rivers flow across political boundaries and require international negotiation and coordination for their management. In the following sections, we first critically examine some of the commonly held myths about shared waters. We then present a tool for understanding the nuances of transboundary cooperation and conflict, applying this framework to questions of power and diplomacy in the example of the Mekong River basin. But we begin with a provocative question: has a bullet ever been fired over so-called "water wars"?

The Myth of Water Wars

The narrative of "water wars" is like James Bond: it never dies. Read a news story about transboundary rivers or groundwater basins, and the idea that states will incite conflict and "fight" over shared waters – rather than negotiate or cooperate – is a common description or entry point (Ramsay 2020). The Nile and Jordan rivers in the arid Middle East region are prone to such characterizations. Often, the "water wars" discourse is used as a shorthand for prophecies of violence over scarce resources – where goes oil, so shall water.

To date, the evidence proves otherwise. The myth of water wars was debunked by scholars of **hydropolitics**, the study of politics over shared waters. Across the world, there are 153 countries that share the 286 transboundary river basins and some 592 transboundary aquifers (UN-Water 2021). The map of transboundary waters is stunning: more fresh water is shared than not (Figure 7.2). Historically, states do not go to war over water. Studies have quantitatively assessed instances such as verbal disputes and expression of disagreements that characterize conflict (Yoffe et al. 2003). In fact, states tend to engage in low-level conflict and their cooperative behavior is also equally restricted to low levels (Wolf et al. 2003).

Furthermore, the water wars rhetoric distracts us from a critical understanding of why water resources are political. Crucially, the water wars hypothesis does not distinguish how water directly or indirectly plays a role. Sentiments are strong but do not necessarily manifest in an acute military conflict between states (Wolf et al. 2003). More insidiously, a Malthusian assumption of resource scarcity underpins the water

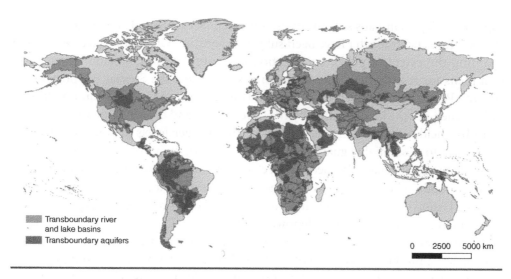

Figure 7.2 World map of transboundary rivers, lakes, and aquifers. Source: UN-Water 2018a, p. 14 / United Nations / CC BY-SA 3.0 IGO.

wars discourse: the idea that resource scarcity is inevitable and will result only in conflict (see Chapter 1 for a critique). Malthusian thinking is much too reductionist (and racist) to appreciate the varied ways water is accessed, allocated, and used. Moreover, the water wars hypothesis does not consider the ways that scarcity is constructed – a core plank of this book. The geography of water production and distribution is mediated through structural relations, not only mere hydrological or climatic forces.

What is fought over in shared waters? Let's return to the Mekong River basin to find an answer. In July 2018, villagers in southern Laos unexpectedly lost their lives, homes, and livelihoods when a dam collapsed and flooded the region with a reported 5 billion cubic meters (m³) of water. The Xe Pian-Xe Namnoy hydropower project, a major enterprise of USD 1.02 billion, was near completion when one of the saddle dams broke (Figure 7.1).[1]

The dam collapse was an unprecedented accident in Laos and across the Mekong basin region. The Lao government reported the death toll at 71 people (Inclusive Development International and International Rivers 2019). A further 7000 people in 19 villages were affected, many requiring resettlement. With little to no time for advance warning about the flooding, villagers found themselves without their possessions and in temporary camps. Crops, livestock, and farmland were lost. Damage extended 100 km further downstream in northeast Cambodia. As a result of this transboundary accident, 15 000 people were affected in Cambodia and their livelihoods were at risk (Inclusive Development International 2020).

The struggles over shared waters constitute a chance for a prosperous life and just treatment of communities. After all, water is life. Even after two years, there are 5000

[1] A saddle dam is the term given to a water barrier built into a topographic depression or gap in the rim of a reservoir. It is typically built as an auxiliary feature to increase water storage capability (Jansen 2003).

displaced villagers in Laos still awaiting permanent settlement. The Lao government and its actions post-disaster have been subject to scrutiny. Criticisms have been made regarding the extent of public consultation and information provision about long-term plans for resettlement. The situation is so dire that the Special Rapporteur on Extreme Poverty and Human Rights of the United Nations Commission on Human Rights called out the lack of action by the government and urged the private developers to be "held fully accountable," otherwise the meager levels of financial support only work to "essentially guarantee people will live in poverty" (Alston 2019). Cambodian victims have also been neglected and continue to suffer hardships from lost livelihoods. They have given evidence in a letter of allegation submitted to the Working Group on Business and Human Rights and other business enterprises of the UN Human Rights Council (Peoplepower21 2019).

Hydrosocial thinking shows us that the mainstream and tributary rivers of the Mekong are a product of social, ecological, and spatial relations concerning international cooperation, infrastructure development, and livelihoods. This relational perspective sheds light on the private sector and their role in producing Mekong waters. As a typical build-operate-transfer project in Laos, the Xe Pian-Xe Namnoy project will be managed by international private companies and handed over to the Lao government after 27 years. A Korean company, SK Engineering & Construction, was responsible for building the dam. Along with this company, a Thai company called Ratchaburi Electricity Generating Holding, Korea Western Electric Power Co., and Lao Holding State Enterprise make up the consortium of developers. To date, none have provided a full explanation of the cause nor provided adequate compensation for damage to survivors in Laos and Cambodia (Inclusive Development International 2020). This significant disaster has been attributed to poor construction, which immediately calls into question the responsibility of project developers.

The events surrounding the Xe Pian-Xe Namnoy hydropower project are not those of water war. Military power was not exercised over the tributary and no "war" was waged. The situation is hardly what one would imagine as war over scarce water resources. However, this example represents the production of water and its serious contestations that expose systemic dimensions of water conflict. The project reflects the deep divides between the state and private companies that develop the river – those who reap the benefits from generated hydropower – and the villagers relying on river resources for their livelihoods and now bearing the burden of the disaster.

Limits to Water Cooperation

We now understand that "wars" are not fought over water – but what does a critical approach to cooperation look like? A crucial point in understanding the disaster of the Xe Pian-Xe Namnoy hydropower project is in recognizing the limits of international water cooperation. Problematically, the Mekong Agreement is silent on the management of tributaries. While there are various procedures for mainstream use, the Agreement contains no specific articles that deal with tributary development. Tributary projects are effectively treated as national projects outside the bounds of the Mekong Agreement, even if these projects have potential transboundary effects. Consequently, accountability for this disaster falls outside of the scope of the existing Mekong arrangement. Despite the disastrous effects felt in both Laos and Cambodia, the Mekong Agreement can do little to address the concerns of those affected in the two countries. Indeed, taking water cooperation for granted risks overlooking the causes and implications of the Xe Pian-Xe Namnoy Dam disaster.

This case study illustrates that water cooperation is not a panacea – it is a process, a relational struggle. The mere existence of a transboundary water agreement does not eliminate grievances or tensions over the way water is used and allocated. "Cooperation" itself necessitates critical scrutiny. The establishment of official agreements and founding of river basin organizations are only partial indicators of cooperation. Indeed, their existence and creation can also hide (or narrowly define) an entire set of relations, reflecting the hegemony of the cooperating states.

For example, the Colorado River Compact of 1922 (and its amendment of 1944) established water rights allocation between upper basin and lower basin states of the United States, as well as between Mexico and the United States. Nowhere in this agreement were the water rights of Native Nations and Indigenous communities, despite Indigenous presence that pre-dated imperial settlement in Mexico and the United States, and tribal status in the United States as sovereign nations (Curley 2019a).

Meanwhile, in the Jordan River – a basin often misrepresented as a water conflict "hotspot" – an agreement has existed since 1987 between Jordan and Syria. Nonetheless, the agreement lacks key clauses for a robust international treaty and misses out on a major element of flexibility: an important feature required to manage water equitably, especially when there have been significant volumes of migration into Jordan (Zeitoun et al. 2020).

Water is thus a site of both **conflict** and **cooperation**. A major advance in the scholarship of hydropolitics is to theorize conflict and cooperation as *coexisting* key features of shared waters. This nuanced approach considers how and why dynamics change instead of seeing politics as made up of singular events of conflict and cooperation. A critical approach situates such dynamics within the broader backdrop of political, economic, and social drivers that influence the allocation and use of water. What is important here is the *quality* of cooperation: how equity is considered; how responsibility and accountability for governing water are determined; how transparency and participation in decision-making regarding use and access are assured. Answers to these questions explain the power relations and politics that produce water.

This approach is broadly known as the **London school** of hydropolitics, best represented by the work of Naho Mirumachi (2015), a co-author of this book. The London school questions the normative implications of orthodox understandings of conflict and cooperation – that conflict is unconstructive and problematic, and that it needs to be averted or replaced with productive cooperation (Zeitoun and Mirumachi 2008; Zeitoun et al. 2011; Warner et al. 2017). A linear progression from conflict to cooperation is widely seen as desirable.

The London school argues that moving automatically from conflict to cooperation is, in fact, an apolitical and standardized approach (Zeitoun and Mirumachi 2008). This is because the very political nature of contesting access and allocation to water does not follow a linear progression. When basins and their contexts are highly varied across the world, there are no uniform steps that can ensure less conflict and more cooperation. Any intervention needs to be adapted to the context.

Moreover, conflict can be *productive* in raising issues of debate and contention, so that they are recognized as items on the agenda for deliberation. Cooperation, such as the examples of the Mekong Agreement, the Colorado River Compact, or the Jordan–Syria Water Agreement and similar transboundary agreements, can obscure historical injustices, grievances, and structural inequalities. Cooperative agreements can function as a status quo of unresolved issues, rather than act as a "settling" of problems and conflict (Box 7.1).

Box 7.1   Transboundary Cooperation and the SDGs

The binary view "conflict" or "cooperation" has implications for policy development. The lack of overt militarized conflict and low levels of cooperation have encouraged policymakers to seek measures to enhance cooperation. While there has been an increase in transboundary water agreements, the details and scope of such agreements have been varied (Giordano et al. 2014).

A prominent example is the Sustainable Development Goal (SDG) Indicator 6.5.2 which assesses the "[p]roportion of transboundary basin area with an operational arrangement for water cooperation." Rather than simply tally the number of agreements, the Indicator examines whether arrangements are operational. In other words, it looks at the existence of a joint governing mechanism, regular meetings, data and information exchange, and joint water management plans or objectives, all of which need to be fulfilled to be considered operational.

The combination of these four criteria reflects the scholarship that has analyzed the role and function of institutions, such as river basin organizations (Kittikhoun and Schmeier 2021). While data and information sharing are fundamental in theory, incentivizing these fundamentals in practice has proven to be challenging, with questions about the quality of exchange and compliance (Mukuyu et al. 2020). A range of case studies have shown the actual process of joint water management plans, particularly in the case of large infrastructure developments such as in the Nile (Wheeler et al. 2018) or when environmental flows are considered in adaptative management plans, as in the case of the Glen Canyon Dam, part of the Colorado river (King et al. 2019).

Thus far, progress of the SDG transboundary cooperation indicator has been reported by 61 countries out of all 153 countries that share transboundary water bodies. Results show that operational agreements cover less than 60% of transboundary basin area. Moreover, coverage is patchy, with only 17 countries establishing operational agreements in all basins within their territory (UN-Water 2018a). Clearly, there is more work to do.

A sticking point in this process is the lack of recognition of the quality of cooperation. For example, the 6.5.2 Indicator is focused on procedural aspects and is missing out on substantive features such as rights and obligations as per international water law (de Chaisemartin 2020). In addition, the geographical coverage of agreements presents biases in gauging the actual practice of cooperation (McCracken and Meyer 2018).

A Spectrum of Intensities

If conflict and cooperation are not binary, what tools and frameworks can help us make sense of shared waters? Applying the idea of coexisting conflict and cooperation can be done using the matrix developed for the Transboundary Waters Interaction NexuS framework, known as **TWINS** (Mirumachi 2015). The TWINS framework visualizes how interactions change over time using four quadrants: low conflict-low cooperation; high conflict-low cooperation; low conflict-high cooperation; high conflict-high cooperation (Figure 7.3). Different intensities of conflict and cooperation coexist, comprising varying characteristics of interactions over water (Mirumachi 2015). This approach provides a compelling insight as to why, for

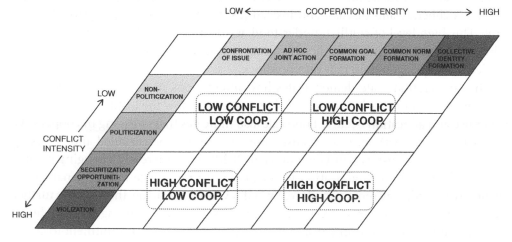

Figure 7.3 Transboundary Waters Interaction NexuS (TWINS) matrix. Source: The Authors.

example, a joint study may be carried out by governments but fail to resolve tensions over water infrastructure development. Such a situation would be one where there are stable relations of moderate levels of cooperation coexisting with non-overt conflict.

In other instances, technical water management committees might continue to function in shared basins despite the political fallouts of central governments. Here, the interaction may be unstable and volatile, made up of high conflict and low levels of cooperation. Low levels of conflict and cooperation coexisting suggest very little meaningful interaction. Understanding the coexistence of low and high levels of conflict and cooperation respectively gives an indication of why relations might be at a stalemate, or at times rapidly changing.

The TWINS framework is helpful in unpacking who makes the conditions of shared waters "scarce" or in "crisis." The framework focuses on the **hydrocracy**, or state elites who decide how water should be governed. The framework typifies the actors responsible for investing in and managing water, the associated infrastructure, and services. These actors are from various ministries and departments within the government and a key feature is their knowledge and technical expertise. These actors are in privileged positions of decision-making because they are based within the central government and possess accumulated know-how to shape agendas on river planning and water governance (Mirumachi 2015). The TWINS framework presents a different starting point for analysis than those studies using a binary notion of conflict and cooperation because it does not treat nation-states as monolithic entities. Rather than discussing "basin states" being in conflict (or cooperation) through generic terms, the TWINS framework specifies a precise set of actors who wield influence over state interests in interstate negotiations. This delineates where centers of power lie in deciding priorities over water resources.

A notable feature of the conflict scale is that it allows for the discursive effects of contentions to be examined. In other words, while there may be no violent acts, the TWINS framework accounts for discourses that shut down the usual means of engagement to deliberate and discuss, and instead justify extraordinary, exceptional measures. This high level of conflict, understood as a form of securitizing water, explains why unilateral action might occur, why usual protocols of decision-making are disregarded, or why negotiation falls into deadlock.

The cooperation scale demonstrates how interactions might evolve through the creation of common goals and norms, thus suggesting how the institutionalization of water use and allocation occurs. This scale also indicates that engaging in cooperative interaction is not without risk because future action may be curtailed and reversing arrangements is a costly affair (Zeitoun et al. 2020). Put differently, changing institutionalized water use and allocation is not easy. Oft times, once international agreements are signed, they remain in place. Even if revisions are made, they require further investment of time and resources. In the circumstances of changing basin political economy, not to mention climate and hydrological factors, the commitment to cooperative interaction involves risk-taking. This point is not highlighted in the existing literature, especially when cooperation is presented as normatively good. Evaluating cooperation is more complicated than observing joint action in the present and requires consideration of the extent to which future uncertainty features in deliberations.

The TWINS framework was originally developed to analyze interstate transboundary water interactions. But the TWINS framework could be applied to subnational actors, not limited to state elites engaged in interstate negotiations. In other words, interactions can focus within a state amongst a range of actors: (i) between different governmental agencies, industries, and civil society groups; (ii) between different communities along the river; and (iii) between sectors with vested interests in water resources such as agriculture or energy. The TWINS framework examines the motivations, drivers, and demands for water and the power relations between actors. By tracing how intensities of coexisting conflict and cooperation change over time, the matrix provides the analyst with a refined grasp of how nodes of power and influence change, with alterations to the constellations of influential actors. In this regard, while the exemplification of TWINS has been based on dyads of actors, there can be multiple actors that can be examined. Clearly, the politics of water do not occur only between two sets of actors, and multiple actors contribute to decision-making. Multi-actor water interactions within and across the basin scale make up dynamic coexisting conflict and cooperation.

Power and Hydro-hegemony

The way the hydrocracy produces water is situated within **hydro-hegemony**: conditions of asymmetric power relations that determine the control and access of transboundary waters (Zeitoun and Warner 2006). Basin states sharing the same river, lake, or aquifer are equal in that they are recognized as sovereign states. However, a single hydro-hegemon has overwhelming control over water resources through the application of power. This state not only sets the rules of negotiations but also rebuffs any challenges to them.

The framework of hydro-hegemony, developed by Zeitoun and Warner (2006), is fundamental to the body of work in the London school. Factors such as the geographical location of the basin state or technical capacity and expertise to secure water are advantageous to control water resources. States also exercise power, which is multidimensional. Hydropolitical power can be used to physically divert and store water for unequal access and promote narratives about the "best" way to utilize water resources or the "natural" way of managing the river.

Hydro-hegemony is based on a set of strategies and tactics to consolidate control over water resources as expressions of power. Such power is not always coercive but used to establish legitimacy (**bargaining power**) or to influence mindsets (**ideational power**). Within the framework of hydro-hegemony, bargaining and ideational power are more effective than material power. **Soft power**, as it is described, works to maintain consent to the hydro-hegemonic order. Soft power enables the hydro-hegemon to exercise its control over waters without it being constantly challenged; instead, this flex is accepted as the way of things by other basin states (Zeitoun et al. 2011). The hydro-hegemony framework recognizes that knowledge is power. Knowledge construction is one significant way of ensuring compliance, drawing on not only science but also normative beliefs or "received wisdom" (Zeitoun and Warner 2006). Marshaling these dimensions of power enables a hydro-hegemon to seek compliance from others.

The hydro-hegemonic state consolidates control through the establishment of institutions as they install normative agreements, such as water sharing treaties or legal agreements on basin development. For example, Egypt has long been considered the hydro-hegemon in the Nile River basin. Through a series of agreements set during British colonial rule, it has had overwhelming control of the river and has actively developed projects to utilize the water. Notably, Egypt is the downstream state, meaning despite being geographically distanced from the water sources, it has been able to command control over water through a combination of bargaining and ideational power (Cascão and Nicol 2016).

Hydro-hegemony is not fixed or immutable. Upstream on the Nile River, Ethiopia has constructed the **Grand Ethiopian Renaissance Dam** (GERD) to produce hydropower, an example of the "dam fever" pursued by developing nations, explained in Chapter 6. Ethiopia has exercised its power to generate project financing from foreign banks (notably from China) and to negotiate many rounds of talks with Egypt to ensure that downstream water flow will be unaffected and guaranteed with a legally binding agreement. The massive GERD project, with a potential for generating more than 6000 MW of electricity, will have basin-wide implications. GERD will change the hydro-hegemonic dynamics of the region, from reliance on decades-old colonial agreements to requiring multilateral cooperation (Yihdego et al. 2018). To date, much of the focus on water security for upstream and downstream states has been more of a technical matter concerning dam management (Wheeler et al. 2018). But as our analysis reveals, GERD is imbued with power relations, made even more complicated and delicate through the involvement of external actors such as the United States and the African Union.

Returning to the Mekong River basin, the issue of **sovereignty** spotlights another important dimension of power among shared waters. The 1995 Mekong Agreement set out measures for prior notification and consultation among nations when utilizing the shared waters. The Agreement also created the Mekong River Commission (MRC) – an international river basin organization represented by Laos, Thailand, Cambodia, and Vietnam – to guide policy and plans for basin development. However, long-standing tensions have simmered between maintaining national sovereignty and allowing policies and restrictions on water use to coordinate transboundary basin development. This issue was seen as problematic for upstream Thailand. As a hydro-hegemon, Thailand was able to leverage its soft power such that water use in wet and dry seasons was agreed upon as non-legally binding protocols (Mirumachi 2015).

A state not within the bounds of the Mekong Agreement is a hydro-hegemon: China. As we wrote in Chapter 6, China is equipped with the financial and technical expertise to develop the basin, exerting its influence through the construction of large dams. China has constructed 11 major dams on the main stem of the Mekong River in the upper basin, leading to concerns for downstream states. Studies show that upstream dams have restricted flows during an already drought-prone wet season in recent years (Eyler et al. 2022).

China is an observer to the MRC (along with Myanmar) and does not take part in or have influence over official transboundary decision-making. However, China continues to exert major influence over downstream states by extending dam building opportunities in the lower basin through the Belt and Road Initiative: a major program of infrastructure investments that aims to enhance global trade and economic development (Oliveira et al. 2020; Sidaway and Woon 2017; Sum 2019). Consequently, the production of water happens both through transboundary water institutions and beyond, mediated by hydro-hegemony.

Waterscapes of transboundary rivers are shaped with tightly intertwined ideas of economic progress and modernity. The case of the Nile shows that while Ethiopia may be finally exercising power to control water upstream, there are critiques that dam development is a tool for the political ambition of the ruling party, the Ethiopian People's Revolutionary Democratic Front. The developmental state model has facilitated the rapid expansion of hydropower with leaps in economic growth but also the grip of the party on state control. The overtly technical and economic debates of hydropower development overshadow urban–rural divides over water access, impacts on communities within project sites, and household burden – often on women – from irrigation or hydropower schemes (Müller-Mahn et al. 2022).

The top-down development in the Mekong (both in the upper and lower basin) portrays the river as an object of development, facilitated by the hydrocracy. Ideational power was effective in shaping a geographical imagination of a large river with high prospects for spurring economic development. The Mekong Agreement was negotiated by the hydrocracy to prioritize basin development. The Belt and Road Initiative also reveals the expectation of states in utilizing the river for development. Critically, the Mekong discourses that do not conform to state-oriented narratives of development (and the knowledges that contribute to shaping and deploying them) are ignored, marginalized, and disregarded. With businesses invested heavily in infrastructure projects, contestations over the Mekong question who owns the river (Figure 7.4). The perspectives of local communities and the impacts on ecology are missing from the debate (Fox and Sneddon 2019; Sneddon and Fox 2006). Their knowledge does not "count" in the eyes of the hydrocracy.

Critical hydropolitics provides sharper analysis to distinguish groups of people who face struggles over water. This analysis extends to those who are excluded from discourses or who pose fundamentally very different discourses than those perpetuated in the mainstream debate and channels of decision-making. Hydro-hegemony begins to answer questions of where water is and how its conditions are produced through power relations and

Figure 7.4   Activists ride a boat along the Mekong River, Thailand. Credit: International Rivers.

resisted in the social reproduction of peoples' lives. This analysis is the first step in identifying how states develop dominant discourses, such as those around economic development. By illuminating how these discourses shape and are shaped by the social, economic, and ecological relations in the basin, we can analyze how the diverse actors entrench themselves through multiple levels of decision-making.

Water Diplomacy as Praxis

In recent years, peacebuilding and security have become associated with governing transboundary waters. By building up trust between states, water acts as a catalyst for geopolitical stability and the observance of the rule of law (UNECE 2015). Consequently, one of the advantages of water diplomacy is argued to be the gains toward peace and security. The proponents of this view argue that water diplomacy is related to but distinct from those technical discussions of water cooperation or transboundary water management (Molnar et al. 2017). While transboundary water management may have existed within the purview of environmental ministries, those entities are regarded as having less influence compared to other ministries in state decision-making. Instead, security actors such as governmental agencies involved in foreign policy can act preventatively for instability and conflict. The introduction of the ministry of foreign affairs means that water issues assume a coveted position within the agenda of a prominent state agency. The involvement of foreign policy-makers can give more authority to decisions taken in international negotiations and persuade the public (Pohl et al. 2014).

Foreign policy experts provide useful knowledge on regional priorities that can be leveraged for out-of-the-basin or "problemshed" solutions compared to government agencies tasked solely with environmental management (Box 7.2). However, water can be "securitized" such that it is taken out of the public domain of usual decision-making. Because water is relational and concerns social and ecological relations, water diplomacy

---

Box 7.2   Problemsheds, Not Watersheds

Simply looking within the hydrological boundaries of transboundary waters does not tell us the full story of coexisting conflict and cooperation. Rather, we need a focus that reaches beyond the basin. The notion of the **problemshed** became critical to the analysis of transboundary waters in the pioneering work of geographer J.A. Allan on the international political economy of water. He argued:

> Explanations are not to be found by narrow analysis at the catchment and water budgets at the national level or even at the regional level. Catchment hydrologies are not a complete source of explanation because they are not determining the options available to those managing a national economy. If national hydrological systems restrict economic options then politicians have to find remedies in systems which do provide solutions. [...] national economies operate in international political economic systems – in problemsheds – and not just in hydrological systems – or watersheds. (Allan 2001, p. 19)

The focus on the international political economy enabled a new focus on local agriculture development to produce food for global export and how that impacts states' water use, and thus their interest in transboundary water allocation. The context is a complex relationship of food producing states, food importing states, agribusinesses, and farmers. As we explored in Chapter 5, the analysis of international political economic systems shows how food has a hidden geography of water that sheds light on the significant role of virtual water in intensifying and abating water struggles. Consequently, the analysis on conflict and cooperation would be amiss if it did not extend to the global institutions of food and trade connecting basin states.

The problemshed is a way to think about the spatiality of water that highlights the social relations between states, people, businesses, and other actors. This is advanced by geographers such as Wescoat and Halvorson (2012, pp. 90–91), who highlighted how examining different kinds of problemsheds reveal the "real everyday context of regional water struggles and management practice." For example, the links between water, food, and energy could characterize one problemshed. Another might be about water, hydroclimatic variability, and security. The analysis of a problemshed shows that water struggles are found in the multiple social, spatial, and ecological relations which defy the hydrological basin scale. Mollinga et al. (2007) describe the problemshed as an issue network in which actors identify, shape, and decide on water management issues. Water management is at once political and social. The boundaries of water management are a social and political construct. Problemsheds reject that "boundaries are pre-defined spatially, sectorally and analytically through the primacy of 'water'" (Mollinga et al. 2007, p. 707).

Importantly, utilizing the problemshed as an analytical lens helps further uncover the answer to who produces water. The spatiality of water points to multiple organizations, various institutions, and plural actors that are involved in water, food, energy, and other issues that affect water struggles. This analytical lens illustrates the polycentric nature of decision-making at multiple scales and across sectors. The problemshed questions the effectiveness of international river basin organizations determined by the physical boundaries and siloed in their attention to the water sector through a narrow representation by water ministries.

of this mode does not address the struggles of individuals and communities. Water diplomacy is bound up with the national interests of elite decision-makers, representing a limited and skewed view. A critical reading of shared waters requires an examination of the assumptions about water diplomacy resolving conflict – for whom and at whose expense?

Securitization of water results in decision-making passed on to opaque, exclusive processes in the name of sovereign interests and state survival (Mirumachi 2015). Foreign policy stakeholders are particularly well-positioned to determine how and when sovereign interests and state survival matter. In such cases, scrutiny by the public – which requires polycentric, deliberative decision-making – may pose security risks to central authorities. Consequently, the struggles of people dependent on the transboundary water resources and their ecosystems can be disarticulated from state security agendas and foreign policy goals. Securitization renders a situation with no alternatives and the hydro-hegemon can discursively shape the single option as valid, urgent, and essential, while constructing objections as unviable. Securitization of water poses challenges to the transparency and accountability of the state, as well as changing the dynamics of transboundary water relations. Cooperation may emerge as a result of this opaque process, but generally results in outcomes that favor the powerful state (Mirumachi 2015).

Recent policy discussion promotes water diplomacy because of the possible gains in peace. As Pohl et al. (2014, p. i) note, "transboundary waters constitute a promising entry point for diplomats aiming for high peace dividends." Policy support tools are aimed to "use water to generate peace dividend[s]" (Strategic Foresight 2021). However, the academic literature on the link between environment and peacebuilding requires careful analysis of causal relations. For example, the link between environmental cooperation and peace is nuanced and highly context dependent. While environmental cooperation can nurture non-violence, there are many other factors at play for peacemaking and no single cause for peace (Ide 2019). This means that there are potential risks in overclaiming peace dividends at scale due to water diplomacy. Furthermore, scholarship has begun to posit that environmental peacebuilding may have adverse effects, not least of which can be an acceleration of environmental degradation and displacement of populations, especially when large-scale projects are implemented under peacebuilding aims (Ide 2020).

As we have established in this book: camp is everywhere. The study of hydropolitics is overtly focused on the state. Struggles over water are not limited to the level of international negotiations – a topic we examine in Chapter 8. Water diplomacy can be another tool in a suite of approaches that community leaders, concerned water users, civil societies, and individuals can employ for alternatives to existing arrangements that affect equity. While formal water diplomacy may be represented by state officials, there are multiple tracks to diplomacy. If formal water diplomacy is considered Track 1, then informal water diplomacy makes up Track 2, 3, and more. There are different definitions and classifications of these tracks (Diamond and McDonald 1996), and studies on multi-track diplomacy demonstrate that a wide range of actors – from citizens to religious leaders to businesses – can be involved. Within transboundary water governance, Track 2 is used regularly to bring together non-governmental representatives to support the formal negotiations and is regarded as useful when there is a deadlock in Track 1 (Barua and Vij 2018). From a critical perspective, water diplomacy can question, resist, and reaffirm worldviews over transboundary water resources.

Crow and Singh (2003) assert an important point on the possibility of multi-track water diplomacy to generate alternatives. Whereas large-scale developments in transboundary rivers are often debated, they suggest that multi-track water diplomacy allows for smaller projects that are decentralized, based on bottom-up demand, and flexible to the local context. Large-scale projects tend to have steep stakes and trade-offs that affect more individuals and communities than small-scale ones. Multi-track diplomacy confronts top-down megaprojects based on ambiguous national interests. The multi-track approach creates a platform to discuss how equity can be achieved through context-appropriate development interventions. Potential harm may be more easily identified and addressed by diplomatic dialogue between a wider set of stakeholders who know the local context well.

The fact that there needs to be political will to deal with water issues is recognized within scholarship and policy. Water diplomacy is one distinct way in which such political commitment can be achieved, especially in formal, Track 1 processes. But it can be the stage of ontological struggles as well. With multiple sectors and stakeholders engaged in the use, allocation, and access of water, multi-track diplomacy is relevant to deliberating trade-offs and harms.

The Future of Shared Waters

For a just and sustainable future of transboundary rivers, lakes, and aquifers, what can be done? One approach is to challenge hydro-hegemony. **Counter hydro-hegemony** is a mode of practice that contests existing institutions shaped by a combination of power and narrow decision-making. Tacking back to concepts we introduced in Chapter 1, counter hydro-hegemony promotes alternative discourses and governance models that center water struggles. This approach does not merely reform, adjust, or revise agreements and policies – it recognizes the ontological politics of water, with the goal of fundamentally *transforming* the production of shared waters.

Importantly, to make such changes, it is vital to begin with a focus on the kinds of harm that people experience. By focusing on the production of **harm** to individuals, the community, and ecosystems, a transformative approach presents a method to overhaul priorities and practices around water use (Zeitoun et al. 2020).

What is "harm" in water? Harm is widely recognized in the governance of internationally shared freshwater bodies. Perhaps the most notable usage is the principle of "do no harm" espoused in a landmark convention of international water law, the 1997 UN Convention on the Law of the Non-navigational Uses of International Watercourses. As a key principle, along with "equitable and reasonable use" and "prior notification," the recognition of harm is central to regulating state behavior and peaceful engagement over shared waters.

In practice, the definition of harm is vague and comes with various interpretations. Much of the legal debate has focused on what "no harm" might entail. Tignino and Bréthaut (2020) summarized that the process to develop the legal understanding of "no harm" has covered the degree or scale of effects, damage, and impacts that might occur from one state's use of water. The object of harm can be sovereignty and territorial claims as well as environmental concerns.

Harm is largely understood in a material way, connected to environmental damage or the loss of territorial integrity. But harm can also be experienced and manifest in discursive and emotional ways. Drawing on feminist and political ecology perspectives, geographers have unpacked how **suffering** occurs from various interventions of water management. These studies show how suffering shapes the flows of relations between water and society, impacting the dynamics of water conflict. Farhana Sultana (2011) shows that conflicts over access to or use of safe water in Bangladesh result in uneven degrees of suffering, particularly for women, the traditional managers of household water. These women experience embodied and emotional pain of securing clean water and mediating relationships with the owners of safe wells and the wider community.

There are hardships from risking unsafe, arsenic-contaminated water, including poor health. But conflicts over access, use, and claims to water rights also produce suffering and emotional labor, making these already vulnerable women rely on polluted sources. The stigma of those ill from arsenicosis not only presents difficulties in employment and loss of economic opportunity, but also implies such people are less valued within the community than those who are healthy (Sultana 2012). Discursive harms distinguish between those people who are affected by arsenic and those who are not; and moreover, these discourses devalue the former. Such struggles for water occur daily and the sufferings aggregate if not addressed. Water struggles and suffering shed light on the need for alternative water governance to protect and value affected people and their communities.

A focus on harm informs an alternative perspective on the production of water. A critical approach might ask: who (or what) produces harm in shared waters? Such an approach questions who is involved in decision-making about water and how they exercise their authority. Take the example of Basra, Iraq, positioned downstream of the transboundary Tigris–Euphrates rivers. There is no basin-wide agreement between Turkey, Syria, Iran, and Iraq. The hydraulic missions of upstream countries, as well as changing climatic conditions, have contributed to significant declines of river flow into Iraq. Problems of water quality have compounded supply issues, causing problems in Iraqi cities like Basra, heavily affected by war.

What produces harm? In Basra, water provision is hampered by **clientelism** – defined as the capture of public resources for private gain (Herrera 2017). Clientelism in Basra is a kind of politically "sanctioned" corruption bound up in oil revenues that impedes progress in extending and updating critical water infrastructures (Mason 2022). As a result, the state does not have the sufficient capacity to provide water to a growing population; nor do Basra residents trust the water quality, instead relying on private water sources. Many residents are not connected to the public water network, exposing those without appropriate housing arrangements to contaminated water (Mason 2022). Harm is experienced by poor and marginalized people, while elite actors benefit. In thinking about the Tigris–Euphrates River basin with a critical lens, we begin to see how hydrosocial relations are experienced unevenly in different parts of the basin, an approach that pushes our understanding beyond the limits of nation-states and questions about upstream development altering water flows.

While hydro-hegemony is often mediated through river development projects, the process is not simply about control over water. Once again, the Mekong River provides a pertinent illustration. Blake and Barney (2018) point out that there is coercion and violence directed at individuals and communities by the state and development project authorities. Drawing on the example of dam building in Laos, they explain that the Laos state implemented a strategic plan of hydropower development to take advantage of its Mekong basin water resources. However, the plan operates as authoritarian state control over the Laotian people. Authority is not merely exercised in dam construction but extends to the involuntary resettlement and the alternative livelihood means offered to affected communities (Blake and Barney 2018). People have no choice but to give up their existing customary rights and accept these options, amounting to indirect coercion and state violence.

Here, **slow violence** degrades livelihoods and erodes bestowed rights to resources of the affected communities. Slow violence is not immediate or spatially even, often manifesting over time and exacerbating precarity and vulnerability (Blake and Barney 2018). Slow violence extends existing suffering and harm. In their study, while the Laotian state was the focus as the source of slow violence, they also point to private sector entities involved in the business of investing and developing infrastructure projects (Blake and

Barney 2018). The coalition of state and capitalist enterprise contributes to the production of slow violence. Indeed, the interplay between state and non-state actors is important to locate the ways that harm, violence, and resistance are produced and experienced. Understanding the agency of state and non-state actors offers insight into how inequality is bound up in the access and allocation of shared waters.

Summary and What's Next

Shared waters are a common feature of planetary life, with hundreds of international transboundary river basins and aquifers. This chapter examined the relational nature of transboundary waters as they flow into multiple states and shape their interaction in material and political ways. While the water "wars" thesis is popular in mainstream accounts, researchers have found no evidence of waters wars or direct conflict. At the same time, the notion of "cooperation" as a straightforward solution to water disputes is a fallacy because of the political production of water. International water agreements cannot fully accommodate a heterogeneous basin with contested worldviews of shared waters.

This chapter offered a different set of intellectual tools to understand shared waters. Water struggles in transboundary contexts reveal relations of power between states and actors, how they mediate which state has access to water, when, how much, and for what development goals. The inequality of water access and allocation is observed between states but also within states, whereby individuals and communities experience conditions of water scarcity or abundance differently. Water diplomacy can be a way to mediate the ontological politics of water. A relational view of shared waters situates questions of state interests and sovereignty in tandem with broader concerns of people and the environment. For example, the focus on harm and slow violence experienced by individuals and communities testifies that there are no universal, global measures to enhance water cooperation or address water conflict.

A critical approach to shared waters examines and elicits the multi-scalar dimensions of geopolitics – not simply confined to the scale of the nation-state. In the next chapter, we pick up the thread of bodily harm and slow violence to explore how and why the experience of water is an intimate and fundamental part of citizenship – demarcating who gets to belong (and who does not) to the social fabric. Where do you belong? Find out next.

Further Reading

Transboundary water governance
Allan, J.A. (2001). *The Middle East Water Question: Hydropolitics and the Global Economy*. London: I.B. Tauris.
Bréthaut, C. and Pflieger, G. (2020). *Governance of a Transboundary River: The Rhône*. Cham, Switzerland: Palgrave Macmillan.

Earle, A., Cascao, A., Hansson, S., Jägerskog, A., Swain, A., and Öjendal, J. (2015). *Transboundary Water Management and the Climate Change Debate*. Abingdon, UK: Routledge.

Transboundary aquifers

Burchi, S. (2018). Legal Frameworks for the Governance of International Transboundary Aquifers: Pre-and Post-ISARM Experience. *Journal of Hydrology: Regional Studies* 20: 15–20. https://doi.org/10.1016/j.ejrh.2018.04.007.

Linton, J. and Brooks, D.B. (2011). Governance of Transboundary Aquifers: New Challenges and New Opportunities. *Water International* 36 (5): 606–618.

Villar, P.C. (2016). International Cooperation on Transboundary Aquifers in South America and the Guarani Aquifer Case. *Revista Brasileira de Politica Internacional* 59 (01). https://doi.org/10.1590/0034-7329201600107.

Hydro-hegemony

Rudolph, M. and Kurian, R. (2022). Hydro-hegemony, Water Governance, and Water Security: Palestinians Under Israeli Occupation in the Jordan Valley, West Bank. *Water Alternatives* 15 (1): 73–92.

Warner, J., Mirumachi, N., Farnum, R., Grandi, M., Menga, F., and Zeitoun, M. (2017). Transboundary 'Hydro-hegemony': 10 Years Later. *WIREs Water* 4 (6): e1242.

Zeitoun, M., Cascao, A., Warner, J., Mirumachi, N., Matthews, N., Menga, F., and Farnum, R. (2017). Transboundary Water Interaction III: Contesting Hegemonic Arrangements. *International Environmental Agreements* 17 (2): 271–294.

International water law

Leb, C. (2020). *Data Innovations for Transboundary Freshwater Resources Management: Are Obligations Related to Information Exchange Still Needed? Brill Research Perspectives in International Law 4 (4)*. Leiden: Brill.

McCaffrey, Stephen C. (2019). *The Law of International Watercourses*, 3rd edition. Oxford, UK: Oxford University Press.

Rieu-Clarke, A., Moynihan, R., and Magsig, B-O. (2012). United Nations Watercourses Convention: User's Guide. IHP-HELP Centre for Water Law, Policy, and Science (under the auspices of UNESCO), United Kingdom. https://unece.org/fileadmin/DAM/env/water/meetings/Water_Convention/2016/10Oct_From_Practitioner_to_Practitioner/UN_Watercourses_Convention_-_User_s_Guide.pdf (accessed 13 June 2022).

River basin organizations and regional institutions

Huitema, D. and Meijerink, S. (2014). *The Politics of River Basin Organizations: Coalitions, Institutional Design Choices and Consequences*. Cheltenham, UK: Edward Elgar.

Ingram, H. and White, D. (1993). International Boundary and Water Commission: An Institutional Mismatch for Resolving Transboundary Water Problems. *Natural Resources Journal* 33 (1): 153–175.

Zikhali-Nyoni, T. (2021). The Role of SADC in Transboundary Water Interactions: The Case of the Incomati International River Basin. *Journal of Southern African Studies*, 47 (4): 703–718.

Water diplomacy

Carmi, N., Alsayegh, M., and Zoubi, M. (2019). Empowering Women in Water Diplomacy: A Basic Mapping of the Challenges in Palestine, Lebanon and Jordan. *Journal of Hydrology* 569: 330–346.

Islam, S. and Smith, K. (2020). *Interdisciplinary Collaboration for Water Diplomacy: A Principled and Pragmatic Approach*. New York: Routledge.

Mirumachi, N. (2020). Informal Water Diplomacy and Power: A Case of Seeking Water Security in the Mekong River Basin. *Environmental Science and Policy* 114 (1): 86–95.

Part 3

Water is Life

Chapter 8

Intimate Waters

Citizenship and the Toilet

Picture England in the 1960s. The "Swinging Sixties" produced color television, pirate radio, contraception pills, The Beatles, Notting Hill Carnival, anti-war protests, and a pivotal World Cup victory in men's football [soccer], the 4–2 win over Germany in 1966. The decade was a time of tumult, change, promise, and hope.

This era also revealed stark legacies of social and spatial inequality, including our most intimate water-related activities. A 1967 English Housing Survey (EHS) found that 25% of homes in England lacked an indoor toilet, bath or shower, and/or a sink with hot and cold running water (UK Ministry of Housing, Communities & Local Government 2017) – a situation we call **plumbing poverty** (Deitz and Meehan 2019). Across England, some 2.5 million homes did not have an indoor flush toilet and 5 million homes lacked at least one of the three basic facilities. Hard to imagine today, but some of our UK-based friends and neighbors recall growing up with an outside toilet (also called a "privy" or "outhouse") and filling up tin baths in the "wet room."

In England, the transition from outdoor water to indoor plumbing required state muscle, massive infrastructure investment, new technologies, targeted legislation, and public campaigns to change norms and behavior. In the mid-nineteenth century, when housing stock boomed in Britain's industrializing cities, indoor plumbing was rare and limited to wealthy homes. Outbreaks of water-related diseases, like cholera,

Water: A Critical Introduction, First Edition. Katie Meehan, Naho Mirumachi, Alex Loftus, and Majed Akhter.
© 2023 John Wiley & Sons Ltd. Published 2023 by John Wiley & Sons Ltd.

were common. The 1858 summer heat wave amplified the smell of untreated sewage in London's Thames River. City life came to a standstill and the odor chased people out of Parliament. The "Great Stink" helped to galvanize public officials to approve funding and designs to modernize London's sewage system.

The infrastructural rebirth of London was ambitious, reflecting new relationships between people, nature, and the state. Engineer Joseph Bazalgette's plan involved building 1100 miles of drainpipes under London's streets, feeding into 82 miles of new brick-lined sewers, and carrying sewage to six intercepting sewers (Ackroyd 2008). Many of these sewer outlets had once been open streams; "lost rivers" like the Fleet River were brick-lined and covered over to become hidden flows of waste (Ackroyd 2008). A boom in infrastructure construction swept the island, introducing modern sewage disposal methods, treatment techniques, and improvements in filtered water and sanitation provision. The co-benefits supported public health fights against cholera and water-borne diseases. However, these co-benefits were concentrated in the growing middle classes, leaving the working class (especially in regional cities) reliant on communal water sources and outdoor toilets, as pictured in Figure 8.1.

By the 1960s, these conditions remained largely the same. Mr. Frank Allaun, the elected Member of Parliament for Salford East (a working-class community near Manchester), reported that 5 out of 10 houses in his constituency had no bath or inside toilet (*The Guardian* 1960). The government made grants available to homeowners to provide missing plumbing, but serious gaps in infrastructure coverage remained, especially for working-class renters. "Owner-occupiers were taking advantage of the Government scheme for installing bath and inside toilets in approved houses," *The Guardian* (1960) noted, "but it was being largely ignored by private landlords."

Fundamentally, the toilet is an *intimate* experience. At the same time, the toilet is a direct connection between you – the subject doing your private business – and the broader public realm of water utilities and providers, environmental sources and sinks (i.e. rivers), local and national government, and social norms and expectations. Consider Figure 8.1, a 1960s indoor plumbing campaign poster by the UK Ministry of Environment. Dressed in a pink bathrobe, the "wife" of the home waits angrily in the rain, while "Fred" (presumably her husband) uses the dilapidated toilet. Outdoor toilets are backwards, the image says. She is a modern white woman on the verge of upward mobility, the poster implies, and he is lazy. Sexualized humor and gendered norms do the work here. Nameless women in bathrobes are the responsible catalysts of household water, but Fred does the finances. The audience might chuckle at this cartoonish class scene. But the poster draws on gendered norms to send a clear, "logical" message. "Don't just sit there Fred," the Ministry gently chastises, winking to the viewer, "Get a house renovation grant."

What is revealed in this image? And, more importantly, what remains absent or hidden? Toilets have everything to do with water, race, gender, class, citizenship, and the public sphere. In previous chapters, we examined many of the "big" water issues in the hydrosocial cycle: agriculture and food production, law and policy, large

Figure 8.1 "Don't Just Sit There Fred, Get a House Renovation Grant." This poster, produced by the Ministry of the Environment, was part of a nationwide effort in the 1960s to provide "complete plumbing" (running hot/cold water, indoor shower/bath, and a flush toilet) to English houses. Source: Katie Meehan (author).

dams and hydropower projects, the economy. In this chapter, we focus on more *proximate* relations, what we call **intimate waters**. A critical understanding of intimate waters requires a focus on the relations that shape the most personal aspects of our lives: our bodies, labor, households, health, subjectivity, and belonging to – and exclusion from – the wider social fabric. Intimate waters are not singular or private, but collective achievements. This chapter will develop an understanding of how Fred, sitting "alone" on his outdoor toilet, is not isolated but in fact deeply embedded in the wider hydrosocial fabric.

Margaret Thatcher, the Prime Minister of the UK (1979–1990), was notoriously cautious when it came to privatizing water. Her concern was that something as intimate as water, central to people's everyday existence, would not be easily framed as a commodity to be sold on a market and bought by consumers. Nonetheless, the UK government privatized water utilities in England and Wales under her administration (Chapter 4). Paradoxically – or perhaps not – in the parts of the world where water has been privatized, it has helped to reframe water-using citizens as "consumers." Politics is practiced and goes unnoticed through our intimate relations with water, reproduced daily in our homes and neighborhoods with every flush of the toilet. Individuals as part of social groups are made through relations with those who buy and sell water, with those who guarantee their right to water, and with those who work to ensure ongoing access to water. Everything from a gendered division of labor to ongoing problems of water access and security – even in the places we consider to be "modern" and "developed" like England – can be seen through our intimate relations with water.

If we are what we eat, then taking the hydrosocial cycle seriously means we are also what we drink. This chapter argues that water is a fundamental part of being a citizen – demarcating who gets to belong (and who does not) to the social fabric. By taking a relational approach to intimate waters, we explore the uneven conditions of citizenship and belonging: the gendered and racialized divisions of labor in securing water; the shifting subjectivities in new market models of water; and the emotional and corporeal dimensions of water, including connections between water, stress, stigma, and mental health.

What do we mean by **citizenship**? A starting point is T.H. Marshall's (1949) influential theorization of citizenship as an aspirational ideal associated with three types of rights. Citizenship, on paper, is a legal relationship with a state that comes with rights and responsibilities. Marshall's analysis paves the way for us to distinguish between *de jure* (formal) and *de facto* (informal) rights, and to understand that citizenship rights do not emerge ready-made but are struggled for and claimed over time. "Formal" citizenship – symbolized by a passport – is as much about exclusion as inclusion. But Marshall's model now seems limited in important ways, opening useful lines of critique for thinking about citizenship in the context of water. First, the model is rooted in the idea of the nation-state (and the history of Britain specifically). To what extent can this model account for political belonging that doesn't easily map onto the nation-state? Second, Marshall was mainly concerned with defining citizenship as a formal status. What if, alternatively, we foreground the substantive dimensions of citizenship as *practice* – as an ongoing set of struggles to claim rights, reflecting different group identities and positionalities?

Citizenship, as Hannah Arendt describes, is the "right to have rights." Sara Smith (2020, p. 17) reminds us, "this means citizenship will *always* fail to protect some people." Nikhil Anand (2017) explores this failure in the waterscapes of Mumbai, India, where people living in informal settlements struggle to gain access to the city's potable water infrastructure network. Anand (2017, p. 16) demonstrates how this process of claiming citizenship – "the ability of residents to be recognized by city agencies through legitimate water services" – is an incremental, contingent, and even reversible form of political subjectification. Similar dynamics of claiming citizenship are at play in Buenos Aires, Argentina, where residents of a low-income neighborhood rejected temporary "high-tech" toilet solutions. Instead, they demanded the extension of integrated, universal, state-managed sewerage systems or the "kind you flush and forget like in the rich countries" (Morales et al. 2015; see also Sultana 2020).

Citizenship has never been neutral or universal (Smith 2020, p. 17). At its core, citizenship is a negotiation for **belonging** – to the nation-state, but also to cities, communities, and wider social worlds. Water is an important and intimate marker of belonging. The humble toilet is just the tip of an iceberg: the tip of a material culmination of power, class struggle, social norms, ecological modernity, and social relations. Think about that the next time you flush.

The Work of Water

Water is hard work. In Tlalpan, a borough of Mexico City, Lupe wakes up early to check the tap and cistern in her modest house. The pipe is empty. Lupe feels the whoosh of air against her cheek. While her house is connected to the municipal network, water only comes once a week, part of a distribution schedule called the *tandeo* (distribution by turn). The *tandeo* periodically doles out water to "connected" customers in this working-class hillside neighborhood, perched on the side of a steep volcano.

Water service is unpredictable, Lupe explains, so she always keeps the tap open. She buys expensive bottled water for drinking and purchases water from *pipas* (mobile water trucks) to fill the cistern. For Mexico City residents without piped connections to the public water network, local governments often subsidize trucked water delivery. However, Lupe must pay private rates because she is technically "connected" to the grid. During the rainy season, Lupe captures and filters rooftop rainwater for nonpotable purposes – laundry, cleaning, and bathing her pet dog (Figure 8.2). All this work takes time, physical labor, and money. Still, it's better than when we used to haul water to the house, Lupe recalls, water is heavy.

Who does the work of water? And what creates these situations in the first place? As we analyze citizenship through its margins, edges, complications, and failures – such as households without running water or toilets – we gain a stronger grasp of the institutional structures and dynamics that actively *produce* conditions of water marginalization (Ranganathan and Balazs 2015; Jepson 2012; Meehan et al. 2020a, b). In Mexico City, the lack of water is no accident, nor is it the fault of residents who do not "work" hard enough. As Lupe's story shows, water is very hard work and the labor disproportionately falls on the shoulders of the most marginalized: a situation of uneven citizenship.

Water marginalization is not unique to Mexico City. Lupe's situation reflects what experts call **household water insecurity**: the insufficient supply and quality of water to achieve a healthy, productive, and thriving life (Jepson et al. 2017). Household water insecurity is more than a lack of physical access to water. Experts point to key

Figure 8.2 A rainwater harvesting system in a Tlalpan home, Mexico City. This system, designed by the non-governmental organization Isla Urbana, features a "first-flush" device (the barrel) designed to expunge and filter initial rains. In this home, the storage cistern is located underground. Source: Katie Meehan (author).

issues of water quality, safety, reliability, affordability, and acceptability that influence whether a household is "secure" or not (Box 8.1; Jepson et al. 2017; Meehan et al. 2020b, 2021; Wutich 2020). For example, unreliable water service means that Lupe must rely on expensive sources (bottled water, private *pipa* water) or seasonal supply without centralized quality control (rainwater). Lupe's house has access to the grid. Insecurity, in this case, is more about reliability, affordability, and quality. A focus on these myriad factors shows that household water insecurity is driven not simply by technical failures but by social relations. These social relations differ across space and place, and are institutionalized in policy and practice, linking water insecurity to broader political, economic, and cultural systems of social inequality, marginalization, and exclusion (Deitz and Meehan 2019; Loftus 2015; Jepson 2014; Jepson et al. 2017; Wutich and Ragsdale 2008).

Infrastructure absence and malfunction occur for a variety of reasons. But, as Tatiana Acevedo-Guerrero (2019) and others argue, infrastructure malfunction does not necessarily indicate the "absence" or "failure" of the state. In the Colombian city of Barranquilla, Acevedo-Guerrero (2019, p. 480) excavates the specific state laws and regulations that contribute to systematic malfunction and details the practices that "lay bare the ways in which malfunction is enforced and tolerated through institutional channels." Indeed, a critical engagement with household water insecurity reveals the state's *active* role in the production of marginality – through regulations, zoning, funding, decisions, programs, policing, and infrastructure (Acevedo-Guerrero 2019;

Box 8.1 Unaffordable Water in the United States

Household water insecurity involves more than just piped connections. Most households in the United States have piped running water (see Deitz and Meehan 2019 and Meehan et al. 2020a for exceptions) but rising costs and high-profile crises have brought increasing attention to the **affordability** of water and sewer service as a potential driver of household water insecurity in the United States, one of the world's wealthiest countries (Teodoro 2019; Pierce et al. 2021; Swain et al. 2020).

Unaffordable water disproportionately burdens low-income households. Manuel Teodoro (2019, p. 1) found that "low-income households must spend an average of 9.7% of their disposable income and/or work 9.5 hours at minimum wage to pay for basic monthly water and sewer service" – a figure that varies considerably across the country (see also Teodoro and Saywitz 2020). In Teodoro's (2019) analysis, San Francisco, California, had the least affordable rates, whereby low-income households must work 13.6 hours at minimum wage to cover their monthly bills for water and sewer service.

The COVID-19 pandemic revealed: (i) the uneven nationwide regulation of water tariffs, (ii) the lack of comprehensive support and effective intervention programs, and (iii) the sheer degree of a water service shut-offs – punitive measures targeted at ratepayers who are delinquent in their water bills. While several US cities like Detroit and Phoenix passed emergency measures to temporarily suspend their water shut-off programs during COVID-19, many US residents were outraged at the ubiquity of water shut-offs in the first place (Lakhani 2020).

What does it mean to transform water "citizens" into "consumers"? In some cases, the options are debilitating, causing both short-term and long-term health and economic consequences (Swain et al. 2020). A Cleveland resident, who owed more than USD 5000 in charges (including late penalties and interest), told reporter Nina Lakhani of *The Guardian*, "I've done two payment plans, but I'm still in foreclosure, it's like they're trying to make me homeless. There is no way I'm using the amount of water they're charging me for but I'm in a no-win situation. I don't want to lose my home, so I have to keep finding the money" (quoted in Lakhani 2020).

Gregory Pierce and others illuminate the human stakes of unaffordable water. As they explain (2021, p. 2), "low-income households unable to afford water service face trade-offs that may harm their health and welfare." People may cut back or forego household drinking and sanitary needs, risking their health (Rosinger 2020). Households may be forced to make welfare-harming decisions when choosing between paying for the minimum amount of water services to satisfy basic needs and other essential goods and services (Rockowitz et al. 2018).

New research argues that affordability in the United States should be normatively guided by the core tenets of the human right to water (Goddard et al. 2021).
Additional research is needed to comprehend the roots and drivers of the water
affordability crisis in one of the world's wealthiest countries, and to develop
mitigating policies and practice.

Adams 2018a, b; Adams et al. 2018b; Kooy and Bakker 2008; Kooy 2014; Meehan 2013;
Ranganathan 2014a; Ranganathan and Balazs 2015). For example, in many informal
settlements across the global South, the extension of drinking water infrastructure is
prevented or complicated by state refusal to grant legal tenure and land title to settlers
(Acevedo-Guerrero 2019; Adams 2018b; Anand 2011; Kooy 2014; Ranganathan 2014b).

Marginality is produced and policed by the state, even at the corporeal level.
In Tijuana, Mexico, water officials ignore unauthorized water connections by
"conventional-looking" households in irregular settlements (*colonias*), while policing
and disrupting the same kind of unauthorized water usage managed by unhoused
(homeless) people and deportees. This exercise of state power delineates citizenship
and belonging through techniques of **biopower** (Meehan 2013). Citizenship is thus
the "embodied intersections of sociospatial differences" that collide with the felt geographies of infrastructural materiality (Sultana 2020, p. 1408).

Struggles for secure and safe water can be seen as making space for citizenship
(Sultana 2020). Malini Ranganathan (2014a) interviewed property owners in
Bangalore, India, who lacked formal water access and land tenure – groups she refers
to as the "peripheralized middle class." Owners agreed to pay high costs to connect
their households to the municipal network, despite the expense. She argues that such
practices are acts of claiming citizenship:

> [F]ar from reflecting an internalization of a "willingness to pay" or "stakeholder" ethos
> celebrated by development practitioners today, payment for water provides an insurgent
> means to bargain for greater symbolic recognition, respectability, and material benefit
> from the state. In particular, payment for pipes enables peripheral dwellers to strengthen
> their claims to secure land tenure in an era of exclusionary and punitive spatial policies.
> Payment comprises a terrain of contested meaning-making and political struggle, at the
> heart of which lie the stakes of urban citizenship. (Ranganathan 2014a, p. 590)

Ranganathan's analysis shows the importance of considering "everyday" acts of political agency through seemingly ordinary and mundane practices. Paying for a pipe, in
this case, is a bargaining chip to win further rights from the state.

Insecurity reveals "cracks" in the system, and who performs the gendered and
racialized **labor** of managing life among these cracks. Labor includes both the waged,
formal labor of production in modern society and the reproductive and often unwaged
labor required to produce life and show up to a paid job. Building on the rich insights

of feminist theory and scholarship (Ahlers and Zwarteveen 2009; Dickin and Caretta 2022; Harris et al. 2017; Sultana 2009; Truelove 2011), a critical approach recognizes the existence of overlapping systems of water marginalization based on multiple markers of social identity and difference, such as gender, ethnicity, race, sexuality, class, caste, and (dis)ability (Harris et al. 2017).

Who does the work of water? Worldwide, the **gendered division of labor** helps to determine women's role as the primary domestic caretakers, responsible for water-related management and tasks within households and smallholder agriculture (Bennett et al. 2008; Cleaver 1998a, b; Crow and Sultana 2002; Meinzen-Dick and Zwarteveen 1998; O'Reilly et al. 2009; Sultana 2009; Truelove 2011). Gender is a social construction, not a biological category. The central role of women in water – reflected in the third plank of the Dublin Statement (Chapter 4) – is due to institutionalized social norms that designate who performs "caretaking" work and reinforce a hierarchy that punishes deviation or difference.

The gendered division of labor has evolved through space and time, but such modifications do not necessarily reflect a systemic transformation or societal change. Lupe once manually fetched water daily from a community well. Now, she is responsible for managing the irregularities around a piped water service and finding alternative financing and sources, like trucked deliveries and rainwater harvesting. Even though women shoulder a disproportionate burden of water work at the household and community levels, they are routinely marginalized from decision-making and excluded from formal water governance – yet another effect of the gendered division of labor (Crow and Sultana 2002; Meinzen-Dick and Zwarteveen 1998; Zwarteveen 1997, 2010).

The gendered division of labor results in a hierarchy of citizenship. In other words, *who* gets to establish the "right to have rights"? As Ellis Adams and colleagues show in the case of Lilongwe, Malawi, despite valiant efforts to promote "gender equality" and participatory decision-making in local water management, it is exceedingly difficult to transform hierarchies, social norms, and power relations. In Lilongwe, the patriarchy of water is expressed in the male-dominated make-up of local water user associations. The patriarchy of water suppresses women's agency and reinforces existing gender-based inequalities in governing community water systems (Adams et al. 2018a).

More broadly, the labor of water can be understood as a form of **social reproduction**. There is a range of regenerative and unwaged work practices such as caretaking, provisioning, cleaning, feeding, and watering required to produce life and show up to a paid job (Katz 2001; Strauss 2013; Meehan and Strauss 2015). In this way, social reproduction is necessary not only for the individual but to reproduce entire societies through time and space (Katz 2001). Tracing the origins of the gendered division of labor, Maria Mies (2014) theorizes how capitalist society possesses a parasitic dependence on the free appropriation of nature – such as rivers and groundwater – and the bodies and work of women, especially women of color. Capitalist social order is thus characterized by **exploitation** as it extracts value (for example, workers and labor power) from the unpaid labor of social reproduction. In the case of Lupe, her relentless work of securing water for the household is unpaid but essential labor. In Mexico

City, there are many thousands of Lupes, performing similar work. As we mentioned previously, a critical take on insecurity can reveal "cracks" in the system and who shoulders the burden. A critical feminist perspective further illuminates how our system takes advantage of such "cracks."

Our working relations with water are both intimate and enmeshed in broader structures of belonging. In a famous turn of phrase, Cindi Katz (2001, p. 711) describes social reproduction as the "fleshy, messy, and indeterminate stuff of everyday life," drawing attention to its ordinary and earthly nature. At the same time, because of the sheer necessity of providing life-sustaining resources (e.g. water) at scale to working populations to maintain production, the social reproduction of water has been intricately bound up with the state (Katz 2008; Meehan and Strauss 2015).

In high-income countries like England and the United States, the state assumed many of the functions and responsibilities of social reproduction at a mass scale: developing systems for public health care, education, mass transit, pensions, housing, water, and sewerage. Water was one of the first broad-scale resource provision systems to fall under the control and management of the nation-state, as we explain in Chapter 4. In the past several decades, however, the rise of **austerity** policies has resulted in declining levels of state support and investment in the infrastructures of social reproduction – including water, sewerage, and flood defense. In post-Katrina New Orleans, decades of austerity policies and state disinvestment in public services like water infrastructure, flood control, health care, education, and housing has led to extreme spatial and racial disparities in livelihoods and wellbeing – a scenario Katz (2008) calls the "scoured landscape" of social reproduction. Within this scoured landscape, households and communities must knit together to provide vital resources, an extra burden on low-income households. "In many places, these shifts have had a particularly chilling effect on women, who for the most part continue to fill the gap between state and market in ensuring their households' reproduction and well-being" (Katz 2001, p. 713).

A feminist perspective draws attention to how the **body** is an important site to understand water insecurity, marginalization, and citizenship (Truelove 2019a; Truelove and Ruszczyk 2022). Thinking relationally about water means thinking about how humans internalize relations with water, and how they internalize relations with the broader practices and technologies through which water is produced and distributed. For example, Elizabeth Shove (2002) shows how new water-related technologies influence ideas (and practices) around cleanliness. She adopts washing machines as an example. "At first sight," she admits (2002, p. 3), "laundering is a curiously low-profile, even boring, example."

New technologies such as washing machines, while not disrupting a gendered division of labor, do reduce the time spent on washing clothes. At the same time, Shove (2002) argues, these "labor saving" devices begin to influence ideas of what constitutes "clean" clothes, resulting in clothes being washed *more* frequently and using more water overall. Similarly, the increasing prevalence of showers within the modern home began to shift perceptions of what is (and is not) a clean body (Kaika 2004).

Technologies don't simply change practices or vice versa; rather, they unfold in specific ways in specific contexts (Shove 2002). Within a given context, technologies, ideas, and changing relations of production all come together to produce a topic we turn to next: subjectivity.

From Citizen to "Consumer"

What does it mean to transform the "citizen" into a "consumer" or customer? What are the effects of this shift in relations? In South Africa, a country that has undergone profound social, political, and economic transformations over the past three decades, the emergence of a new, market-based **subjectivity** around water is particularly evident.

Throughout the 1980s, boycotting payment for services was one of the tools used by South African anti-apartheid activists to bankrupt institutions such as the local municipality. In the dying days of apartheid, Black Local Authorities were imposed. These institutions were accurately regarded as illegitimate props for a racial capitalist project. Refusal to pay for water became part of a "militant subjectivity" and a rejection of the apartheid institutions.

When apartheid ended in the early 1990s, democratic elections transformed the national government to support a municipal rollout of water services to those who had previously lacked a connection. The new services included those who weren't able to exercise such militant subjectivities through non-payment. At the same time, and as we discussed in Chapters 3 and 4, international financial institutions heavily promoted neoliberal ideas around privatization and corporatization within the water sector.

In South African municipalities, this turn to neoliberalism often meant a brutal focus on ensuring that household bills covered the cost of providing municipal water services, a policy principle known as **full-cost recovery**. Municipalities were unable to draw on general taxation to finance such services. As a result, they focused on overturning what they perceived to be "a culture of non-payment." The prevailing discourse was that the militant subjectivities of the 1980s endured, an assumption overlaid with racial and class-based stereotypes about morality and a work ethic. Consequently, public education programs focused on developing good consumer behavior – essential for a good neoliberal subject – and instilling recognition of the fourth Dublin Principle that "water is an economic good."

Two competing rationalities converged within the early 2000s. On the one hand, South Africans were citizens who had a right to water. Following the example of Durban, many cities were able to provide that constitutional guarantee through a free basic allowance and a progressive ratepaying system of increasing block tariffs. On the other hand, South Africans were rescripted as neoliberal subjects, described as "consumers" and disciplined by "a culture of payment" if they embraced that right.

By the early 2000s, Durban's approach appeared to be ensuring full cost recovery for water services. Municipal bureaucrats disciplined poor households into paying for water. Market-based subjectivities appeared to override citizenship within the rainbow nation. While education campaigns provided one way of instilling market rationalities, new technologies provided further support. The municipality installed devices that limited household supplies, either by shutting off the flow of water when a minimum allowance had been reached or, more commonly, reducing the diameter of the pipe to a trickle of water. These technologies regulated when people could wash, what time they might get up in the morning, and when daily activities like cooking, cleaning, and going to the shops commenced. They were one of the main tools through which neoliberal rationalities could be imposed on citizen subjects.

In trying to make sense of what was happening in Durban, Alex Loftus (2006) suggests that we should understand technologies like the **water meter** as an artifact of relations. In post-apartheid South Africa, water embodied all the contradictions of the post-apartheid project – one marked by profound efforts to achieve greater equality and social justice, albeit within the confines of neoliberal economic policies with an emphasis on fiscal discipline. These broader political economic tensions were exacerbated by a series of poor decisions made by the bulk water provider to the city, which raised the cost of water by around 25%.

The municipality installed meters in households and source points. With each rise in the unit cost of water, the meters gained greater power over the lives of residents and became a greater force in shaping market subjectivities. The slogan "Destroy the Meter! Enjoy Free Water for All!" began to circulate readily within South Africa's townships and informal settlements. Despite the temptation, as Loftus (2006) concludes, destroying a water meter does not necessarily alter the social relations that make it so powerful. Changing those relations is a re-evaluation of water as a commodity.

The market-based subjectivities in early 2000s South Africa share similarities with, as well as having major differences from, those in England and Wales, where water supplies have been privatized and now financialized. From the vantage point of capital, while financialized water utilities require regular paying consumers, equally important are predictable consumption patterns. Water meters are crucial in ensuring such predictable patterns.

Universal metering of domestic water supplies has been discussed for many decades in England and Wales. Evaluating the first large-scale compulsory metering program in England, Ornaghi and Tonin (2021) suggest that metering reduces household consumption by a remarkable 22%. Given the pressure on supplies, the future effects of climate change, and the relative balance between costs and benefits of reducing demand as opposed to increasing supplies, water metering begins to look like a win-win solution.

Thus, the relatively slow rollout of metering in England and Wales is perplexing. By 2009, just one-third of households in England were charged for water and sewerage according to a metered bill (Loftus et al. 2016, p. 325). An initial lack of interest in the private sector is understandable if we think of their economic model based around the sale of a commodity. The more you sell, the more you are likely to generate profits. Concurrently, the less likely you are to install a costly technology that might dissuade people from consuming your commodity.

In recent years, water utilities such as Thames Water have become receptive to metering households (Loftus et al. 2016). For the water companies, water meters – and especially smart meters – ensure far greater predictability in how much water is consumed. Metering has focused on reducing fluctuations in household demand throughout the day. This shift restructures the relationship between water users and utilities:

> Traditional "dumb" meters enable companies to better secure the water network by more easily identifying leaks and tracking flows of water. However, buried in the ground outside of the home, these meters are largely invisible to households and have a limited influence on governing household behaviors. (Loftus et al. 2016, p. 328)

Smart meters, meanwhile, provide granular usage data which is used to monitor household demand (as in South Africa) and convey information to (and about) British households to "produce more 'responsible consumers' in a way that minimizes the risks of revenue volatility" (Loftus et al. 2016, p. 329). In England and Wales, the water meter has been enrolled in a suite of practices aimed at ensuring that household revenue streams remain as predictable (and stable) as possible.

The water meter is one of the crucial socio-technical mediators through which broader forces and institutions – companies, capital, the state – touch the lives of individual subjects (Loftus et al. 2016). Water and the technologies through which it is provided come to shape differing subjectivities. In the case of Durban, water embodies all the contradictions of the post-apartheid project. The right to water is written into the country's constitution and citizenship is defined in relation to water access. Nevertheless, in delivering water through a corporatized and fundamentally neoliberal model, South Africans are disciplined into being passive neoliberal subjects, as opposed to the "militant" subjects who challenged apartheid by non-payment of water services.

In the example of England and Wales, the political economic rationale takes a more supple form. The shift to universal metering is not necessarily the drive to "pay for every last drop." Indeed, given that the financial model relies on income from household bills, a rapid reduction in consumption would be a disaster for different utilities. Instead, water metering is intended to achieve more predictable patterns of consumption, a subjectivity that better matches the needs of a financialized utility. What is common between the two examples is evidence of how human practices and subjectivities are shaped in relation to the practices and new articulations of political rule, embodied through our most intimate relations with water.

"Here's to Flint"

Writing this book in London, our daily life is saturated with the quiet and often hidden presence of networked infrastructure: water, electricity, waste, communication, and transport. Water infrastructure sustains life – the social reproduction of societies – and demarcates an important site of citizenship and belonging. When public water infrastructure breaks down, unravels, or collapses, these "crises" reveal the weaknesses of technology and underlying priorities and inequalities in social and institutional arrangements (Kaika 2005).

The **Flint water crisis** has become synonymous with environmental injustice in the United States. In 2014, Flint, Michigan, city managers – searching for a quick fix to ailing finances – switched the municipal water source from Lake Huron and the Detroit River to the less expensive Flint River, full of corrosive chemicals from its industrial past (Pauli 2020). "Here's to Flint," said Mayor Dayne Walling as he lifted a glass of tap water, toasting the switch (Carmody 2019).

> At first, Flint's nearly 100 000 residents complained their tap water was undrinkable: cloudy, foul smelling and tasting of chemicals or worse. The system also suffered *E. coli* outbreaks. Eventually, the city acknowledged it was in violation of the Safe Water Drinking Act. (Carmody 2019, n.p.)

Elevated levels of lead in Flint residents, especially in children, terrified the city. Neighbors forged alliances and demanded new pipes and water, civic accountability, and institutional change (Pauli 2019, 2020). Nearly five years after the initial crisis, thousands of Flint families still purchased bottled water for drinking and cooking. Trust in the municipal government was shattered. "The trust is gone," a resident said in a news interview, "The trust is gone for everybody" (Shapiro 2018, n.p.).

Flint is one of the poorest cities in the United States, where poverty is racialized. Over 40% of Flint residents live below the federal poverty line, and the city is home to more than 50% Black residents. A woman broke down in a National Public Radio (NPR) interview as she struggled to find an explanation: "Honestly, I feel like it was done on purpose because Flint is predominantly Black. And who cares? I feel like it's pretty much where the nation is right now. You see young Black boys getting murdered by white police officers all across the nation. So what do I think as a Black mother raising Black boys? How do I think a government that's predominantly white – they showed me what they feel about me and us here in Flint. They showed us" (Shapiro 2018, n.p.).

What implications does the Flint crisis have for understanding citizenship? As the woman noted in the NPR interview, "They showed us." In her analysis, Laura Pulido (2016) argues that instead of seeing the Flint crisis as an unfortunate "anomaly" or accident, we should view it as "business as usual" in the context of austerity politics and capital abandonment.

> Flint was abandoned by capital decades ago, and as it became an increasingly poor and Black place, it was also abandoned by the local state. The abandonment can be seen in shrinking services, infrastructure investment, and democratic practices. Such treatment, including deliberate poisoning, is reserved for those who are not only racially devalued but considered incapable of contributing to accumulation. (Pulido 2016, p. 2)

Specifically, Pulido argues that the **racialized division of labor** – a corollary to the gendered division of labor but based on constructed racial and ethnic differences – explains why Black lives are broadly treated as "surplus" in the United States and routinely neglected by the state. In the case of Flint, state neglect culminated in actions to switch to a cheaper water source, putting thousands of mostly Black lives at risk. "My argument is that the people of Flint are so devalued that their lives are subordinate to the goals of municipal fiscal solvency," Pulido writes (2016, p. 2), "This constitutes **racial capitalism** because this devaluation is based both on their blackness and their surplus status, with the two being mutually constituted. It is no accident that US surplus populations are disproportionately nonwhite."[1] In treating the Flint water crisis as routine, and not accidental, we are able to "see" the systemic and institutionalized conditions that make hydraulic citizenship and belonging unequal in the first place. And we can examine these dynamics through space and time. Indeed, as Pulido (2016, p. 2) urges, "the situation in Flint is of concern to all of us, not only because of its tragic nature, but because as a racially devalued, surplus place, it is a testing ground for new forms of neoliberal practice that will become increasingly common."

What about the other Flints? In 2017, nearly 460 000 US households – some 1.1 million people nationwide – experienced "plumbing poverty" or the lack of piped running water in their homes (Deitz and Meehan 2019; Meehan et al. 2020a; see Figure 8.3).

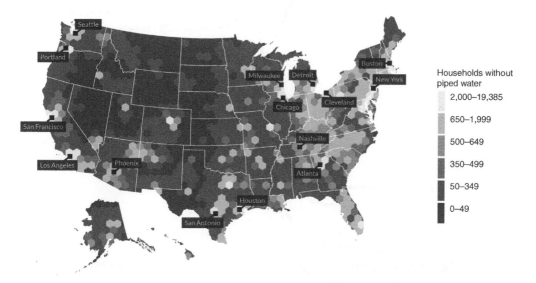

Figure 8.3 Households without piped running water in the United States, 2013–2017. Lighter shades (in yellow) depict higher numbers of unplumbed households, with notable concentrations in West Coast cities, the urbanized Eastern seaboard, and cities in the Rust Belt and Sun Belt. Data sources are from the US Census Bureau and base map design is derived from Meehan et al. (2020a). Source: Meehan et al. (2021) / King's College London / CC BY 4.0.

---

[1] For elaborations on theories of racial capitalism and surplus populations, see Ruth Wilson Gilmore's *Golden Gulag* (2006) and Robinson's *Black Marxism* (2000) – two brilliant and definitive texts.

Most of these unplumbed households (73%) are in urban areas and nearly half (47% or an estimated 514 000 people, a population the size of Sacramento, California) are in the 50 largest metros (Meehan et al. 2021). For Katie Meehan and collaborators, cities have become key sites to understand the institutionalized dynamics of racial capitalism, spatial inequality, and social reproduction – especially in high-income countries like the United States, where water service is assumed to be "universal."

San Francisco, California, is a prime example of worsening trends (Figure 8.4). From 2000 to 2017, the number of households without running water increased

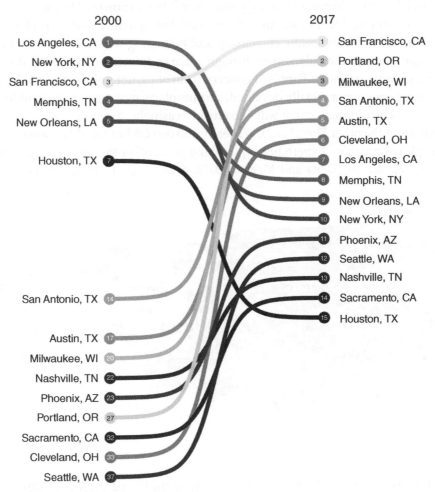

Figure 8.4   Changes in urban plumbing poverty rank in the United States from 2000 to 2017. This diagram ranks select US metro regions by share (%) of the population without piped running water, what we call "plumbing poverty" (Deitz and Meehan 2019). At left is the metro ranking in the year 2000; the right column displays the 2017 rank. Following the Great Recession of 2008, San Francisco became the leading metro (in terms of unplumbed households) by 2017. Los Angeles, the leader in 2000, "improved" its rank to seventh on the list. Source: Meehan et al. (2021) / King's College London / CC BY 4.0.

by 1600 – a relative increase of 12% (Meehan et al. 2021). An estimated 14 900 households in the San Francisco metro area lack running water: the highest per capita figure of all 50 US metros (Meehan et al. 2021). Of this unplumbed population, 90% are renters and 17% are Black/African American people, despite making up just 9% of the metro population.

Unlike Flint, San Francisco is one of the wealthiest urban areas in the country, home to tech firms, expensive housing, and the third-most billionaires of all global cities (Meehan et al. 2021). Yet the dynamics of racialized and working-class abandonment – central to the Flint water crisis – are present in San Francisco. Renters without running water are a growing subclass in allegedly "progressive" San Francisco. A typical city resident in 2017 spent 32% of their monthly household earnings on a home with piped water, while the average unplumbed renter in San Francisco spent 44% of their monthly income to live in a home *without* piped running water (Meehan et al. 2021).

San Francisco is one of the few top-50 metros where the Black population has shrunk in the past two decades, from 418 000 in 2000 to 393 000 in 2017. Yet the number of Black people without piped running water *increased* by nearly 50% over that same period, from 3200 to 4600 individuals (Meehan et al. 2021). As San Francisco grows in wealth and prestige, its Black population is more likely to experience water insecurity or to leave the metro region entirely. Household water insecurity is not limited to the global South. As we evaluate the social relations that underpin water infrastructure provision – from Mexico City to San Francisco to Lilongwe – we can acknowledge the institutions and conditions that enable belonging in modern society, as well as the ways that citizenship fails us.

Emotional Geographies of Water

Have you ever felt like your bladder would burst? Or worse, you urgently needed to poop? Where did you go? How did you *feel* in those final moments? Even reading these sentences might prompt an intimate sense of anxiety and panic. In 2020, during the height of the global COVID-19 pandemic, emergency lockdown policies closed public water and toilet access in many cities around the world, revealing fundamental linkages between access, citizenship, and the **emotional geographies** of water.

In London, pandemic policies in spring 2020 led to the temporary closure of many public toilets, drinking water fountains, and taps, especially in parks (Meehan et al. 2022). Crowd-sourced "loo maps" sprung up online, inviting people to post the locations of toilets with details of opening hours, cleanliness, and charges. By the summer, lockdown restrictions remained and city parks were filled with human excrement. Londoners gathering in outdoor spaces resorted to using their green surroundings – bushes, trees, grass – as a toilet. There was nowhere else to "go."

Normal human body functions were policed and shamed. In the London borough of Hackney, local officials issued so many fines to public urinators on a single May weekend that they reportedly ran out of citation paper (Meehan et al. 2022). Hackney

Figure 8.5   A sign in a Hackney park in London says, "Go home if you need to go." What happens if you are without a home and need water and sanitation access? Source: Katie Meehan.

used messaging of public shaming and disgust to dissuade potential offenders (Figure 8.5), akin to shame-based techniques used by Community-Led Total Sanitation (CLTS) programs to discourage open defecation (Brewis et al. 2019a; Galvin 2015). Banners posted on park gates by Hackney Council boldly declared: "It's a park, not a toilet. Go home if you need to go" (Figure 8.5).

What are the ramifications of putting up barriers to the places where we perform our intimate needs? We all drink water – often, and without substitute. We all need the toilet. On average, women use the toilet eight times per day and men six times per day. Yet public toilets and drinking water fountains are absent or rapidly *closing* in many countries (Lowe 2019). In industrialized countries like the United States and Canada, we have *less* access to public water facilities than ever before (Lowe 2019).

Scarcity is produced – a core plank of this book – and in tandem with the closure of public water facilities, market-based outlets increasingly provide for essential human needs. For example, bottled water has cornered a market once occupied by free tap water. Private businesses, such as Starbucks or McDonald's, have become the *de facto* "public bathroom" but at a price (the purchase of a product). A trip to the toilet in London's Greenwich Park now requires a bank card and costs 20p (Figure 8.6). This practice of ring-fencing is a potential barrier to marginalized people – such as unsheltered and houseless people – without a permanent address and bank account (Meehan et al. 2022).

A crisis, like a global pandemic, reveals a lot about the unassuming but powerful relations between water, society, and the market. The London toilet closures revealed

Figure 8.6 Barriers to entry in Greenwich Park. Since 2015, Royal Parks charges "customers" to use the bathroom in Greenwich Park. Source: Katie Meehan (author).

how we may feel the stress of insecure water and sanitation as individuals ("I need the toilet!"): the *production* of these water-related emotional geographies – like all relations – is social and collective.

Anthropologists and global health experts Alex Brewis and Amber Wutich (2022, p. 3) define **stigma** as "the process by which people become classified within society as less valuable, undesirable, or unwanted" (2019, p. 3). Stigma is a collective human norm and practice, though it serves to identify and label particular people or places. "Stigma is best viewed as a process rather than a thing," they write (2019, p. 207), "Though the process of being *stigmatized*, people become socially stained and discredited because they hold a characteristic that is classified as unacceptable or undesirable."

In the case of London, Hackney's tactical use of shame and stigma is on full display in Figure 8.3. Unwittingly, this tactic raised some awareness of a population that is routinely marginalized and stigmatized: people experiencing homelessness. While the COVID-19 toilet closures were likely many Londoners' first encounter with public barriers to water, such conditions are a daily reality for unhoused (homeless) people, who regularly experience water, toilet, and sanitation (WaSH) insecurity and overt barriers (DeMyers 2017; Meehan et al. 2021; Neves-Silva et al. 2018). When the institutional policy is "go home if you need to go," what happens if you do not have a stable or adequate "home"?

Stigma and its companion, shame, tend to fail as techniques of inducing mass behavior change (Brewis and Wutich 2022). More nefariously, as Brewis and Wutich explain, stigma exacerbates bad feelings and worse behaviors. Alex Brewis and co-authors (2019a, p. 12) describe the "walk of shame" and how the Community-Led Total Sanitation (CLTS) approach – discussed briefly in Chapter 1 – uses emotions of shame and disgust to label so-called "hygiene violators" and spaces:

> Triggering strategies include the so-called walk of shame, where facilitators work with village members to identify the defecation sites and who is using them, including which parts of the village are the "dirtiest." As the handbook recognizes "transect walks are the single most important motivating tool. The embarrassment experienced . . . results in an immediate desire to stop open defecation and get rid of these areas. Even though every-one sees the dirt and shit every day, they only seem to awaken to the problem when out-siders force them to look at and analyse the situation in detail." (Brewis et al. 2019a, p. 12)

Growing evidence points to the collateral damage from community sanitation inter-ventions that use stigmatizing labels (and "other indices of contempt") for hygiene violators, a cycle that "reinforc[es] stigmatized identities in ways that can drive social or economic marginalization" (Brewis et al. 2019a, p. 12). The popularity of disgust-based programs is expected to continue and expand across the global South for sev-eral reasons: these programs are low-cost and preferred by donor agencies; the campaigns are described as "participatory" and "empowering" for communities; and experts in anthropological social science have provided "science-based arguments" in favor of the efficacy of such interventions (Brewis et al 2019a, p. 13).

The **stress** of water insecurity is real: physical, emotional, and mental. Experts have long documented strong connections between insecure water and impaired physical human health (Rosinger and Brewis 2020; Rosinger and Young 2020). Mental health is equally bad (Wutich et al. 2020). In Cochabamba, Bolivia, Amber Wutich first developed systematic methods to fully capture water-related mental health condi-tions, including psychological distress, perceived stress, depressive symptoms, and anxiety symptoms (Wutich and Ragsdale, 2008). Wutich et al. (2020) trace the con-temporary origins of water and mental health scholarship in Ennis-McMillan's 2001 ethnographic study of "**suffering from water**" in a Mexican town. The study found that people who lacked sufficient water experienced negative emotions including worry, anger, anguish, frustration, and bother. Importantly, this work found that suf-fering from water insecurity was concentrated among people from lower and middle social strata, experienced collectively, and disproportionately affected women.

Water struggles are mediated through bodies, spaces, and emotions – a corporeal realm that extends far beyond a single human body (Sultana 2011). In examining the lived realities of arsenic contamination of drinking water in Bangladesh, Farhana Sultana (2011, p. 165) ties together the intimate with the political in compelling detail: "[G]iven the gendered division of labor in water where women are responsible for obtaining domestic water (men [in Bangladesh] do not participate in the feminized activity), the water crisis has made women's everyday life more difficult." Women and girls must weave their way "through labyrinths of red [do not use] and green tubewells to fetch water on a daily basis for their families."

The struggle to obtain safe water in Bangladesh reveals the stakes of claiming citizenship in an unequal world. "Most people in my study noted that to sustain access to a safe water source, it was generally important to maintain a good relationship with the owners," Sultana explains (2011, p. 166), "often pay a fee, clean the area, give free labor in exchange for water, or pay hired labor to get water." Such stress and anxiety, Sultana argues (2011, p. 167), is embodied not only in Bangladeshis' testimonies of arsenic poisoning ("*panir theke koshto*" or "suffering from water") but also in their hardship, struggle, and pain of insecure access ("*panir jonno koshto*" or "**suffering for water**"). A fieldwork anecdote illustrates this pain in vivid detail:

> Rahman's family obtains drinking water from the safe deep tubewell located in the courtyard of the *bari* (household) of Monir, a wealthy family nearby. . . . [Rahman's wife] Halima sometimes takes the water quietly and tries not to draw any attention, but often she is rebuked or humiliated. Halima says she doesn't like that feeling, but she has no choice if she wants any water. She lamented, "I have a thousand things to do as it is, I can't spend all day just getting water." But Rahman and Halima cannot afford a tubewell of their own, so she says she will manage somehow. She sighed and looked away, saying "We have to suffer a lot for water." (Sultana 2011, p. 166)

Take Back the Tap

What might "camp" look like in the case of intimate waters? In 2010, students at the University of Portland (UP) in Oregon, USA, voted in favor of banning the sale of single-use plastic water bottles on campus. Prior to the ban, UP sold roughly 53 000 single-use water bottles per year, through catering services or the 15 vending machines owned by Pepsi. In its place, the students demanded the installation of *more* drinking water fountains – in the cafeteria, classrooms, dormitories, and sports facilities.

These actions by the UP students push back on a growing trend of **bottled water**. Beverage companies earn massive profits from bottled water (a "cooperative" commodity) and clever variants of water-based drinks. In 2014, bottled water was the second largest beverage category in the United States with an annual per capita consumption of 34 gallons and total consumption of almost 11 billion gallons (Hawkins 2017). While the beverage industry presents a dizzying array of brands – what the market calls "choice" – maneuverings behind the scenes suggest otherwise. In many US schools, corporations such as Pepsi-Co, Nestlé, and Coca-Cola have quietly negotiated exclusive-rights contracts to sell their products in vending machines, with stipulations that prevent outside competitors and discourage (or prohibit) the presence or construction of drinking water fountains. School districts agree to such contracts, often as a desperate way to generate revenue when austerity policies mean public funding is on the decline.

Around the world, dozens of university campuses feature student-led campaigns to "Take Back the Tap." In 2008, the University of Leeds was the first British university to ban bottled water from its bars, cafes, and shops. The National Hydration Council called the Leeds vote "a shame, removing the right to choose" (Wainwright 2008).

But for the students, opposition to bottled water is far greater than a menu debate. This opposition represents a clear refusal of market pressures to transform us into "consumer" subjects, rather than citizens. Here, the personal and intimate – what we drink – are very political. "These politics foreground the effects of representing and delivering drinking water as an individualized good versus distribution via reticulated networks in which consumers gain access to a collective service," writes Gay Hawkins (2017, p. 1). Students offer a viable choice: taking back the public tap, with improvements in access, reliability, service, and quality. Water for all.

Like students everywhere, the Leeds and Portland students are at the vanguard of struggle. But camp, as we have learned, is a difficult path. As we write this chapter, England is in the grip of a severe seasonal drought, the worst since 1976. Tawny fields, low rivers, and talk of water restrictions raise questions about the future of this island's water. A major football [soccer] victory for the England women's team capped an otherwise dry summer.

The artful goals and confetti belie a long struggle. England's Football Association (the FA) banned women's football from 1921 to 1971. The game was first run by volunteers, later by a handful of professionals who tirelessly built the sport. Sheila Parker, the first England women's captain (1972–1983), worked a day job at the reception desk at Leyland Motors to pay for her travel to tournaments (Beaty et al. 2022).

Today, we still push. Since 2007, the women's team has qualified for every major tournament and reached the semi-finals in 2015, 2017, and 2019. And in the dry summer of 2022, England broke the soccer drought and won the European Championships, a 2–1 victory over Germany, bringing the international trophy home for the first time in 56 years. Much work is left to accomplish, notably to open more pathways for people of color. Camp is never easy. The impossible will take a little while. But still, we push.

Summary and What's Next

Intimate waters are not singular or private, but collective and relational. Water, as we argued in this chapter, is an intimate and fundamental part of citizenship – a means to demarcate who gets to belong (and who is marginalized from) the community, city, and nation. A critical understanding of water requires a focus on the proximate hydrosocial relations that shape the most personal aspects of our lives: our bodies, labor, households, subjectivity, and belonging to – and exclusion from – the wider social and ecological fabric. This chapter traveled across the global North and South to explore key manifestations of intimate waters including the struggle to define subjectivity, the gendered and racialized labor of water, and the emotional distress and physical impacts of a life without secure water.

Household water insecurity is more than a lack of physical access. Insecurity touches on aspects of reliability, affordability, quality, acceptability, trust, and risk. If you have never thought about water – how heavy it weighs, how much it costs, where you might find your next drink – then we hope this chapter has changed your mind. Struggles for water justice have proliferated worldwide despite considerable odds, evidenced in the case of Flint and beyond. In the next chapter, we pry open the black box of justice to examine struggles around the human right to water.

Further Reading

Household water insecurity

Adams, E.A. (2018). Thirsty Slums in African Cities: Household Water Insecurity in Urban Informal Settlements of Lilongwe, Malawi. *International Journal of Water Resources Development* 34 (6): 869–887.

Household Water Insecurity (HWISE)-Research Coordination Network (RCN). (2022). Research network includes many resources and scholars who work in this area, funded by the National Science Foundation: https://hwise-rcn.org/ (accessed 23 June 2022).

Jepson, W., Budds, J., Eichelberger, L., Harris, L., Norman, E., O'Reilly, K., Pearson, A., Shah, S., Shinn, J., Staddon, C., et al. (2017). Advancing Human Capabilities for Water Security: A Relational Approach. *Water Security* 1: 46–52.

Meehan, K., Jepson, W., Harris, L.M., Wutich, A., Beresford, M., Fencl, A., London, J., Pierce, G., Radonic, L., Wells, C., et al. (2020). Exposing the Myths of Household Water Insecurity in the Global North: A Critical Review. *Wiley Interdisciplinary Reviews (WIREs): Water* 7 (6): e1486.

Wutich, A., Budds, J., Eichelberger, L., Geere, J., Harris, L.M., Horney, J.A., Jepson, W., Norman, E., O'Reilly, K., Pearson, A.L., et al. (2017). Advancing Methods for Research on Household Water Insecurity: Studying Entitlements and Capabilities, Socio-cultural Dynamics, and Political Processes, Institutions and Governance. *Water Security* 2: 1–10.

Wutich, A. (2020). Water Insecurity: An Agenda for Research and Call to Action for Human Biology. *American Journal of Human Biology* 32 (1): e23345.

Water and mental health

Brewis, A., Wutich, A., du Bray, M.V., Maupin, J., Schuster, R.C., and Gervais, M.M. (2019a). Community Hygiene Norms Violators are Consistently Stigmatized: Evidence from Four Global Sites and Implications for Sanitation Interventions. *Social Science & Medicine* 220: 12–21.

Brewis, A., Choudhary, N., and Wutich, A. (2019b). Household Water Insecurity May Influence Common Mental Disorders Directly and Indirectly through Multiple Pathways: Evidence from Haiti. *Social Science & Medicine* 238: 112520.

Rosinger, A.Y. and Brewis, A. (2020). Life and Death: Toward a Human Biology of Water. *American Journal of Human Biology* 32 (1): e23361.

Sultana, F. (2011). Suffering for Water, Suffering from Water: Emotional Geographies of Resource Access, *Control and Conflict. Geoforum* 42 (2): 163–172.

Wutich, A. (2009). Intrahousehold Disparities in Women and Men's Experiences of Water Insecurity and Emotional Distress in Urban Bolivia. *Medical Anthropology Quarterly* 23 (4): 436–454.

Wutich, A., Brewis, A., and Tsai, A. (2020). Water and Mental Health. *WIREs Water* 7 (5): e1461.

Informality and water

Ahlers, R., Cleaver, F., Rusca, M., and Schwartz, K. (2014). Informal Space in the Urban Waterscape: Disaggregation and Co-production of Water Services. *Water Alternatives* 7 (1): 1–14.

Kooy, M. (2014). Developing Informality: The Production of Jakarta's Urban Waterscape. *Water Alternatives* 7 (1): 35–53.

Meehan, K. (2013). Disciplining De Facto Development: Water Theft and Hydrosocial Order in Tijuana. *Environment and Planning D: Society and Space* 31 (2): 319–336.

Schwartz, K., Luque, M.T., Rusca, M., and Ahlers, R. (2015). (In)formality: The Meshwork of Water Service Provisioning. *Wiley Interdisciplinary Reviews (WIREs): Water* 2 (1): 31–36.

Ranganathan, M. (2014). 'Mafias' in the Waterscape: Urban Informality and Everyday Public Authority in Bangalore. *Water Alternatives* 7 (1): 89–105.

Truelove, Y. (2019a). Rethinking Water Insecurity, Inequality and Infrastructure through an Embodied Urban Political Ecology. *Wiley Interdisciplinary Reviews (WIREs): Water* 6 (3): e1342.

Truelove, Y. (2019b). Gray Zones: The Everyday Practices and Governance of Water beyond the Network. *Annals of the American Association of Geographers* 109 (6): 1758–1774.

Feminist political ecology

Adams, E.A., Juran, L., and Ajibade, I. (2018). 'Space of Exclusion' in Community Water Governance: A Feminist Political Ecology of Gender and Participation in Malawi's Urban Water User Associations. *Geoforum* 95 (1): 133–142.

Crow, B. and Sultana, F. (2002). Gender, Class, and Access to Water: Three Cases in a Poor and Crowded Delta. *Society and Natural Resources* 15 (8): 709–724.

Dickin, S. and Caretta, M.A. (2022). Examining Water and Gender Narratives and Realities. *Wiley Interdisciplinary Reviews (WIREs): Water* 31 (2): 319–336.

Harris, L. (2008). Water Rich, Water Poor: Intersections of Gender, Poverty, and Vulnerability in Newly Irrigated Areas of Southeastern Turkey. *World Development* 36 (12): 2643–2662.

Truelove, Y. (2011). (Re-)Conceptualizing Water Inequality in Dehli, India through a Feminist Political Ecology Framework. *Geoforum* 42 (2): 143–152.

Truelove, Y. (2019). Rethinking Water Insecurity, Inequality and Infrastructure through an Embodied Urban Political Ecology. *Wiley Interdisciplinary Reviews (WIREs): Water* 6 (3): e1342.

Gendered labor, irrigation, and rural water

Birkenholtz, T. (2013). "On the Network, Off the Map": Developing Intervillage and Intragender Differentiation in Rural Water Supply. *Environment and Planning D: Society and Space* 31 (2): 354–371.

Cleaver, F. (1998a). Choice, Complexity, and Change: Gendered Livelihoods and the Management of Water. *Agriculture and Human Values* 15 (4): 293–299.

Cleaver, F. (1998b). Incentives and Informal Institutions: Gender and the Management of Water. *Agriculture and Human Values* 15 (4): 347–360.

Meinzen-Dick, R. and Zwarteveen, M. (1998). Gendered Participation in Water Management: Issues and Illustrations from Water Users' Associations in South Asia. *Agriculture and Human Values* 15 (4): 337–345.

Chapter 9

The Right to Water

Victory or Defeat?

On July 28, 2010, the United Nations General Assembly explicitly recognized the human right to water and sanitation (Figure 9.1).[1] The Canadian water justice activist Maude Barlow (2012) described the jubilation erupting on the floor of the General Assembly when the resolution was passed by majority vote – a vote that was far from a forgone conclusion, with several powerful countries lining up in opposition. Confirming Barlow's sense of the importance of this moment, two months later the right to water and sanitation would be further formalized in a second resolution at the UN Human Rights Council. For many non-governmental organizations (NGOs), scholars, and activists such as Barlow, these two resolutions are viewed as a major victory and a clear turning point in the struggle for water justice:

> These historic resolutions present an incredible opportunity for other groups, communities, and Indigenous peoples around the world suffering from water shortages, unsafe drinking water and poor or non-existent sanitation services. It is not often that a new right is recognized at the United Nations, especially around an issue as increasingly

---

[1] Our focus in this chapter will be on the human right to safe drinking water. The United Nations recognized a distinct right to sanitation in 2015.

Water: A Critical Introduction, First Edition. Katie Meehan, Naho Mirumachi, Alex Loftus, and Majed Akhter.
© 2023 John Wiley & Sons Ltd. Published 2023 by John Wiley & Sons Ltd.

10 YEARS OF THE

THE HUMAN RIGHTS TO WATER AND SANITATION

2002

The General Comment No. 15: The Right to Water is adopted at the 29th Session of the Committee on Economic, Social and Cultural Rights. The General Comment outlines that everyone is entitled to sufficient, safe, acceptable, physically accessible and affordable.

2008

United Nations Human Rights Council appoints **Catarina de Albuquerque** as Independent Expert on the human rights to water and sanitation

2010

United Nations General Assembly officially recognises the human right to water and sanitation in resolution 64/292

2011

The United Nations Human Rights Council extends the mandate of the human rights to water and sanitation as a **Special Rapporteur**

2015

The human right to safe drinking water and the human right to sanitation are referred to as separate rights by the United Nations General Assembly

In his first report as the new Special Rapporteur, **Léo Heller** sets out guidance on levels of services that comply with the human rights

2030 Agenda for Sustainable Development is adopted, including SDG 6 on ensuring the availability and sustainable management of water and sanitation for all

2018

The Special Rapporteur launches the first initiative to follow-up official country visits

2019

The Special Rapporteur is mandated to organise a year-long campaign to celebrate the **10-year anniversary** of the recognition of the human rights to water and sanitation and to raise awareness of the rights

2020

The Special Rapporteur releases monthly materials to promote the human rights to water and sanitation, and to celebrate the work of the mandate

✉ srwatsan@ohchr.org   🌐 www.ohchr.org/srwaterandsanitation   🐦 @srwatsan

HUMAN RIGHTS TO WATER AND SANITATION

Figure 9.1   Timeline of key events related to the human rights to water and sanitation. Source: From 10 Years of the Human Rights To Water And Sanitation, © 2020 United Nations. Reproduced with permission from the United Nations.

political and urgent as the global water crisis. The right to water and sanitation are living documents waiting to be used for transformational change around the world. (Barlow 2012, pp. xvi–xvii)

In sharp contrast, only one day after the General Assembly vote, the publisher of *Global Water Intelligence* (GWI) wrote that "the adoption of the human right to water has got to be a massive defeat for the Global Water Justice Movement" (Gasson 2010). The column's author argued that the adoption of the human right to water amounted to no more than "a statement of good intentions by governments" and in no way undermined the ability of the private sector to serve as one of the main providers of water services (Gasson 2010).

To confuse things a little more, Gasson quoted Catarina de Albuquerque, the United Nations' independent expert on the human right to water (later to become the UN Special Rapporteur on the human rights to safe drinking water and sanitation), as evidence. In her role as Independent Expert, de Albuquerque wrote, "[H]uman rights are neutral as to economic models in general, and models of service provision more specifically." Gasson's message to the private sector was clear: they should feel reassured that the human right to water did not in any way undermine the ability to profit from providing water. Not simply a defeat for water justice, the vote was interpreted by Gasson as a victory for the private sector.

The opposing positions of Gasson and Barlow indicate not only the radically different hopes that individuals invest in the human right to water but also the ways in which that right has been struggled over amongst states, peoples, and businesses. Despite the efforts of various UN experts to solidify this human right, the meaning of the right to water has shifted through time and across contexts.

Ten years after the General Assembly's historic vote, de Albuquerque's successor in the role of UN Special Rapporteur, Léo Heller, authored a report in which the role of the private sector was interrogated more intensely (UN OHCHR 2020). For many people, Heller's report promised clarity on the question of whether the right to water is a victory for water justice or a defeat. Following a forensic process of data gathering alongside intense diplomatic efforts to ensure balance with "privatization debates," Heller concluded that economic models do matter and that processes of privatization, municipalization, and financialization (Chapter 4) can influence whether or not people are able to access clean drinking water.

By tackling the question of private sector participation head on, Heller's report generated a vicious backlash from private sector representatives, concerned that the UN was taking an anti-privatization stance. Water justice activists, on the other hand, rallied to support the UN Special Rapporteur. We will examine this confrontation in more detail at the end of this chapter. For now, we illuminate how different groups interpret the human right to water and how they articulate it with and through different sets of struggles. For some, such as Barlow, the human right to water remains one of the greatest opportunities for rendering democratic, just, and equitable access to water. For others, the human right to water presents an opportunity for the private sector to become even more involved in the provision of water.

As we analyze these sharply divergent views on the right to water, we build on the idea that law is relational (Chapter 3) to explore the human right to water as one site where struggles over the law are particularly acute. And we position the right to water within the long arc of struggles for water justice, considering its material consequences for people and the environment. The first step in building such an understanding involves clarifying what is meant by a human right to water.

What is the Human Right to Water?

A "human right" can refer to several different things. Generally, a **human right** refers to a set of legal judgments, texts, and frameworks that seek to minimize harm and maximize freedoms for individuals. Combined, these judgments, texts, and frameworks restrict the power of sovereign authorities, such as states, to delimit the freedoms of the individual. NGOs such as Amnesty International protect this vision of human rights through the defense of prisoners of conscience and the freedom of individuals against violent and oppressive regimes. Responding to the criticism that such an understanding of human rights is too narrow, Amnesty International has broadened the range of its activities in recent years, including anti-racist activities within the liberal democracies that claim to be bastions of the freedom of the individual.

Written constitutions often serve as one of the most direct expressions of such a conception of human rights. These constitutions express what Thomas Hobbes (2017 [1651]), an English seventeenth-century philosopher, referred to as a **social contract** between citizen and state. Individuals, in Hobbes's view, surrender certain limited freedoms in return for the protection of other liberties from the state. Hobbes had a somewhat negative view of "human nature," seeing human rights as necessary protection from "the state of war" into which humans would naturally descend without such protections. Other philosophers, such as Jean Jacques Rousseau, perceived rights as facilitating the development of an underlying moral potential among humanity.

The roots of these ideas of human rights extend back to Enlightenment conceptions of the human subject, captured in arguments for both the American War of Independence and the French Revolution. Since then, human rights have evolved primarily through the 1948 UN Universal Declaration of Human Rights. Signed after the horrors of two world wars, the UN Declaration elevated human rights to a universal status above and beyond the particularities of individual social contracts between citizens and states. At the same time, the UN Declaration framed human rights as indivisible and interdependent, meaning that individual rights should be viewed as part of a functioning and inseparable whole.

Nevertheless, the universality of this human rights framework raised questions about the abilities of some liberal democracies (largely in the West or the global North) to define the content of human rights. And of more concern, could liberal democracies apply those frameworks in ways that deepen neo-colonial power imbalances? The fact that the United States provides around 22% of the entire UN budget heightens concerns around what Alston (2017) refers to as "the power of the purse in multilateral institutions" and the ability of the powerful to influence international relations and human rights policy to fit with their own interests.

But to criticize human rights as somehow based on "Western principles" opens the possibility of cultural relativism that, for some, would permit states to deny their obligations to certain social groups, or to inflict harm under the guise of "culture" or "religion." Countering the claim that the human right to water is an imperial or Eurocentric framework, we note that some of the more obvious abstentions in the 2010 UN vote on the human right to water and sanitation were wealthier countries in the global North. Perhaps more importantly, the resolution was proposed by the Bolivian ambassador to the United Nations, Pablo Solón, a statesman who has written widely on both grassroots environmentalism and Indigenous rights. In sum, how the human right to water is understood and interpreted by the powerful points to differing perspectives on water–society relations. As with the Dublin Principles (Chapter 4) and in the opposing positions at the beginning of this chapter, the human right to water can be seen as a set of individualist principles that support the economic valuation of water. But the human right to water can also be interpreted as a form of protection for communal rights. In such contrasting perspectives, the human right to water can either be support for or a challenge to settler colonial frameworks in water governance.

Teasing out these different perspectives, Maude Barlow (2012) distinguishes between what are commonly referred to as "first-generation" human rights – more closely associated with the civil and political liberties laid out at the start of this section – and "second-" or even "third-" generation rights that might prioritize economic and social change or communal rights respectively. She poses the question as to whether the right to water will be understood in the limited sense of a first-generation right, or defined along the lines of the second or third, with the last coming closest to Pablo Solón's proposition in the UN General Assembly vote. Fulfilling second-generation or economic and social rights requires money. Fulfilling third-generation rights may involve addressing some of the injustices of settler colonial relations, such as the construction of an oil pipeline through Indigenous lands – something most powerful countries would be unwilling to do.

As a leading scholar of human rights law and as the UN Special Rapporteur on extreme poverty and human rights between 2014 and 2020, Philip Alston (2017) has focused on the relationship between different generations of human rights and the best strategies for achieving those rights (Box 9.1). Alston deconstructs the position often expounded by wealthy states: civil and political liberties will enable citizens to actively pressure governments to achieve economic and social rights. Prioritizing first-generation over second-generation rights serves wealthier interests well, benefiting those who may already have access to the services they require, access to decent health care, a pension, etc. However, such prioritization can result in undermining the needs of the vast majority of people, whether in the global North or the global South, who lack economic and social rights.

In contrast, Alston (2017) argues, *both* sets of rights should be pursued alongside each other. He finds lessons in the experience of struggling for civil and political rights. Recognition, institutionalization, and accountability are crucial in achieving

Box 9.1 The United Nations Special Rapporteurs

Who are UN Special Rapporteurs? *Rapporteur* is taken from the French word meaning investigator, and the UN Special Rapporteurs are essentially investigators charged to compile reports on a specific aspect of human rights. They are not paid by the United Nations in the role and many work for universities or think tanks and rely on their university or NGO role for their main salary. At present, the UN Human Rights Council has appointed Special Rapporteurs to cover 44 thematic areas within the Declaration of Human Rights.

With the recognition of the human right to water and sanitation in 2010 (and subsequent recognition as two distinct rights in 2015), a UN Special Rapporteur on the human rights to safe drinking water and sanitation has compiled annual reports on specific aspects of these rights. The first UN Special Rapporteur on the human right to safe water and sanitation was Catarina de Albuquerque, a Portuguese lawyer and human rights activist. The second, Léo Heller, served in the role from 2014 to 2020 while serving as Professor in the Department of Sanitary and Environmental Engineering at the Federal University of Minas Gerais, Brazil. The role was then passed to Pedro Arrojo-Agudo, a Professor of Ecological Economics at the University of Zaragoza, Spain, who also represented the Spanish Parliament as an elected member with a particular interest in human rights. As can be seen from the response to Heller's final report on privatization, states and businesses can take great offense at the reports these Special Rapporteurs author.

both generations of rights. Responding to the specific concern that governments might agree to the principle of economic and social rights (such as the right to water) without any intention of paying for the fulfillment of those rights, Alston argues that a commitment to such economic and social rights may not induce *immediate* change. However, it does mean a commitment on the part of state institutions to recognize and begin institutionalizing universal access to water. Developing mechanisms through which greater accountability might be achieved represents a further step. In the next section, we explore how these different arguments played out in a South African context.

Alston's argument for the intertwining of civil and social rights counters the claim that struggles for the human right to water fail to grapple with the underlying causes of unequal access to water. For Alston, both the recognition of the right to water and fairer access to water need to be fought for together. Furthermore, economic and social rights build greater solidarity among populations. If a population has equal rights to health care, water, and housing, the people are bound together in common causes. Writing in 2017, Alston was looking for a counterpoint to the divisive language of right-wing populism. He called for greater solidarities to be built between different social groups around human rights principles. The geographer David Harvey (2000) would seem to offer qualified support for such a position. Harvey distances himself from those who dismiss human rights as a distraction from "structural" change to the economy. Instead, he argues, exploring the UN Universal Declaration of Human Rights points to fundamental contradictions that might themselves become the starting point for broader struggles for social and economic justice.

Our account provides a broad overview of human rights, and we find it vital to position the more recent (and more rapid) evolution of the human right to water within such an overview. While various agreements within the United Nations acknowledged the human right to water throughout the 1990s, the UN General Assembly finally recognized the right to water at the international level in a 2010 resolution (for a detailed review of prior legislative frameworks, see Langford and Russell 2017). There was mention of water in association with other rights (for example the right to development and the rights of the child), but 2010 was the first moment in which the right to water was fully recognized. The General Assembly vote was not unanimous and many powerful states abstained, including the United Kingdom and the United States – even though the United States supported the second resolution at the UN Human Rights Council two months later.

Once the resolution had been signed into international law, every member state of the United Nations was obliged to prepare a Plan of Action. Specifically, action must include three key state obligations: to respect, fulfill, and protect the right to water for their citizens. This is the "plan" that Gasson mocks as a major defeat for the water justice movement in his GWI column, even if, as Barlow (2012) points out, there is much more involved in implementation than a simple statement of good intentions.

For de Albuquerque (2014) such a framework provides tools through which citizens as right-holders can pressure states to meet their obligations and ensure equitable access to water, a similar argument to Alston's. The UN model ostensibly relies upon Hobbes's notion of a social contract between citizen and state, with the latter having obligations to the former that must be fulfilled. If human rights obligations can be exceeded by different states, international law stipulates that those same states must achieve at least a basic minimum service, and that there must be no retrogression in water provision. The human right to safe drinking water therefore states that water should be sufficient, safe, acceptable, physically accessible, and affordable. While these basic details are clear, some have argued for a more expansive interpretation of the human right to water. Christy Clark (2019) notes the importance of the right to participation within the human right to water, a point echoed by Gosling and Tobin (2020) when they argue that the right to water must embody principles of participation, transparency, accountability, and sustainability.

Portions of these different positions are echoed in two books by Farhana Sultana and Alex Loftus, *The Right to Water* (2012) and *Water Politics* (2019). In both volumes, Sultana, Loftus, and the participating authors frame the human right to water as a starting point for efforts to achieve water justice. Rather than dismissing the human right to water as vacuous, unachievable, or a tokenistic commitment on the part of governments who have no real interest in bringing about change, the authors consider the different ways in which meaning is invested in struggles for the human right to water, and how that meaning might be turned into action.

A similar argument is developed in Christy Clark's work (2019) on how water justice activists mobilized around a conception of the human right to water in Flint, Michigan, a case we will explore in more detail shortly (see also Chapter 8). While early studies on the human right to water predominantly focused on the global South, many now recognize how important the right to water can be for ensuring fairer access to water in the global North. Crucially, studies of – and activism in – the global North have learned from experiences in the global South. Of those experiences, none is perhaps as iconic and interesting as that of South Africa.

Struggles in South Africa

For much of the twentieth century, people of color in South Africa endured and struggled against one of the most brutal state-sanctioned forms of racism. Apartheid, an Afrikaans word meaning "separateness" and implying a policy of "separate development," stipulated that people of different races should live in geographically distinct spaces. The racial categories constructed by the architects of apartheid further demonstrate the racism at the heart of the project, as an attempt to divide people on the basis of spurious difference. Apartheid sought to split South Africa into 20 different homelands, or Bantustans, according to racial constructs. As the geographer David Smith (1982, p. 1) wrote, apartheid came to be one of "the most ambitious contemporary exercises in applied geography." Defeated through internal political protest, external sanctions, and the ending of the Cold War, the spatial legacies of apartheid are still evident across South Africa. Nevertheless, the ending of apartheid in 1990 and the first free elections of 1994 permitted a democratic government to come to power and usher in a period of rapid change within the country. Unjust legacies could finally be addressed.

Signed into law by Nelson Mandela on December 10, 1996, the South African Constitution is one crucial plank in this process of addressing past racial injustices. A remarkable document, the Constitution conceptualizes a "rainbow nation" that challenges the legacies of apartheid, while charting a progressive future for all citizens based on civil and political liberties alongside guarantees of social and economic justice. As part of this progressive vision, the Constitution was one of the first (if not *the* first) documents in the world to acknowledge that all citizens had the right to water. Section 27:1 of Chapter 2 of the Bill of Rights states that "everyone has the right to have access to . . . sufficient food and water." While there might be some slippage in the initial term "to have access to" (e.g. does that simply mean that you have the right to a tap, but not necessarily the water within?), Section 27:2 bolsters the statement further by emphasizing that "the state must take reasonable legislative and other measures within its available resources to achieve the progressive realization of these rights."

As part of these legislative measures, the Water Services Act of 1997 (later clarified in 2001) confirmed the right "of access to basic water supply and sanitation," specifying, quite precisely, a "minimum quantity of 25 l per person per day or 6 kiloliters per household per month . . . within 200 m of the household" (RSA Government Gazette 2001). The legislation would appear to be watertight, even down to the basic minimum that each person and household should receive. Therefore, it was a surprise to find by the early 2000s, many thousands of households were disconnected across the country from their municipal water networks for non-payment of bills. Despite the raft of legislation from the Constitution onwards, racial inequalities continued to manifest insidiously around water access. The inability of many to pay for water – despite the fact they now had access to the potable water network – had become the proximate cause for these inequalities, even if that proximate cause must be positioned alongside the legacies of apartheid.

The city of Durban provides a telling illustration. In Durban, the executive director of the city's water utility estimated that roughly 4000 households were being disconnected per week during the early 2000s. A single household could well have been disconnected on multiple occasions, skewing official figures, but a cautious estimate would suggest hundreds of thousands of disconnections took place per year in one municipality alone. Assuming a modest household size of four individuals, the number of Durban residents affected by such measures was sizeable. As if to prove our point that human rights (and the law more broadly) are defined in different ways at different times, the right to water in South Africa meant an increased focus on the question of payment.

Perhaps the best-documented case – written up by Durban-based scholar-activist Ashwin Desai (2001) and presented in several documentary films on Durban – is that of Christina Manquele. A resident of Chatsworth, Manquele was one of many individuals who found herself in serious arrears to the municipal water provider around the late 1990s. Unable to pay those arrears, Manquele was disconnected from the water supply and forced to turn to neighbors for water, or to unsafe, unprotected supplies, such as streams outside of her home – a perverse situation in which the state continued to pledge a constitutional guarantee of her right to water. When activists formed a Concerned Citizens group in Manquele's neighborhood of Chatsworth, they took her case to court demanding that the municipality reinstate her supply. Initially, the court ruling went in Manquele's favor. However, the ruling was later overturned with the municipality's lawyers claiming that in not paying her bill Manquele had *foregone* her right to water.

Prior to Manquele's court battle with Durban municipality, South Africa had also been forced to deal with a cholera outbreak in rural KwaZulu Natal. According to one important study (Deedat and Cottle 2002), the reason for the outbreak was the implementation of pre-payment water meters, forcing poorer residents to turn to unsafe sources of water. Pre-payment meters work in a similar way to a mobile phone top-up or a "pay as you go" plan. Residents are expected to charge up a card that gives them access to water until they have fully expended the pre-paid amount. When these same pre-payment water meters were installed in parts of Soweto, Johannesburg's largest township, a further legal challenge was presented by that community, a legal challenge seen as a test case on the human right to water. Disappointingly, the community's legal challenge was overturned by the courts on the grounds that pre-payment meters merely "suspended" access to a household's free basic water allowance (Clark 2012). Given that such an allowance was made available monthly, water supplies would resume within at most a 30-day period.

While the South African courts appeared to consider it possible for a municipality to meet its obligations around the human right to water even if a household is without water for 29 days every month, such a judgment went against a similar case presented in the UK in the 1990s where a judge concluded that it was "common sense" that pre-payment meters disconnected a household from the water supply. Aside from these legal intricacies, the crucial point was that in the 1990s and early 2000s, whether in Durban, Soweto, or rural KwaZulu Natal, the inability to pay for water continually eclipsed a person's right to water. Worse perhaps, the need to ensure that user payments would finance a free basic water policy meant an even greater focus on full cost recovery: water disconnections were the inevitable result. Classed, racialized, and gendered inequalities were exacerbated by apparently well-meaning policies meant to address such inequalities. Perhaps, the right to water was as much a defeat as a victory.

South Africa embodies some of the best and worst practices when it comes to water policy (Box 9.2). If these three examples provide a somewhat negative overview, we are not seeking to diminish the lengths to which many have gone to develop a free basic water policy for all, as well as a pricing system for water that might foster social and environmental justice. Nevertheless, and here we return to the point made in Chapter 4, the need to pay for the pipes has led to the erection of unfair and unnecessary barriers that prevent people from accessing water. The need for municipal water utilities to ensure all costs are recovered has meant that people are denied access to water in a context in which some of the most progressive policies have been tested. Grassroots struggles in places such as South Africa exemplify ways in which activists give new meaning to the human right to water. And they often challenge the very governments who have pledged to provide such rights. Crucially, these struggles are in no way confined to the global South. Indeed, struggles for the right to water have multiplied within the global North in recent years, in part learning from the experiences of activists in places like Durban and KwaZulu Natal. Intriguingly, calls for the right to water are now almost as loud in the global North as the global South. Maintaining the idea that law is relational, these calls have invested yet new meaning in the right to water and represent a new moment in the arc of the struggle for water justice.

Box 9.2   Free Basic Water in South Africa

One of the novel policy strategies used by South African municipalities to provide a human right to drinking water has been through the provision of a free basic allowance to all households. Although there are other examples of this having been implemented around the world, the city of Durban was a leader in South Africa. The free basic water policy in Durban was initially a response to the difficulties of providing water to informal settlements (e.g. shack settlements declared "illegal" under apartheid but whose status was often ambiguous in the transition to democracy). From the 1970s onwards, all households on the water network in Durban had been metered for their consumption, so the free basic allowance granted to shackdwellers could be easily made universal. A monthly 6 kl was the initial free basic allowance (based on the 200 l tank that shackdwellers relied on). Later, this increased to 9 kl per month. The free basic allowance was then subsidised by charging more in progressive blocks for those who consumed higher volumes of water. By charging more for higher usage, such a scheme has the apparent benefit of discouraging overconsumption, but the allowance has been criticized for ignoring household size and for leading to a greater focus on limiting household supplies to the basic minimum.

From South to North

On May 3, 2016, the United Nations Office of the High Commissioner on Human Rights posted a press statement:

> Speaking ahead of President Barack Obama's upcoming visit to Flint [Michigan] on May 4, three United Nations experts have called for immediate action to address the serious human rights concerns brought upon by the contamination of Flint's water supply and the devastating consequences for its residents.
>
> The Flint case dramatically illustrates the suffering and difficulties that flow from failing to recognize that water is a human right, from failing to ensure that essential services are provided in a non-discriminatory manner, and from treating those who live in poverty in ways that exacerbate their plight, said the UN experts on extreme poverty, water and sanitation, and housing. (UN OHCHR 2016)

The three Special Rapporteurs urged President Obama to exercise global leadership by acknowledging the importance of a human rights framework. Philip Alston, as Special Rapporteur on extreme poverty and human rights, continued by noting the underlying racism and declining levels of federal funding being spent on water and sanitation infrastructure, that had given rise to the situation in Flint, a water crisis and lead poisoning of the city's water supply, as introduced in Chapter 8 (Pauli 2019; see also Further Reading section). Crucially, Léo Heller, as the UN Special Rapporteur on the human rights to safe water and sanitation noted, "the fact that Flint residents have not had regular access to safe drinking water and sanitation since 2014 is a potential violation of their human rights."

The water crisis in Flint serves as a clear demonstration of how the human right to water is not merely a question of "under-development" to be addressed by NGOs in the global South. Nor is it simply a "target" to be rolled into Sustainable Development Goal 6 on Clean Water and Sanitation. As seen in the statement from the three Special Rapporteurs, the human right to water also matters for citizens of countries in the global North. The actions of social movements – often in conjunction with elected representatives – have sought to use the human right to water as a means of ensuring better access to clean drinking water both in Flint and beyond (Clark 2019).

The water crisis and human rights struggles are not limited to Flint. Between 2013 and 2020, the city of Detroit, Michigan, had shut off an estimated 137 165 water accounts – including homes and businesses – under the direction of a debt recollection plan and state-appointed emergency manager (Meehan et al. 2020b). By 2019, water service was disconnected for more than 23 000 accounts, three-fifths of which were still without service as of late January 2020, as the COVID-19 pandemic started to take hold (Lakhani 2020). Thus, as disconnections were beginning to soar in Detroit, and people found themselves without running water to meet water, hygiene, and public health needs, a cross-party group of House representatives sought to introduce a bill around water as a human right.

Two Congresswomen from Michigan (Rep. Rashida Tlaib and Rep. Debbie Dingell) introduced the "Emergency Water is a Human Right Act" in the US House of Representatives, citing disconnections in Detroit and other US cities as the justification for banning water shut-offs through the course of the pandemic. Back in 2016, the three UN Special Rapporteurs were present in Flint at the invitation of social movement activists within the city working with (and often against) legislators, thereby seeking to bring about greater material change when it comes to the human right to water. Such actions are an evocative example of how struggles around the human right to water have moved from sites that we think of as "the global South" to sites within the global North – places we assume have universal access to safe drinking water.

For example, global North examples of struggles emerging around the human right to water are evident in the European Union, including Ireland, Italy, and Greece. As in Flint, social movements in these countries have challenged environmental injustices through a human rights framework. Such struggles can be seen as part of a broader response to the politics of austerity that had reconfigured state–society relations across the global North following the financial crisis of 2008 (Ranganathan 2016). In the case of the European Union, a range of social movements and unions worked collectively to achieve the first successful European Citizens' Initiative around the human right to water. This remarkable achievement sought, in the words of van den Berge et al. (2019), to shift the European Union from a market-oriented framework to a rights-based one, when it comes to water provision.

The provision for the European Citizens' Initiative was relatively new, drawing on promises within the Lisbon Treaty of 2007 which legally obliges the European Commission to consider legislative proposals if one million signatures could be gathered on a particular proposal. These signatures required a minimum number from at least seven EU countries. Right2Water formed as a coalition to gather sufficient signatures and call upon the European Commission to ensure all member states guarantee

Figure 9.2 Demonstrations held by Right2Water in Spain. Source: Campaign for the fulfilment of the Human Right to Water and Sanitation, Sevilla, Spain, 7 June 2016.

safe, sufficient supplies of water and sanitation services. Mobilizing around the tagline, *Water and sanitation are a human right! Water is a public good, not a commodity!*, the group secured over 1.8 million signatures in support of the proposal, which led to a hearing within the European Parliament (Figure 9.2). Following the hearing, the European Commission reaffirmed the commitment to ensuring water quality, improving access to water, and increasing transparency. A further consultation on the EU Drinking Water Directive was also proposed, which "foresees inter alia an obligation for Member States to improve access to water and ensure access for vulnerable and marginalized groups."

In some respects, the response from the European Commission deflected attention from the underlying issues that motivated activists in the first place. Activist opposition to privatization is largely ignored, with the Commission maintaining its commitment to "neutrality" when it comes to the question of whether water should be provided by a public or private provider. Nevertheless, the fact that Right2Water was the first ever successful Citizens' Initiative served to inspire other campaigns for equitable access to water across the European Union. As we have argued in this chapter, the right to water is a living thing. We understand that this living thing is to be struggled over and defined by relations between social groups. Right2Water plays a distinctive role by giving new meaning to the right to water in the global North.

Van den Berge and co-authors (2019) powerfully demonstrate the ways in which Right2Water influenced the campaign to overturn plans for the privatization of water services in Thessaloniki, Greece, in 2014. Opposition to water privatization in the port city cohered around the social movement "SOS for water," which "was set up by the same water workers that made the link between Right2Water and the fight against privatization of the water companies in Greece" (Van den Berge et al. 2019, p. 166). Following a successful referendum campaign (one that was at first declared illegal by the Greek Parliament), the water privatization program was overturned and, following the collapse of the government, the President of the Greek Parliament pledged her support for the human right to water. While the legislative change was not as ambitious as most of the movement's supporters had hoped for, the influence of the human right to water on people's broader consciousness was immense.

Elsewhere in the European Union, movements for the human right to water reflected similarities to the Michigan movement. In Ireland, struggles for the right to water emerged to challenge the simultaneous introduction of user charges for water and of legislation enabling water utilities to be able to disconnect consumers (Bieler 2019, 2021). And in Italy, citizens campaigned in a referendum opposing water privatization and articulating this struggle in the language of human rights (Bieler 2019, 2021). In each of these cases, as in South Africa, opposition to social movement struggles was framed as the prudent and necessary response to economic crisis. As Clark (2019) argues, such "financialized logic" exerts a remarkable power across several different contexts. With the Constitutional Court of South Africa upholding the legality of Johannesburg's installation of pre-payment metering, the court appears to focus on "the technical and financial complexity of the City's policy dilemmas rather than the rights of Soweto residents" (Clark 2019, p. 179). While recognizing the frustrations of such struggles, Clark remains attentive to the ways in which water justice activists "have embedded their struggles for the right to water within a broader campaign to reclaim power from below" (Clark 2019, p. 185).

A Right to the Hydrosocial Cycle

How might a victory be a defeat? Or a defeat a potential victory? In 1888, the English utopian socialist William Morris wrote that "men [and women] fight and lose the battle, and the thing that they fought for comes about in spite of their defeat, and when it comes turns out not to be what they meant, and other men and women have to fight for what they meant under another name." Although written about struggles in the Peasants' Revolt, six hundred years prior to the United Nations' recognition of the human right to water, Morris deftly captures the dual nature of struggle. Was the right to water a victory, after all (Angel and Loftus 2019)? Transposing Morris's insight to the present moment, it becomes crucial to identify the efforts of scholars and activists doing this work and recognize that meaning is just as likely to be defined by those, such as Gasson (2010), for whom the United Nations' recognition of the human right to water is a green light for privatization.

In their perspective of the right to water, Sultana and Loftus (2012, 2019) suggest that it might be possible to think of the right to water as akin to what the French philosopher, sociologist, and taxi driver Henri Lefebvre referred to as **the right to the city**. For Lefebvre, the right to the city was not about existence within or usage of the city (Merrifield 2006). Instead, he wrote about the right to *shape* the city by meanings, forms, and functions. The right to the city was about working with other social groups to define the practices that make cities what they are (Merrifield 2006). With water, this approach could signify the right to create a more just and equitable water network, one with a fairer distribution of water. To put things in the language of this book, this approach means the right to shape the hydrosocial cycle more democratically. If, currently, the hydrosocial cycle is influenced as much by the business of water and by the need to ensure profitable returns on infrastructural investments, the right to water defined in this way would ensure democratic decision-making and participation in the hydrosocial cycle itself.

All of this probably sounds utopian, and such a discussion needs to be grounded in the pipes, in the practices, and in the institutions through which water is provided. Perhaps most important among the latter is the institution of the state, which is framed as both the guarantor of the human right to water and as overseer of the business of water (Chapter 4), in addition to its role in the development of global water networks. Catarina de Albuquerque, the former Special Rapporteur on the human rights to safe water and sanitation, is justified in viewing the state as the necessary link in the chain between the United Nations' development of a legislative framework and those citizens or "rights holders" who live within the jurisdiction of individual states. This prioritizing of the state resonates with our earlier discussion of Thomas Hobbes's social contract. However, as we argued, framing the state in this way can result in the exclusion of various others. The resulting framework is a Eurocentric one, based on an understanding of "first-generation" human rights over the communal "third-generation" human rights, including Indigenous framings of natural resources. Writing on the Bolivian experience, Rocio Bustamante, Carlos Crespo, and Anna Walnycki (2012, p. 223) argue:

> If we are to consider how rights can be recognized and employed by the state, we are recognizing and justifying the state as responsible for ensuring compliance. A rights-based approach means that other institutional and organizational forms are not recognized, even though they may occupy spaces for interaction and rights that don't necessarily originate from the state.

Picking up this thread, James Angel and Alex Loftus (2019) note the paradox that an institution is often seen to undermine people's ability to access basic resources – or, at least, often seen to be responsible for reproducing the unequal relationships that undermine people's ability to access such resources. In this paradox, the state is also the main institution responsible for guaranteeing the human right to water. Nevertheless, rather than a simple refusal to engage with state institutions, the authors propose ways of working in-against-and-beyond the state or with-against-and-beyond the human right to water. As in the case of South Africa, this implies: (i) working to invest meaning in the human right to water; (ii) opposing the state's framing of the human right to water (especially, the imposition of full cost-recovery on impoverished populations); and (iii) looking beyond the human right to water for broader democratic struggles of water access. In short, Angel and Loftus (2019) put William Morris's dual nature of struggle into practice. This kind of process-based understanding of the human right to water can learn from Léo Heller's report, which questioned the private sector's role in relation to the human right to water. Heller posed questions around the previously "agnostic" position of the United Nations and shifted attention from outcomes to the underlying processes that produce and reproduce such outcomes.

Rights Beyond the Human?

Thus far, we have discussed the right to water through the lens of human rights and the relationship between the state and its citizens. Yet, struggles over water have transcended the concerns of humans, related to cosmologies, non-human entities, and the "rights of nature" to exist and thrive. Once again, the struggle over rights to water assumes different meanings and relationalities – this time, through the prism of rights for "nature."

For example, in the Andean region, prominent struggles for the right to water have been rooted in Indigenous ontologies, values, and relationalities – which extend the subject of rights beyond the human figure, to include the rights of Mother Earth, or *Pachamama*. In a landmark move, Ecuador established a new constitution in 2008 that adopted the Rights of Nature, based on the notion of *sumak kawsay*, or living well. In the Ecuadorian Constitution, *Pachamama* is regarded as "where life is reproduced and exists, has the right to exist, persist, maintain and regenerate its vital cycles, structure, functions and its process in evolution" (Article 71).

In Bolivia, the Law of the Rights of the Mother Earth (*Ley de Derechos de la Madre Tierra*) was passed in 2010, which included the right to water. Established under Evo Morales, the first Indigenous president of Bolivia, this law specifies inherent rights for Mother Earth, human communities, and ecosystems. Nature is not subject to humans. The more recent 2012 Framework Law of Mother Earth and Integral Development for Living Well (*La Ley Marco de la Madre Tierra y Desarrollo Integral para Vivir Bien*) signifies that the health and integrity of Mother Earth is fundamental, otherwise the "*vivir bien*" [a thriving life] is not possible. These Andean ontologies present how water is life in a broad sense. Rights are a way to defend life in which humans are *part of* nature.

The right of nature is increasingly recognized in other places as well. In New Zealand, the Whanganui River was given legal personhood with human rights in 2017. This was the result of more than 160 years of struggle by the Māori peoples, who traditionally recognize their wellbeing as inherently tied to the wellbeing of the river. Māori cosmology sees water as life in which humans and nature are inseparable. In India, there have been efforts by a state High Court to recognize the Ganges and Yamua rivers and the Gangotri and Yamunotri glaciers as living entities (though the rivers were later denied such right by the Supreme Court). In Bangladesh, the High Court recognized the Turag River as a living entity with legal rights in 2019. In these efforts, the idea that "water is life" moves beyond seeking justice for only humans.

The rights of nature are steeped in contentious politics and debates about development and the neoliberal legal orders and economic policies that abstract nature for profit. At first glance, the rights of nature might seem counter-hegemonic to the human–nature hierarchy, a clear rejection of Western notions of capitalist development. Yet, as Mabel Gergan (2017, p. 490) reminds us, "[P]eople's relationship with a sacred, animate landscape is not easily translatable into the clear goals of environmental politics." In research with anti-dam activists and Indigenous Lepcha youth in the Indian Himalayan region, Gergan exposes the cultural and political anxieties expressed by participants, who cope with the layered effects of loss brought by climate change, colonialism, hydropower development, and economic change. Gergan (2017, p. 495) argues that "[C]ritical engagements with indigenous environmentalism must be in dialogue with diverse interpretations and registers of loss and erasure beyond those addresses by environmental politics" – a claim and cautionary tale echoing Rachel Arsenault, Jody Inkster, Deb McGregor, Teresa Montoya, Nicole Wilson, and others (Chapter 1).

Experience from Ecuador shows that claims made through the *Pachamama* are complex, with outcomes that are not straightforward. The rights of nature can be used by Indigenous communities to expand support for their claims to protect the *Pachamama* and watersheds. But the sacred concept has also been used by the state to facilitate mining plans where water is polluted and transfer of watershed management to the private sector. Here the right of nature is paradoxically instrumental to continuing destructive practices of mining but with regard to water and environmental conservation, thus setting on a path of "market-oriented sustainability that is similar to neoliberal or green environmentalism but with social redistribution as a crucial difference" (Velásquez 2018, p. 166). Such redistribution has not been taken seriously enough by the state, and the ideas of development stemming from *sumak kawsay* have contributed to further cementing inequalities between peoples in Ecuador (Radcliffe 2012). The rights of nature certainly have not been the panacea to water struggles. But the debates invite us to ask critical questions about the fate of nonhumans in the hydrosocial cycle, and what conditions are necessary to ensure water "justice" for all.

The Right to Water as a Process

The "agnostic" position that Heller's report set out to interrogate is perhaps best seen in some of the statements of his predecessor, Catarina de Albuquerque, in her role of Special Rapporteur. For de Albuquerque, the practical outcomes – such as whether people have access to safe, sufficient supplies of water – are far more important than debates over who should provide that access. In an interview with *The Guardian*, de Albuquerque goes further than this agnostic position, suggesting that it is a "no brainer" that the private sector will have to be involved in water provision (Purvis 2016). But for Heller this isn't such a "no brainer." Instead, the role of the private sector should be open to question. This question became the basis for Heller's final report to the UN General Assembly, submitted on July 21, 2020. Heller's conclusion was clear and unambiguous: privatization is a risk to the human right to water (UN OHCHR 2020).

To arrive at this conclusion, Heller engaged in a forensic process of data collection, including expert consultations and public hearings with several interested groups, as well as responses from NGOs, think tanks, member states of the United Nations, and the private sector (all data can be accessed through a UN webpage).[2] Based on this evidence, Heller's report challenges the narrative that human rights are neutral when it comes to the type of water provider. The report goes on to identify specific risks relating to privatization: risks associated with profit maximization; the fact that water is a natural monopoly; and power imbalances associated with the introduction of private actors. For the human right to water, this risk-based analysis identifies six areas of concern, ranging from questions of affordability to a lack of accountability. Perhaps the key message is that processes matter in defining outcomes: "[P]rocesses underlying water and sanitation service provision are not neutral and shape the social, political and economic environment in which human rights are realized" (UN OHCHR 2020, p. 4).

This last statement captures some of the themes we have emphasized throughout this book. Indeed, by adopting a relational approach to the hydrosocial cycle, we hope you are now attuned to thinking in terms of processes and how they shape people's ability to access water. Heller transposes such an understanding of the human rights to water and sanitation. Unlike his predecessor, Catarina de Albuquerque, Heller is adamant that public and private providers are not equivalent. Rather than being a "no-brainer," questions of ownership are crucial in realizing the human right to water.

The response to the UN report from the private sector was swift and aggressive. Neil Dhot, executive director at AquaFed, the International Federation of Private Water Operators, argued that Heller's report was "anti-private-sector and pre-judged" (quoted in Root 2020). AquaFed wrote directly to Heller, threatening (in what Heller describes as a "very aggressive email") to "escalate the matter" (Heller 2022, p. 121). Prior to Heller's presentation of his report to the UN General Assembly, the group wrote to senior officials, including the High Commissioner, in the UN Human Rights Office of the High Commissioner. They accused Heller of having violated his mandate (Root 2020). To even question the role of the private sector, it would seem, should not be permitted. A wide range of responses to Heller's consultations were published on the UN webpage (see footnote 2), including a long document from AquaFed itself. Nonetheless, Aquafed claimed a "lack of published meeting notes and critiques" (quoted in Root 2020). Clearly

[2] Refer to the UN Human Rights Office of the High Commissioner report: https://www.ohchr.org/en/calls-for-input/reports/2020/privatization-and-human-rights-water-and-sanitation-report (accessed 20 June 2022).

frustrated by the personal attacks, Heller hit back, claiming "AquaFed pressed all the time to subvert this standard procedure and to publish in advance documents that they produced" (Root 2020). Civil society groups rallied to Heller's defense, issuing a statement of support. As for Aquafed, the group argued:

> This interference [from Aquafed] is a transparent and unacceptable attempt to protect the industry's profits from exposure to the reality of the lived experience of far too many who have had their human rights violated under privatization. (Transnational Institute 2020)

We have come full circle. We started this chapter by arguing that different groups struggle over the human right to water and invest different meanings, and Heller's forensic report is the latest document to become embroiled in the complex politics around the human right to water. These struggles only serve to reinforce Heller's claim that processes matter in achieving the human right to water and that they are "not neutral" (UN OHCHR 2020, p. 4). Teasing out those processes and relationships is crucial to understand the human right to water in a particular context.

Summary and What's Next

In this chapter, we have discussed how the human right to water has become an important terrain upon which competing visions around water have struggled. The long arc of struggle for water justice has most recently been fought over in this terrain. The UN Special Rapporteur's 2020 report on the human right to water is only the most recent episode to give greater meaning to the right to water, and greater meaning to define the fairest and most equitable means of accessing this basic resource. While some dismiss the human right to water as a distraction from the bigger questions over the ownership of water and broader political economy (Chapter 4), we argue that efforts to shape the human right to water are at the forefront of ensuring greater democratic control over the resource. From the racialized injustices of Flint's water crisis to the debates over how water should be paid for in Ireland and Greece, localized struggles are fundamental to how individuals currently access water and how they are likely to access it in the future. With this nod, our next and final chapter collects the lessons learned in this book – the four planks – and applies critical hydrosocial thinking to the biggest challenge of all: the future.

Further Reading

On human rights
Alston, P. (2017). The Populist Challenge to Human Rights. *Journal of Human Rights Practice* 9 (1): 1–15.

Harvey, D. (2000). Uneven Geographical Developments and Universal Rights. *Spaces of Hope*. Berkeley, CA: University of California Press.

Mégret, F. and Alston, P. ed. (2020). *The United Nations and Human Rights: A Critical Appraisal*. Oxford, UK: Oxford University Press.

On the human rights to water and sanitation

Annual thematic reports authored by the UN Special Rapporteurs on the human rights to drinking water and sanitation. https://www.ohchr.org/en/special-procedures/sr-water-and-sanitation/annual-reports (accessed 18 June 2022).

Baer, M. (2017). *Stemming the Tide: Human Rights and Water Policy in a Neoliberal World*. Oxford, UK and New York: Oxford University Press.

Baer, M. and Gerlak, A. (2015). Implementing the Human Right to Water and Sanitation: A Study of Global and Local Discourses. *Third World Quarterly* 36 (8): 1527–1545.

Heller, L. (2022). *The Human Rights to Water and Sanitation*. Cambridge, UK: Cambridge University Press.

Langford, M. and Russell, A.F. ed. (2017). *The Human Right to Water: Theory, Practice and Prospects*. Cambridge, UK: Cambridge University Press.

Mirumachi, N., Duda, A., Gregulska, J., and Smetek, J. (2021). The Human Right to Drinking Water: Impact of Large-Scale Agriculture and Industry: In-Depth Analysis. Publications Office. European Union, Policy Department, Directorate-General for External Policies. https://doi.org/10.2861/714431.

Sultana, F. and Loftus, A. ed. (2012). *The Right to Water: Politics, Governance and Social Struggles*. London: Earthscan.

Sultana, F. and Loftus A. ed. (2019). *Water Politics: Governance, Justice and the Right to Water*. Abingdon, UK: Routledge.

The state and the human right to water

Angel, J. and Loftus, A. (2019). With-Against-and-Beyond the Human Right to Water. *Geoforum* 98: 206–213.

Bustamante, R., Crespo, C., and Walnycki, A. (2012). Seeing Through the Concept of Water as a Human Right in Bolivia. In: *The Right to Water: Politics, Governance and Social Struggles*. (ed. F. Sultana and A. Loftus), 223–240. London: Earthscan.

On Flint and water/environmental justice

Clark, C. (2019). Race, Austerity and Water Justice in the United States: Fighting for the Human Right to Water in Detroit and Flint, Michigan. In: *Water Politics: Governance, Justice and the Right to Water* (ed. F. Sultana and A. Loftus), 175–188. Abingdon, UK: Routledge.

Pauli, B.J. (2019). *Flint Fights Back: Environmental Justice and Democracy in the Flint Water Crisis*. Cambridge, MA: MIT Press.

Pauli, B.J. (2020). The Flint Water Crisis. *Wiley Interdisciplinary Reviews: Water* 7 (3): e1420.

Pulido, L. (2016). Flint, Environmental Racism, and Racial Capitalism. *Capitalism Nature Socialism* 27 (3): 1–16.

Ranganathan, M. (2016). Thinking with Flint: Racial Liberalism and Roots of an American Water Tragedy. *Capitalism Nature Socialism* 27 (3): 17–33.

Rights of Nature

Boelens, R. (2014). Cultural Politics and the Hydrosocial Cycle: Water, Power and Identity in the Andean Highlands. *Geoforum* 57: 234–247.

Radcliffe, S. (2012). Development for the Postneoliberal Era? *Sumak kawsay*, Living Well, and the Limits to Decolonization in Ecuador. *Geoforum* 43: 240–249.

Salmond, A., Tadaki, M., and Gregory, T. (2014). Enacting New Freshwater Geographies. *New Zealand Geographer* 70: 47–55. https://doi.org/10.1111/nzg.12039.

Tola, M. (2018). Between Pachamama and Mother Earth: Gender, Political Ontology and the Rights of Nature in Contemporary Bolivia. *Feminist Review* 118 (1): 25–40. https://doi.org/10.1057/s41305-018-0100-4.

Chapter 10

Future Waters

At the Confluence of Change

The Okavango Delta in Botswana, south-central Africa, is in flood season. Under the glare of the hot sun, university students giddily clamor onto four-wheel drive trucks and drive into the Delta's swamps and lagoons. They crane their necks for a first sighting of elephants, giraffes, and lions. Students from King's College London, University of New South Wales, and Arizona State University have come together on an interdisciplinary 10-day field trip to observe the sustainability challenges of the Okavango Delta, a complex hydrosocial ecosystem.

Originating in the Angola highlands, the Cubango-Okavango River flows into Namibia and Botswana. The river is a transboundary waterway, 1100 km long and characterized by intense seasonal flooding. Rains from October to March raise water levels. The flood pulse slowly moves downstream over a period of four months. Water creates an alluvial fan or an inland delta in Botswana that never reaches the sea. Surrounded by the Kalahari Desert, the Okavango Delta is often described as a desert oasis that sustains important biological productivity and a multi-million dollar tourism industry. Seasonal flooding contributes to the rich biodiversity of the delta that attracts the "big five game" animals and many hundreds of bird species and fish. Grassland biota has evolved in sync with the flood pulse and timed with seasonal river

Water: A Critical Introduction, First Edition. Katie Meehan, Naho Mirumachi, Alex Loftus, and Majed Akhter.
© 2023 John Wiley & Sons Ltd. Published 2023 by John Wiley & Sons Ltd.

changes. The Okavango Delta is a near-pristine wetland system that drains one of the last free-flowing rivers in the world. Recognized as a site of the Ramsar Convention on Wetlands of International Importance, the Okavango Delta is also the 1000th site on the United Nations Educational, Scientific and Cultural Organization (UNESCO) World Heritage list.

The visiting students survey the river environment on boats and vehicles, examining water quality, flood levels, and the land–water connections supporting animals, birds, fish, and diverse aquatic habitats (Figure 10.1). They set camera traps and underwater cameras to track nocturnal and underwater life. Students test Arduino technology (open-source sensors) in science experiments to assess aquatic community ecology. They learn about the scale and complexity of river basin management. The students listen intently to local guides, whose vast environmental knowledge of the wetlands portrays life on the Okavango in ways the cameras and data could never capture. Local residents coexist with the rhythm of the flood pulse, exemplifying the intimate socio-ecological relations between people and the river (Shinn 2016).

A year later, a new cohort of equally excited students arrive at field camp. One thing is remarkably different, however. Hardly any flood waters are present, and boat surveys must be abandoned. In fact, the Okavango Delta is experiencing one of the driest years on record. Flooding of the delta is highly variable from year to year. Depending on the climatic conditions, the delta swells in size varying from 3500 km^2 to 9000 km^2 (Murray-Hudson and Dauteuill 2019). This spatial and temporal variability – related to precipitation, timing, and depth of flood waters – has shaped the ecosystem and

Figure 10.1   Students surveying the Okavango Delta, Botswana. Source: Naho Mirumachi.

people's wetland livelihoods. Nevertheless, experts are concerned that rainfall and flooding will become less predictable, as analysis by the Intergovernmental Panel on Climate Change (IPCC) suggests more frequent weather extremes in the coming years (Caretta et al. 2022). Climate models present a mixed picture: some models expect the delta to experience more severe droughts; others predict it could be wetter or drier (Shinn 2016). It is, however, noted that climate change will significantly influence the ecology and biodiversity of the delta (World Bank 2019).

The Okavango Delta is undergoing human-induced changes as well. In the past, environmental change in the delta has been both natural and human-induced, though often with more anthropogenic impacts. Human settlements and governance institutions have shaped the quantity and directions of flood waters in the delta. There have been efforts in Botswana to expand irrigation to address food security. Thus far, irrigation has been limited and draws mostly on groundwater and surface water supplies, neither of which are strictly monitored (Masamba and Motsholapheko 2017). Policies have not always worked to prevent such change – and, in fact, even well-intentioned policies have had adverse or unintentional effects (Hamandawana and Chanda 2010).

Hydropower is also on the menu. In Angola, there have been plans to develop hydropower projects on the river since colonial times. While the prolonged civil war made such plans less feasible, hydropower dam prospects have been revived. To date, 28 hydropower project plans have been drawn up; the main dams could have a collective capacity of 391 MW. Additionally, 11 irrigation schemes have been proposed, extending across 279 500 hectares (ha). In Namibia, water scarcity has long been a major concern, especially for its thirsty urban areas. Namibia's capital city, Windhoek, has relied on the Grootfontein-Omatako Eastern National Water Carrier and three interlinked dams for water supply. However, this system is no longer sufficient to support future water demand (Lewis et al. 2019). Namibia has developed a contested plan to transfer water from the Okavango River, which supplies water downstream to Botswana, to the Carrier through a 240 km pipeline. The water transfer will abstract 32 million m^3 of water, potentially up to 67 million m^3 (World Bank 2019).

What does the future hold for the Okavango Delta? Will our students map the same species in a decade's time? Will seasonal floods still sculpt the majestic landscape dotted with islands and lagoons? Who will still fish these waters? Will there be more droughts in the region, changing the way people and ecosystem inhabit the river basin? How will the river system, people, and governing institutions cope with projected impacts of climate change? How will local residents adapt to change in the delta? The hydrosocial cycle of the Okavango Delta begins to reveal the knotty relations – social and ecological – that make up a wetland, and shows the flows, discourses, and ideas that not only produce the existing waterscape but also promise to shape its future.

Without a doubt, the Okavango Delta will face new challenges due to climate and social change. Many of the other watery places mentioned in this book face equally dire pressures, though with fundamentally uneven impacts and implications. Because of this urgency and uncertainty, a critical perspective of **future waters** is more necessary than ever before. To recap, our critical perspective on water and society rests on four core arguments, the four planks that underpin the foundation of this book. They include:

1. Knowledge is power.
2. Scarcity is made.
3. Water is life.
4. Camp is everywhere.

In this final chapter, we will again put these planks to work as tools to probe the narratives, mythologies, and biases embedded within the supposed "solutions" to our water problems. A critical hydrosocial perspective, we argue, cannot take such "solutions" for granted if they do not radically address the relationality of water and society. In thinking critically about water management solutions such as large-scale water transfers or desalination – two of the many options in circulation – we argue that a strict focus on technological promise disguises the fact that water is *produced* in dialectical relation with society. A drought in the Okavango Delta may be part of the "natural" cycle of life. However, the impacts of hydropower development and global warming-induced change on the wetland system are fundamentally bound up with institutionalized dynamics and power asymmetries.

We draw on Erik Swyngedouw's (2013) concept of the **techno-fix** to frame our approach in this chapter. In his analysis of Spain's use of desalination projects to "solve" their water supply crisis, Swyngedouw (2013) argues that a fixation on new technological "solutions" obscures the fact that the Spanish state continued to encourage and *expand* overall water usage. The Spanish state, he argues, uses desalination as a mechanism to sustain its transition toward a market environmentalist governance model.

> Extending the terrestrial management of the hydro-social cycle to include the sea as a "fix" to alleviate the country's uneven geographical distribution of water and for satisfying its unquenchable thirst had been contemplated since the dying days of Fascism. In 1973, for example, the official engineering journal already insisted on the promises of mobilizing the seas: "[W]e should not see the future with pessimism . . . because, after all, the great alternative will be the sea. The sea will be our greatest reserve resource and we have it in abundance." (Swyngedouw 2013, p. 262)

To be clear, we are not haters of technology. People always have devised ingenious technologies to help them adapt and change, but such "fixes" cannot be understood in isolation of the relations and conditions that produce them. The Spanish engineer above insists that we "should not see the future with pessimism," but we argue that a critical approach is the antithesis of pessimism. Remember our adage: skepticism is a superpower. Skepticism is not cynicism or a life without hope, we wrote in Chapter 1. Skepticism is a wariness of institutionalized power and seductive narratives. Skepticism allows us to pause, focus, and interrogate the production and implications of a "solution" like desalination. Skepticism is part of the critical toolbox that we bring to the hydrosocial cycle – not because we are cynical, but because we care deeply about the world and our future.

Stationarity is Dead

Stationarity is dead. This statement was declared by a group of prominent hydrologists in 2008. Paul Milly and colleagues argued that **stationarity** – the "idea that natural systems fluctuate within an unchanging envelope of variability" – was no longer applicable in a heating climate (Caretta et al. 2022). The hydrologists debunked the myth, long held in water resources management and engineering, that anticipated "extreme" weather in the short term but "static" climate conditions in the long term. If water resources management and planning continued to be based on stationarity, they argued, then society would be in big trouble because of the significance of anthropogenic warming of the climate and earth.

Climate change has impacted and will transform major components of the hydrosocial cycle, including precipitation, evapotranspiration, groundwater recharge, and river discharge (Caretta et al. 2022). Historical precedents to cope with water management problems might have looked at developing new water infrastructure or extended drainage. Milly and colleagues argued that these solutions – designed to buffer variability – would no longer be suitable or sufficient in the face of massive climatic and hydrological changes. Stationarity "should no longer serve as a central, default assumption in water-resource risk assessment and planning" (Caretta et al. 2022).

Turbulent changes are already evident in the **world's water towers**: high mountain regions, such as the Himalayas and Andes, with glaciers and snowpack that feed major river systems. Climate change is leading to shrinking glaciers and accelerating the pace of snow melt, making water supply downstream much less reliable – particularly during the dry season – and compounding problems of water quality. There are 78 water towers in the world, which support the water needs of 1.6 billion people, or 22% of the entire population (Immerzeel et al. 2020). Mountains are the site of half of the global biodiversity hotspots. The implications of climate change, glacier recession, and declining levels of meltwater have a tremendous impact on mountain societies (Carey et al. 2017; Carey 2010). "Ice loss has the potential to affect human societies in diverse ways, including irrigation, agriculture, hydropower, potable water, livelihoods, recreation, spirituality, and demography (Carey et al. 2017 p. 350). Coupled with weak institutional governance and tensions over shared waters, it is estimated that 1.9 billion people will feel the adverse impacts of melting water towers (Immerzeel et al. 2020).

Stationarity cannot be revived. For Milly and colleagues, the only option moving forward is to use **non-stationarity** as a principle and starting point to rethink assumptions of water management practices (Caretta et al. 2022). What might this look like? What implications emerge for water management on a heating planet?

First, the constraints posed by existing water infrastructure are massive. The fixity of infrastructure projects creates a kind of **path dependency**, whereby past decisions constrain later actions that might be incompatible with flexibility or rapid adaptation demanded by changing hydro-climatic conditions. Once in place, large-scale water projects such as big dams are difficult to modify or remove – societies become dependent on their irrigation waters or cheap electricity (Chapter 6). Seawalls and riparian dikes can provide flood defense and enable settlement in drainage areas. However, sea-level rise and the increasing severity and frequency of flooding – especially as urban areas become paved over and lose spongy soils – mean that flood risk *increases* over time, "perpetuating investment in escalating levels of water security" (Haasnoot et al. 2020, p. 452). Major infrastructures developed in the past century will need rehabilitation now or in the near future, and the current wave of austerity governance in many countries does not guarantee funds will be provided for maintenance, repair, and replacement (Chester and Allenby 2019; Chester et al. 2019). In the future, dams need not only to store water for the dry season but also to deal with surplus water by providing a buffer to store flood waters (UN-Water 2020). Once these large-scale infrastructures and hydrosocial relations are in place, change seems hard to come by.

Second, non-stationarity requires infrastructure to be robust enough in the face of a larger range of hydro-climatic conditions. This means that investments in new infrastructure will require complex design and higher costs. Decision-making on infrastructure will fundamentally need to embrace uncertainty within a broader strategy of coping with changing conditions (Hallegatte 2009). However, in the case of flood defense, water managers often only have vague guidelines on decision-making under non-stationarity and are not sufficiently equipped with knowledge of flood frequency analysis. This knowledge gap risks new investments in flood defense infrastructure to be either too large or too small in the face of peak flow decrease or increase, leading to high costs to the project, adverse human impacts, and economic losses (François et al. 2019).

Water infrastructure and institutions in the world of non-stationarity should be agile – a seemingly impossible task. But the prospects of establishing smart, flexible, adaptable infrastructure can be considered through a hydrosocial lens. If we understand water as being relational, then we need to revisit existing infrastructures, examine their production and discourses, and explore how flexibility is being sought (or not) and with what implications. We also need to scrutinize "new" infrastructure and technology that supposedly provides solutions for an uncertain future, by questioning the conditions of water that such technologies enable and mediate, its production, and the role of people in it. Drawing on Swyngedouw's (2013) critique of the techno-fix, we turn to four examples.

Techno-Fix 1: Desalination

As climate change puts freshwater sources under increasing pressure, water managers in some parts of the world are turning to **desalination** – the conversion of seawater or brackish groundwater into fresh water through a process of reverse osmosis – as an adaptation response to meet growing demand and buffer against future scarcity (McEvoy and Wilder 2012). Desalination is promoted as a "drought-proof" solution to produce "new" water supply by planners and managers, although existing projects in Spain, Israel, Mexico, Australia, and the United States raise as many questions and unanticipated outcomes as they "solve" problems (Fragkou and McEvoy 2016; McEvoy 2014; Swyngedouw and Williams 2016).

For example, a risk analysis by Jamie McEvoy and Margaret Wilder (2012, p. 353) shows that "while desalination technology can reduce some vulnerabilities (e.g. future water supply), it can also introduce new vulnerabilities by compounding the water-energy nexus, increasing greenhouse gas emissions, inducing urban growth, producing brine discharge and chemical pollutants, shifting geopolitical relations of water security, and increasing water prices." The process is extremely expensive and energy-intensive to produce clean fresh water. Because of this expense, desalination projects tend to be limited to high-income countries. Moreover, the desalination footprint tends to *increase* fossil fuel consumption, actively contributing to negative feedback cycles.

The issue of whether desalination is a "fix" for capital – rather than a necessary or viable solution to climate change – raises serious political implications for water relations. For example, Alex Loftus and Hug March (2016) investigate England's first attempt at desalination, the Thames Water Desalination Plant (TWDP) in Beckton. Expert networks helped make this salty vision a reality. TWDP was built by Acciona, a Spanish infrastructure company with a prominent role in the Spanish desalination industry (March et al. 2014). Built in 2010, the TWDP was intended as an "emergency plant" to supplement water to approximately 400 000 households during drought or exceptional demand.

With a price tag of GBP 270 million, the project won several design awards, e.g. "Most Sustainable Project" in 2009. Nonetheless, operational costs are high and the plant fails to produce a constant supply of water at a viable price. As of writing this chapter in 2022, during England's driest spell since 1976, the TWDP is "out of service" in a serious drought. Experts report the TWDP is ill-equipped to handle the "varying salt water levels" in the Thames Estuary (Weaver 2022).

What the TWDP *did* provide is an opportunity for Thames Water to roll out a byzantine financial model, a lynchpin in the utility's investment strategy and focus on large infrastructure (Loftus and March 2016). Knowledge is power, as this book argues, and flows of capital accumulation were enabled by a network of global finance firms – from Australia to Spain to England – and the speculative promise of the TWDP:

> [O]ne begins to see how the risks entailed in the so-called Australian model have enormous implications for the way in which infrastructure that is geographically rooted in the UK comes to circulate as a financial asset and is "flipped" by different globalize funds . . . Speculation therefore extends well beyond the projected returns on a tightly regulated water market within a given infrastructure network; it also captures the

expected returns on future infrastructural developments, which is where the expectations of profits to be gained from increasingly interlinked infrastructures, across different sectors, begin to emerge. The TWDP is not only linked to Barcelona through
Macquarie's broking role. More importantly, the entire regional raw-water supplies
ATLL (including Barcelona's desalination plant) was leased in 2012 to an international
consortium led by the same Spanish infrastructure company, Acciona, that constructed
the TWDP. Thus, Acciona's technologies and scientific expertise were put to work in
London, at the same time as a financialized model of water provision – in which Thames
Water is surely at the apex – can be seen to have been transported from Australia to
Barcelona via London. (Loftus and March 2016, pp. 57–58)

Where does this leave us? To fully understand the TWDP, Loftus and March (2016,
p. 47) argue, "requires focusing on the scalar interactions between finance, waste,
energy, and water that weave the hydrosocial cycle of London." A focus on the **metabolism** of a "solution" like the TWDP provides us with a fuller view of the relations and
conditions that make desalination a "solution" – and forces us to ask tough questions
about who wins and who "loses" in future waters.

Desalination is a potentially promising alternative to conventional water supply
sources, but a critical approach cannot take hegemonic discourse for granted. "On the
one hand, if we accept the testimony of Thames Water, the answer is simple. Population
demands, a changing climate, and low rainfall place enormous pressures on a water
network that dates back to the Victorian era," write Loftus and March (2019, p. 58).
"[S]uch a response fails to address the question as to why such a large, energy-intensive
infrastructural project provides the best solution to the water needs of London. Why
not invest money in renewing meters, fixing leaks, rainwater harvesting, educational
campaigns, and so on?"

Techno-Fix 2: Moving Water

If water is abundant in one region, why not move it to a water-needy area? The **South–
North Water Transfer project** in China is the world's largest attempt to move water
across vast distances. This project aims to transfer 45 billion m³ of water to the northern regions of Henan, Hebei, Beijing, and Tianjin across a distance of 1273 km
(Figure 10.2). The northern regions have densely populated cities and are the agricultural powerhouses of the country, producing key staples such as wheat, maize, cotton,
and other crops with irrigation, putting surface and groundwater sources in short
supply and at ecological risk (Crow-Miller 2015; Rogers et al. 2020).

Constructed between 2004 and 2014, the project diverts water from the Yangtze
River northwards via canals and tunnels organized along three major water transfer
routes: Eastern, Middle, and Western routes, although the Western route has yet to be
realized (Pohlner 2016; Figure 10.2). This is a **megaproject** of significant proportions,
in terms of geography and cost. Megaprojects usually cost USD 1 billion or more

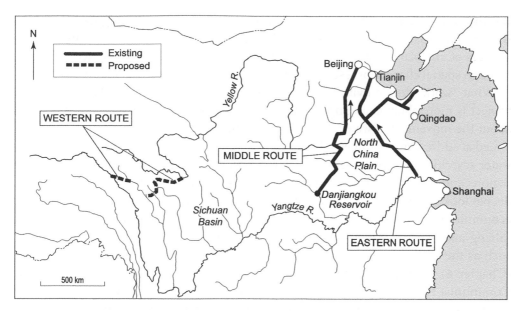

Figure 10.2 Map of the South–North Water Transfer project in China. Source: The Authors, cartography by Philip Stickler.

(Flyvbjerg 2014). This water transfer project is reported as costing just over USD 29 billion (Pohlner 2016), 47 billion, or even close to 80 billion (Webber et al. 2017). While figures vary, there is no denying that this is a costly megaproject in a class of its own.

The grid-like system of water transfer routes laid northbound that interconnects with rivers flowing from west to east exemplifies the material properties of water in shaping this major water supply solution. This megaproject connects the large streams of the Yangtze, Huai, Yellow, and Hai rivers, extending across nearly a third of the country's territory (Sheng and Webber 2019). The transfer project shapes the materiality of river flows of the Yellow, a closed river basin (Chapter 5). Importantly, the project has induced changes to water pricing policy and institutions. The central government has assumed a larger role in determining water pricing, leaving less decision-making power to local governments. Pricing is based on prioritizing the use of transferred water over other alternative water sources, with the effect to secure the financial return of the megaproject and to sustain demand for transferred water (Pohlner 2016).

A megaproject of this scale does not come without resettlement. To realize the Middle route, 340 000 people were resettled when the dam was constructed at Danjiangkou. A further 62 000 residents were removed during the process of canal construction. The Eastern route resettled approximately 8000 people. Furthermore, to ensure the quality of water to be transferred, 88 000 people have been resettled from upstream reaches of the Danjiangkou dam (Rogers et al. 2020). As a result of resettlement, water users in the northern region now have access to high-quality water.

But this access comes at the expense of a burden placed on county governments in source areas that effectively subsidize the cost of pollution control in larger cities (Rogers et al. 2020). The transfer project has created new socio-ecological relations through spatial changes of resettlement and flows of water that are quality assured.

What "work" does moving water do? Arguably, the South–North Water Transfer project is more about changing political relations between people and the state than about the movement of water (Sheng and Webber 2019). At first blush, the project is aimed at addressing the uneven distribution of water within the country and resolving water scarcity. The project delivers water to over 100 cities to sustain growing urban regions and to create jobs through massive construction demand (Pohlner 2016). A deeper look, however, reveals that a motivation behind investment is to ensure China's economic growth and to buffer against global and national economic downturns (Sheng and Webber 2019) – a classic infrastructural justification, used in projects such as the United States' Tennessee Valley Authority in the early twentieth century (Chapter 6). Sustained economic growth is central to the interests of the Chinese Communist Party (Lin 2017). The transfer project helps buttress the political legitimacy of the government by securing continued prosperity, especially in urban areas (Crow-Miller 2015).

Nonetheless, these social and ecological relations shape water flows without giving due regard to future implications of non-stationarity. In the face of the increased likelihood of droughts in the water-supplying southern region and in the north, there is no guarantee that water transfer at this scale can be sustained as planned. Moreover, the project is highly energy intensive. While the Middle Route utilizes gravity to transfer water, the Eastern Route requires pumping stations to lift water more than 65 m. This results in 2.35 billion kWh of energy use for 15.5 billion m^3 of water transfer (Chen et al. 2019). In non-stationary futures, it will be not only the availability of water but also concerns about energy that will determine the effectiveness of infrastructural solutions.

Across China, the future of water is shaped by two powerful discourses identified by geographer Britt Crow-Miller (2015). First, a discourse that naturalizes water scarcity in the northern region is at play. The megaproject is framed as a response to the challenges of a naturally "dry" and drought-prone region – similar to justifications for megaprojects in the US West and British Punjab (Akhter and Ormerod 2015). Even though there have been severe human-induced pressures on water demand, water scarcity is taken as something of a "natural state" irrespective of expanding cities and industries in the region.

Second, while the megaproject supposedly deters the use of limited groundwater sources in the northern region, it is expected that the provision of water will go toward the aims of increasing the irrigation area (Crow-Miller 2015). If the irrigation area expands, then agricultural activity relying on groundwater will continue – a situation where the horse follows the proverbial cart. The demands for water are given so much primacy that solutions for water supply are created to meet such demand. Combined, these two discourses work to naturalize apolitical assumptions as the causes of water scarcity. As a result, options other than supply management are not considered (Crow-Miller 2015).

Will the transfer project deliver on its promise? While these operative discourses are influential practices, they do not offer any insight into how the megaproject might be adaptable or flexible to unpredictable water conditions of the future. For example, the southern region currently imports food from the northern region at an equivalent of 52 billion m^3 of virtual water (Chapter 5). This figure is more than the 45 billion m^3 capacity of the water transfer project itself, raising critical questions as to whether strategic virtual water "imports" might be more appropriate than the actual physical water transfer (Chen et al. 2019). The future hydrosocial relations of the interconnected southern and northern regions are not pre-determined. After all, the conditions of water are constantly produced and regenerated through relations. However, the techno-politics and socio-material impacts of the megaproject are already so significant that there seems little room to (re)design the infrastructure and institutions to anticipate non-stationarity.

Techno-Fix 3: Nature-Based Solutions

So far, we have discussed "hard" (gray) infrastructure and its role in non-stationary futures. In recent decades, planners and environmental engineers have increasingly promoted "green" infrastructures as a way to mitigate environmental problems and deliver ecosystem services. Tree planting and agro-ecological farming practices, for example, could enhance water storage as a solution to problems of flooding or erosion. Designing wetlands to "cleanse" and filter water impurities is another popular approach.

Nature-based solutions (NBS) describes an umbrella category of "natural" fixes to problems, including green infrastructures, ecosystem-based adaption, and natural climate solutions, to name a few (Sneddon et al. 2020, p. 2). Nature-based solutions seek to utilize "nature" and natural processes to provide services for ecosystems and social wellbeing in a climate-uncertain world.

Nature-based solutions emerged out of policy debates in the late 2000s, and the approach has been championed by influential conservation organizations such as International Union for the Conservation of Nature (Cohen-Shacham et al. 2016). The language of NBS has since become widespread, with a growing list of definitions, actions, and types of solutions. Its popularity is such that it is considered "one of the first-class tickets to achieving ambitious social and environmental goals on a mass scale" (Lipton 2020, n.p.).

Nature-based solutions are viewed as promising solutions for water scarcity. For example, the agricultural sector – the largest consumer of water (Chapter 5) – could "solve" droughts and problems of scarce water through agro-ecological techniques, conservation agriculture, or water harvesting. A focus on soil conservation considers both blue and green water. Nature-based solutions are promoted as techniques to reduce societal vulnerability and provide a range of co-benefits (UN-Water 2018b).

What is "natural" about nature-based solutions? And who (or what) benefits? A critical perspective would question the ways that such categories are made and mobilized in policy discourse. As Osaka et al. (2021, p. 2) argue, "defining and presenting a particular course of action or policy as 'natural' can itself be a political act, with consequences for how such policies are interpreted and leveraged in the public sphere." The authors show how NBS frames nature in utilitarian discourse – in other words, as an "instrument" for social gain – such that it becomes yet another example of the valuation or commodification of nature for capital accumulation. The hype around nature-based solutions can work to obscure the intractability of squaring "trickle-down economics, on the one hand, and on decoupling CO_2 emissions (as proxy for environmental harm) from unfettered economic growth through technology and modernization, on the other" (Kotsila et al. 2021, p. 257).

In Chapter 4, we presented a critical view of market environmentalism in nature, including the ways that water is "valued" and managed in pro-market societies. By extension, NBS and its "naturalness" can be seen as "services" provided by nature through a neoclassical and orthodox utilitarian economic lens. Other ways of valuing nature are marginalized when such hegemonic framings are put into place. In a contradictory move, nature-based solutions are enabled by market-based rhetoric and systems that produce the very problems, inequalities, and injustices in the first place (Kotsila et al. 2021).

As neoliberal governmental policies squeeze budgets and shift priorities, payments for ecosystem services (PES) has become a popular policy discourse in the water sector (Pechey et al. 2013). As we explained in Chapter 4, PES is a policy discourse and market instrument that enables buyers of ecosystem services to exchange with sellers who provide stewardship to incentivize pro-environmental water, soil, or land management.

A critical analysis of PES shows how economic and ecological relations are produced and contested. In Colombia, energy and hydropower companies have adopted PES schemes and implemented conservation measures upstream of large dams (Rodríguez-de-Francisco et al. 2019). Analysis of the Colombian case gives insight to the economic and ecological relations that produce conditions ripe for market involvement, and raise questions about the costs and distribution of benefits. In the Hidrosogamoso project, which generates nearly 10% of Colombia's electricity supply, upstream farmers who opt into the PES scheme must change their farming strategies and livelihoods.

At the same time, the Hidrosogamoso dam has transformed riverine ecology and livelihoods. The shift away from agriculture has not been easy and many farmers find the PES regulations onerous and cost-ineffective. Local farmers report declines in their yields and fungus problems from increased humidity caused by the reservoir. As a result of the PES scheme, local communities experience environmental degradation, compounded by a loss of traditional livelihood and survival options (Rodríguez-de-Francisco et al. 2019). A critical lens of the Colombian hydrosocial cycle demonstrates that the uneven production of water has implications for communities' way of life and their means of recourse when their ecological relations are broken: "marking framing silences people's voices who experience environmental degradation as environmental injustice, and instead only refers to it as a mere economic conservation problem to be solved by market forces" (Rodríguez-de-Francisco et al. 2019).

Critiques of nature-based solutions argue that they reflect neoliberal rationalities and techniques of managing the environment and do not fully consider alternative or multiple values of shared resources like water (Osaka et al. 2021). A relational perspective reveals that power – who holds it, who hides it – helps to determine which water policy solutions emerge and are adopted. Power relations mediate supposed solutions; thus, a critical lens is required to question who benefits or loses out from non-stationarity futures.

Techno-Fix 4: Resilience

In recent decades, researchers and policymakers have turned to the notion of **water resilience**. In the face of non-stationarity, socio-ecological systems can deliver on key functions but might also change according to the large range of hydro-climatic conditions – making difficult the "command and control" techniques of past environmental management. Some have argued that water resilience is a new water paradigm, associated with a shift toward thinking about complexity and interconnected sectors and scales that matter to water.

The starting point of water resilience is adapting to change and learning how to cope better with uncertainty (Folke 2003). There are multiple ways that water is connected to resilience, whether through driving humans to resilience via floods and droughts, fluctuating water quality from land-use change, or addressing resilience by supporting biodiversity (Falkenmark et al. 2019). Water resilience aims to focus on the *capacity* of systems to deal with hydro-climatic change (Rockström et al. 2014). In practice, resilient responses to drought call for diversifying water supply, wastewater reuse for potable use (see Further Readings), or even changing livelihoods to rely less on water. Restoring ecosystems to their healthy state is considered a significant step toward resilience (Rodina and Chan 2019).

Water resilience extends beyond the "box" thinking. "Water resilience is no longer just about water, they [the multiple roles of water] link to climate change, dietary choice, trade, consumption, and more. When in the past, water management effectively meant blue water management, water managers of today need to consider vastly cross-sectoral implications and green-blue water interactions" (Falkenmark et al. 2019, p. 10). In this vein, water resilience replaces the paradigm of Integrated Water Resource Management (IWRM), which is considered no longer sufficient by many experts. While IWRM enables flexibility in institutions and legislative systems, resilience calls for a broader focus on adaptive capacity and considers adaptive management and governance (Clarvis et al. 2014), a point which we turn to in the next section. Often, polycentric governance is advocated with a recognition that decision-making across multiple scales is necessary to deal with complexity (Theil et al. 2019). The deliberative feature of decision-making under resilience thinking is coupled with stakeholder participation and an emphasis on social learning that involves different knowledges (Lebel et al. 2006). These features relate to questions of equity and inclusiveness.

Water resilience has been promoted and implemented by a range of actors and governments, from the UK to Jordan to South Africa. For example, in the wake of the Day Zero water crisis, which led to acute water shortages in Cape Town, the city assessed its urban water resilience as part of revisiting its policy and regulatory framework (Rodina 2018). Worldwide, the City Water Resilience Approach is a prominent initiative to encourage resilience actions and strategies, supported by major funders and think tanks such as the Rockefeller Foundation, World Bank, Arup, and the Stockholm International Water Institute. Global companies such as Cargill, Starbucks, and Microsoft are part of the Water Resilience Coalition under the CEO Water Mandate of the United Nations Global Compact, which aims to enhance water stewardship and address water risks to business actors. Resilience is central to global water policy debates involving governments, cities, municipalities, and businesses.

However, when we drill down into the definitions, application, and purpose of resilience, we see a murky picture. In a systematic review, Lucy Rodina (2018) identifies three key problems with the discourse and current state of knowledge. First, while water resilience is widely studied, it lacks a clear definition and is often used as a vague shorthand for an array of policy prescriptions. Second, the application of water resilience is skewed to issues of water supply, water resources management, or drainage/stormwater management, with a notable omission of water, sanitation, and hygiene (WaSH) issues. Third, papers focus on built infrastructure systems with little mention of whom resilience is to benefit.

Resilience, in short, becomes a techno-fix – focused more on expert technical solutions and less on what resilience-oriented governance might look like (Rodina 2018). More than half of the studies examined did not provide any insights on governance or institutional processes. The minority of studies that did consider governance suggested collaborative decision-making and stakeholder engagement, but these concepts remained vague and under-researched. The studies presented resilience as a matter of expertise to be dealt with by water managers, governments, and existing water institutions – leaving out or potentially marginalizing other institutional actors (Rodina 2018). This fuzzy and limited understanding renders issues of equity and inclusion in deliberative decision-making superficial.

These findings are troubling, especially if we see water as relational, as we have done throughout the book. If conventional decision-makers are in the driving seat of water resilience, then what does it mean for those engaged in water struggles as one of life? Water resilience is not a "new" paradigm for people who have survived for generations. Resilience discourse risks normalizing the "business as usual" production of water that runs counter to people's worldviews. Resilience discourse can be an extension of the politics of water that prove incompatible and grounds for strife.

In sum, in taking stock of all four techno-fixes – desalination, water transfers, nature-based solutions, and resilience – their overt emphasis on "solutions" risks **depoliticizing** water, obscuring its relationality and the conditions of its production, and erasing the contentious nature of water issues. Skepticism is a superpower. Skepticism of technological solutions or "new" institutional paradigms does not mean that we hate policy, engineering, or the future. Rather, the future is too precious, and the stakes too high, for us *not* to adopt a critical stance. We cannot nimbly sidestep or erase the politics of water – a critical perspective demands we examine the relational production of water, including its mythologies and hegemonic discourses. We can only go one way: *through* the politics of water.

Adapting to New Worlds

A cornerstone of water resilience is improving adaptive capacity to anticipate future changes, which involves reshaping institutions and governance. Scholarship has focused on seeking key characteristics of **adaptive water governance**. While there are many studies that identify defining features, a useful generalization of attributes is problem-based scale, polycentricity, and collaboration. Focusing on the problem defies a fixed, singular scale of governance. This implies that multiple jurisdictions and sectors are involved, such that authority and decision-making are found in multiple, polycentric nodes. Collaboration is regarded as important to seeking the legitimacy of the governance arrangements and for the inclusion of people who rely on water resources (Cosens and Gunderson 2021). As Chapter 7 explained, the problem-shed extends beyond the water sector and across multiple scales, and decision-making cannot be done by a single entity. Studies indicate that polycentric governance in river basins is vital to deal with climate change. Innovative solutions are found when a diverse set of stakeholders and their knowledges are utilized (Pahl-Wostl et al. 2012).

But this is the theory of what adaptive water governance should look like. How does adaptive water governance work in practice? Let's return to the Okavango Delta. Alongside fishing, keeping livestock, and collecting reeds and grasses of the wetlands, *molapo* farming is an essential livelihood. Highly attuned to the cycle of seasonal flooding, *molapo* farming is an agricultural method utilizing the floodplain. This flood recession-type farming is small-scale, involving plots of approximately 2–3 ha, compared to 25 ha for dryland farming. Many farmers adapt their mode of farming depending on flooding and rains. This means that they shift between *molapo* practices and dryland farming, providing food security amidst water variability (Shinn 2016). However, *molapo* farming is becoming less viable due to institutional changes.

The Botswana government has been concerned with the variability of flooding: for example, they relocated residents from out of the floodplain to newly established dryland plots (Shinn et al. 2014). As a designated Ramsar site, management of the delta requires clear policies to maintain its delicate ecological balance. The government's response has been to devise a strategic plan for adaptive management, including a restriction on development within 200 m of the floodplain. Farming outside of the floodplain can mean insufficient water for growing crops. New policies only allow a farmer to hold one certificate for an agricultural field. Farmers are bound to their dryland plots and can no longer shift to *molapo* field as the seasons and water levels dictate. The government has not universally banned *molapo* farming. However, officials have "suggested" that farmers focus on dryland farming: (i) by providing support such as seed provision, and (ii) by disincentivizing *molapo* practices by ending crop loss compensation (Shinn 2016).

Viewed from a relational perspective, such policies are harmful for inhabitants and run counter to the social reproduction of the delta. Without being able to rely on *molapo* farming, the future for farmers is limited. They expressed a "sense of power-lessness in the face of government requests" and see no other option except dryland farming (Shinn 2016, p. 55). Government "requests" and incentives to relocate contribute to farmers' sense of powerlessness (Shinn et al. 2014). Some farmers have resisted resettlement to drylands, returning to the floodplains and making unauthorized small-scale dams to control the flow of water.

Water is life. But in the face of overwhelming government presence through its diktat on what is adaptive management of the delta, the full expression of sustaining life is curtailed. Clearly, power differentials are at work to determine how adaptation should occur and who should benefit. Even if well-intentioned with no apparent coercion, power relations cannot be overlooked in explaining how adaptive water governance plays out on the ground.

Power is embedded in the waterscape of the Okavango Delta. Changing water levels and the delta environment shape and are reshaped by social, economic, and political relations – a key lesson of this book. For example, customary institutions in the Okavango are underpinned by cultural values of water and decision-making roles held by traditional leaders and district chiefs. This customary practice has been in place for a long time and is regarded as a well-functioning institution to maintain water quality (Gondo et al. 2019b). While there have been attempts to decentralize authority, the effect of policy reform has re-centralized decision-making, further reinforcing elite power and control (Shinn 2016). Government officials tend to dismiss local knowledge as irrelevant or parochial (Gondo et al. 2019c). This results in a situation where customary institutions assume a subordinate role to statutory ones, privileging water as an economic good (Gondo et al. 2019a). Adaptive water governance is not a panacea, but a process, and the struggle to achieve thriving conditions for future waters continues.

A Mudlarkers' Guide to the Future

Low tide in London on a rare sunny day in February, and we are walking the banks of the Thames River. The muddy foreshore near Queenhithe, across the river from the Tate Modern Museum, is open by low tide and perfect for discovery (Figure 10.3). Clay pipes, bone beads, beer bottles, Roman coins, pottery shards, broken tiles, rusty

Figure 10.3 The Thames River, central London. View from the Queenhithe section, looking south to the Tate Modern Museum, formerly a power station. Source: Katie Meehan.

nails, and wooden combs – all shards of London's past, preserved and brought to the surface by the roiling currents of the river. Tides dictate the daily rhythm of the Thames – and, by extension, the flows of London. "It amazes me how many people don't realise the river in central London is tidal," writes author and mudlarker Lara Maiklem (2019, pp. 2–3), "[T]he height between low and high water at London Bridge varies from fifteen to twenty-two feet [and] it takes six hours for the water to come upriver and six and a half for it to flow back out to the sea."

Each tiny discovery is a link to the past. During the Victorian era in England, the term "mudlark" was given to people – often children – who scavenged for usable and sellable debris in the mud of a riverbank. Trash and waste dumped in the Thames reflected London's social and economic connections to the world. "The foreshore too is home to tiny pieces of faraway places," writes Maiklem (2019, p. 199). "I have the shell of a huge Pacific-dwelling barnacle and tiny non-native cowrie shells from as far as Australia, [which] had fallen from hulls or arrived as ballast in the bellies of trade ships." Mudlarks today are hobbyists. Bent at the waist, we peer into the mud and search for treasures at Queenhithe. A mudlarker's eye quickly learns to read the subtle cues of the river; to catch the tidal turn; to find a little piece of joy in the mud. "As I have discovered," confesses Maiklem (2019, p. 5), "it is often the tiniest of objects that tell the greatest stories."

Mudlarking is an apt metaphor for the journey of this book. In searching for gems in the mud of the world, we have sought to excavate the ways that water is fundamentally relational. A critical approach to the hydrosocial cycle, as we demonstrated, enables us to understand how water shapes – and is shaped by – our lives, places, practices, and geometries of power. A relational point of view focuses our analysis on the *conditions* of water and its production – and how we, as people, are part of that production. Trained in the arts of noticing, a mudlarker's sharp eyes will clearly see the forces that produce grand challenges; and also, we hope, a dazzling array of radical possibilities.

The muddy banks of the Thames River are a good place to conclude this book. Think about all the human stories that a river holds – waiting for your (re)discovery and narration. Think about how the waters of the Thames have reflected and shaped the many stories of London – and how much Londoners (and people further afar!) have shaped the Thames. From Georgian halfpennies to champagne bottles, from privatization to financialization, from the Super Sewer to the desalination plant (TWDP) in Beckton, from our home to yours, we have used the Thames to demonstrate the power of a critical approach for understanding water and society. We love this big muddy river; our future is tied to its survival. In honing critical tools to understand the production of water – in a sense, to mudlark – only then can we begin to move toward more generative, resilient, and just futures.

What's next? At the beginning of this book, we wrote about Standing Rock and the invitation to camp. Camp is planetary and frames our collective water futures. Camp necessarily involves critique; and critique is a tricky and beautiful skill, sharpened by practice, reading, listening, debate, curiosity, doubt, risk, and self-reflection. As Holy Elk Lafferty said (in Birkett and Montoya 2019, p. 274), "For me, it's been a continuum. It has never stopped. We're all continuing to fight. Now, camp is the globe. Camp is everywhere." As the winter sun warms the banks of the Thames River, low tide seems like the perfect time to go camping.

Further Reading

Climate change and water

Caretta, M.A., Mukherji, A., Arfanuzzaman, M., Betts, R.A., Gelfan, A., Hirabayashi, Y., Lissner, T.K., Lopez Gunn, E., Liu, J., Morgan, R., et al. ed. (2022). Chapter 4: Water. In: *Climate Change 2022: Impacts, Adaptation, and Vulnerability.* Contribution of Working Group II to the Sixth Assessment Report of the Intergovernmental Panel on Climate Change. Cambridge, UK: Cambridge University Press.

United Nations Water (UN-Water). (2020). *United Nations World Water Development Report 2020: Water and Climate Change.* Paris: UNESCO.

World Bank (2016). High and Dry: Climate Change, Water, and the Economy. Washington, DC: World Bank. https://www.worldbank.org/en/topic/water/publication/high-and-dry-climate-change-water-and-the-economy (accessed 5 August 2022).

Desalination

Fragkou, M.C. and McEvoy, J. (2016). Trust Matters: Why Augmenting Water Supplies via Desalination May Not Overcome Perceptual Water Scarcity. *Desalination* 397: 1–8.

Loftus, A. and March, H. (2016). Financializing Desalination: Rethinking the Returns of Big Infrastructure. *International Journal of Urban and Regional Research* 40 (1): 46–61.

McEvoy, J. (2014). Desalination and Water Security: The Promise and Perils of a Technological Fix to the Water Crisis in Baja California Sur, Mexico. *Water Alternatives* 7 (3): 518–541.

McEvoy, J. and Wilder, M. (2012). Discourse and Desalination: Potential Impacts of Proposed Climate Change Adaptation Interventions in the Arizona-Sonora Border Region. *Global Environmental Change* 22 (2): 353–363.

March, H., Saurí, D., and Rico-Amorós, A.M. (2014). The End of Scarcity? Water Desalination as the New Cornucopia for Mediterranean Spain. *Journal of Hydrology* 519: 2642–2651.

Swyngedouw, E. (2013). Into the Sea: Desalination as Hydro-Social Fix in Spain. *Annals of the Association of American Geographers* 103 (2): 261–270.

Swyngedouw, E. and Williams, J. (2016). From Spain's Hydro-Deadlock to the Desalination Fix. *Water International* 41(1): 54–73.

On wastewater recycling for potable use

Meehan, K., Ormerod, K.J., and Moore, S.A. (2013). Remaking Waste as Water: The Governance of Recycled Effluent for Potable Water Supply. *Water Alternatives* 6 (1): 67–85.

Ormerod, K.J. (2016). Illuminating Elimination: Public Perception and the Production of Potable Water Reuse. *WIREs Water* 3 (4): 537–547.

Ormerod, K.J. (2019). Toilet Power: Potable Water Reuse and the Situated Meaning of Sustainability in the Southwestern United States. *Journal of Political Ecology* 26 (1): 633–651.

Ormerod, K.J. and Scott, C.A. (2013). Drinking Wastewater: Public Trust in Potable Reuse. *Science, Technology, & Human Values* 38 (3): 351–373.

The Cubango-Okavango River basin

Green, O., Cosens, B., and Garmestani, A. (2013). Resilience in Transboundary Water Governance: The Okavango River Basin. *Ecology and Society* 18 (2): 23. https://doi.org/10.5751/ES-05453-180223.

Mogomotsi, G.E.J., Mogomotsi, P.K., and Mosepele, K (2020). Legal Aspects of Transboundary Water Management: An Analysis of the Intergovernmental Institutional Arrangements in the Okavango River Basin. *Leiden Journal of International Law* 33 (2): 391–408.

The Permanent Okavango River Basin Water Commission (OKACOM). (2020). Realising the Benefits of Transboundary Water Cooperation in the Cubango-Okavango River Basin. Gaborone, Botswana: OKACOM. https://unece.org/

fileadmin/DAM/env/water/activities/Benefits_cooperation/OKACOM_Policy_
Document_June_2020.pdf (accessed 5 August 2022).

Saruchera, D. and Lautze, J. (2016). Transboundary River Basin Organizations in Africa: Assessing the Secretariat. *Water Policy* 18 (5): 1053–1069.

Large-scale water transfer and diversion projects

Brite, E.B. (2018). The Hydrosocial Empire: The Karakum River and the Soviet Conquest of Central Asia in the 20th Century. *Journal of Anthropological Archaeology* 52: 123–136.

Gijsbers, P.J.A. and Loucks, D.P. (1999). Libya's Choices: Desalination or the Great Man-Made River Project. *Physics and Chemistry of the Earth, Part B: Hydrology, Oceans and Atmosphere* 24 (4): 385–389.

Mirumachi, N. (2007). The Politics of Water Transfer between South Africa and Lesotho: Bilateral Cooperation in the Lesotho Highlands Water Project. *Water International* 32 (4): 558–570.

Water and adaptation

Allan, C., Xia, J., and Pahl-Wostl, C. (2013). Climate Change and Water Security: Challenges for Adaptive Water Management. *Current Opinion in Environmental Sustainability* 5 (6): 625–632.

Lankford, B. and Hepworth, N. (2010). The Cathedral and the Bazaar: Monocentric and Polycentric River Basin Management. *Water Alternatives* 3 (1): 82–101.

Scott, C.A., Meza, F.J., Varady, R.G., Tiessen, H., McEvoy, J., Garfin, G.M., Wilder, M., Farfán, L.M., Pablos, N.P., and Montaña, E. (2013). Water Security and Adaptive Management in the Arid Americas. *Annals of the Association of American Geographers* 103 (2): 280–289.

Payments for ecosystem services

Fletcher, R. and Breitling, J. (2012). Market Mechanism or Subsidy in Disguise? Governing Payment for Environmental Services in Costa Rica. *Geoforum* 43 (3): 402–411.

Joslin, A.J. and Jepson, W.E. (2018). Territory and Authority of Water Fund Payments for Ecosystem Services in Ecuador's Andes. *Geoforum* 91: 10–20.

Salzman, J., Bennett, G., Carroll, N., Goldstein, A., and Jenkins, M. (2018). The Global Status and Trends of Payments for Ecosystem Services. *Nature Sustainability* 1 (3): 136–144.

Bibliography

Abbey, E. (1968). Water. In: *Desert Solitaire: A Season in the Wilderness*, 112–127. New York: McGraw-Hill.

Abers, R.A. and Keck, M.E. (2013). *Practical Authority: Agency and Institutional Change in Brazilian Water Politics*. New York and Oxford, UK: Oxford University Press.

Acerman, M. (2016). Environmental flows – Basics for Novices. *WIREs Water* 3 (5): 622–628. https://doi.org/10.1002/wat2.1160.

Acevedo-Guerrero, T. (2019). Light is Like Water: Flooding, Blackouts, and the State in Barranquilla. *Tapuya: Latin American Science, Technology and Society* 2 (1): 478–494.

Ackroyd, P. (2008). *Thames: Sacred River*. London: Vintage Books.

Adamczewki, A., Burnod, P., Papazain, H., Coulibaly, Y., Tonneau, J., and Jamin, Y. (2013). Domestic and Foreign Investments in Irrigable Land in Mali: Tensions between the Dream of Large-Scale Farming and the Reality of Family Farming. In: *Africa for Sale: Positioning the State, Land and Society in Foreign Large-Scale Land Acquisitions in Africa* (ed. S. Evers, C. Seagle, and F. Krijtenburg), 159–180. Leiden: Brill.

Adams, E.A. (2018a). Thirsty Slums in African Cities: Household Water Insecurity in Urban Informal Settlements of Lilongwe, Malawi. *International Journal of Water Resources Development* 34 (6): 869–887.

Water: A Critical Introduction, First Edition. Katie Meehan, Naho Mirumachi, Alex Loftus, and Majed Akhter.
© 2023 John Wiley & Sons Ltd. Published 2023 by John Wiley & Sons Ltd.

Adams, E.A. (2018b). Intra-urban Inequalities in Water Access among Households in Malawi's Informal Settlements: Toward Pro-poor Urban Water Policies in Africa. *Environmental Development* 26: 34–42.

Adams, E.A., Juran, L., and Ajibade, I. (2018a). "Space of Exclusion" in Community Water Governance: A Feminist Political Ecology of Gender and Participation in Malawi's Urban Water User Associations. *Geoforum* 95 (1): 133–142.

Adams, E.A., Kuusaana, E.D., Ahmed, A., and Campion, B.B. (2019). Land Dispossessions and Water Appropriations: Political Ecology of Land and Water Grabs in Ghana. *Land Use Policy* 87: 104068.

Adams, E.A., Sambu, D., and Smiley, S.L. (2018b). Urban Water Supply in Sub-Saharan Africa: Historical and Emerging Policies and Institutional Arrangements. *International Journal of Water Resources Development* 35 (2): 240–263.

Agostoni, C. (2003). *Monuments of Progress: Modernization and Public Health in Mexico City, 1876–1910*. Calgary and Mexico City: University of Calgary Press and UNAM Press.

Ahlers, R. and V. Merme. (2016). Financialization, Water Governance, and Uneven Development. *WIREs Water* 3 (6): 766–774.

Ahlers, R. and Zwarteveen, M. (2009). The Water Question in Feminism: Water Control and Gender Inequities in a Neoliberal Era. *Gender, Place & Culture* 16 (4): 409–426.

Ahlers, R., Cleaver, F., Rusca, M., and Schwartz, K. (2014). Informal Space in the Urban Waterscape: Disaggregation and Co-production of Water Services. *Water Alternatives* 7 (1): 1–14.

Ahlers, R., Zwarteveen, M., Bakker, K., and Flyvbjerg, B. (2017). Large Dam Development: From Trojan Horse to Pandora's Box. In: *The Oxford Handbook of Mega Project Management* (ed. B. Flyvberg), 566–576. Oxford, UK: Oxford University Press.

Akhter, M. (2019). Adjudicating Infrastructure: Treaties, Territories, Hydropolitics. *Environment and Planning E: Nature and Space* 2 (4): 831–849.

Akhter, M. (2022). Dams, Development, and Racialised Internal Peripheries: Hydraulic Imaginaries as Hegemonic Strategy in Pakistan. *Antipode* 54 (5): 1429–1450.

Akhter, M. and Ormerod, K.J. (2015). The Irrigation Technozone: State Power, Expertise, and Agrarian Development in the US West and British Punjab, 1880–1920. *Geoforum* 60: 123–132.

Alatout, S. (2009). Bringing Abundance into Environmental Politics: Constructing a Zionist Network of Water Abundance, Immigration, and Colonization. *Social Studies of Science* 39 (3): 363–394.

Aldaya, M., Allan, J.A., and Hoekstra, A.Y. (2010). Strategic Importance of Green Water in International Crop Trade. *Environmental Economics* 69 (4): 887–894.

Ali, I. (2014). *The Punjab Under Imperialism, 1885–1947*. Princeton, NJ: Princeton University Press.

Allan, C. and Watts, R.J. (2018). Revealing Adaptive Management of Environmental Flows. *Environmental Management* 61: 520–533.

Allan, C., Xia, J., and Pahl-Wostl, C. (2013). Climate Change and Water Security: Challenges for Adaptive Water Management. *Current Opinion in Environmental Sustainability* 5 (6): 625–632.

Allan, J.A. (2001). *The Middle East Water Question: Hydropolitics and the Global Economy*. London: I.B. Tauris.

Allan, J.A. (2019). Food, Water and the Consequences of Society Not Valuing the Environment. In: *The Oxford Handbook of Food, Water and Society* (ed. J.A. Allan, B. Bromwich, M. Keulertz, and A. Colman), 859–878. Oxford, UK: Oxford University Press.

Allan, J.A., Keulertz, M., and Woertz, E. (2015). The Water–Food–Energy Nexus: An Introduction to Nexus Concepts and Some Conceptual and Operational Problems. *International Journal of Water Resources Development* 31 (3): 301–311. http://dx.doi.org/10.1080/07900627.2015.1029118.

Allan, J.D., Castillo, M.M., and Capps, K.A. (2021). *Stream Ecology: Structure and Function of Running Waters*. Berlin: Springer Nature.

Allen, J. and Pryke, M. (2013). Financialising Household Water: Thames Water, MEIF, and "Ring-Fenced" Politics. *Cambridge Journal of Regions, Economy and Society* 6 (3): 419–439.

Allouche, J. (2019). State Building, Nation Making and Post-colonial Hydropolitics in India and Israel: Visible and Hidden Forms of Violence at Multiple Scales. *Political Geography* 75: 102051.

Allouche, J., Middleton, C., and Gyawali, D. (2019). *The Water–Food–Energy Nexus: Power, Politics and Justice*. London and New York: Routledge.

Alston, P. (2017). The Populist Challenge to Human Rights. *Journal of Human Rights Practice* 9 (1): 1–15.

Alston, P. (2019). Statement by Professor Philip Alston, United Nations Special Rapporteur on Extreme Poverty and Human Rights on His Visit to Lao PDR, 18–28 March 2019. https://www.ohchr.org/en/statements/2019/03/statement-professor-philip-alston-united-nations-special-rapporteur-extreme (accessed 13 June 2022).

Anand, N. (2011). PRESSURE: The PoliTechnics of Water Supply in Mumbai. *Cultural Anthropology* 26 (4): 542–564.

Anand, N. (2017). *Hydraulic City: Water and the Infrastructures of Citizenship in Mumbai*. Durham, NC: Duke University Press.

Angel, J. and Loftus, A. (2019). With-Against-and-Beyond the Human Right to Water. *Geoforum* 98: 206–213.

Annual thematic reports authored by the UN Special Rapporteurs on the human rights to drinking water and sanitation. https://www.ohchr.org/en/special-procedures/sr-water-and-sanitation/annual-reports (accessed 18 June 2022).

Arendt, H. (1998, 1958). *The Human Condition*. Chicago, IL: University of Chicago Press.

Arsenault, C. (2015). Shrouded in Secrecy, One of Africa's Biggest Land Deals Stalls. *Reuters*. https://www.reuters.com/article/food-africa-land-idUSL5N0XK3KS20150518 (accessed 10 June 2022).

Arsenault, R., Diver, S., McGregor, D., Witham, A., and Bourassa, C. (2018). Shifting the Framework of Canadian Water Governance through Indigenous Research Methods: Acknowledging the Past with an Eye on the Future. *Water* 10 (1): 49.

Arthington, A. (2012). *Environmental Flows: Saving Rivers in the Third Millennium.* Berkeley and Los Angeles, CA: University of California Press.

Baer, M. and A. Gerlak. (2015). Implementing the Human Right to Water and Sanitation: A Study of Global and Local Discourses. *Third World Quarterly* 36 (8): 1527–1545.

Baer, M. (2017). *Stemming the Tide: Human Rights and Water Policy in a Neoliberal World.* Oxford, UK: Oxford University Press.

Baghel, R. and Nüsser, M. (2010). Discussing Large Dams in Asia after the World Commission on Dams: Is a Political Ecology Approach the Way Forward? *Water Alternatives* 3 (2): 231–248.

Bakker, K. (2003). Archipelagos and Networks: Urbanization and Water Privatization in the South. *The Geographical Journal* 169 (4): 328–341.

Bakker, K. (2005). Neoliberalizing Nature? Market Environmentalism in Water Supply in England and Wales. *Annals of the Association of American Geographers* 95 (3): 542–565.

Bakker, K. (2007). The "Commons" Versus the "Commodity": Alter-globalization, Anti-privatization and the Human Right to Water in the Global South. *Antipode* 39 (3): 430–455.

Bakker, K. (2010). *Privatizing Water: Governance Failure and the World's Urban Water Crisis.* Ithaca, NY: Cornell University Press.

Bakker, K. (2013). Neoliberal Versus Postneoliberal Water: Geographies of Privatization and Resistance. *Annals of the Association of American Geographers* 103 (2): 253–260.

Bakker, K. (2014). The Business of Water. *Annual Review of Environment and Resources* 39 (1): 469–494.

Ball, P. (2017). *The Water Kingdom: A Secret History of China.* Chicago, IL: University of Chicago Press.

Banister, J.M. (2014). Are You Wittfogel or Against Him? Geophilosophy, Hydro-Sociality, and the State. *Geoforum* 57 (1): 205–214.

Banister, J.M. and Widdifield, S.G. (2014). The Debut of "Modern Water" in Early 20th Century Mexico City: The Xochimilco Potable Networks. *Journal of Historical Geography* 46 (1): 36–52.

Barbarossa, V., Schmitt, R.J.P, Huijbregts, M.A.J., Zarfl, C., King, H., and Schipper, A.M. (2020). Impacts of Current and Future Large Dams on the Geographic Range Connectivity of Freshwater Fish Worldwide. *Proceedings of the National Academy of Sciences* 117 (7): 3648–3655. https://doi.org/10.1073/pnas.1912776117.

Barlow, M. (2012). *Foreword. In: The Right to Water: Politics, Governance and Social Struggles* (ed. F. Sultana and A. Loftus), xv–xvii. London: Earthscan.

Barry, B., Obuobie, E., Andreini, M., Andah, W., and Pluquest, M. (2005). The Volta River Basin: Comprehensive Assessment of Water Management in Agriculture.

International Water Management Institute. http://www.iwmi.cgiar.org/assessment/files_new/research_projects/river_basin_development_and_management/VoltaRiverBasin_Boubacar.pdf (accessed 10 June 2022).

Bartlett, J. (2022). "Consequences Will Be Dire": Chile's Water Crisis is Reaching Breaking Point. *The Guardian* (1 June). https://www.theguardian.com/world/2022/jun/01/chiles-water-crisis-megadrought-reaching-breaking-point (accessed 6 June 2022).

Barua, A. and Vij, S. (2018). Treaties Can Be a Non-starter: A Multi-track and Multilateral Dialogue Approach for Brahmaputra Basin. *Water Policy* 20 (5): 1027–1041. https://doi.org/10.2166/wp.2018.140.

Bauer, C.J. (1997). Bringing Water Markets Down to Earth: The Political Economy of Water Rights in Chile, 1976–1995. *World Development* 25 (5): 639–656.

Bauer, C.J. (1998). *Against the Current: Privatization, Water Markets, and the State in Chile*. Boston, MA: Kluwer.

Bauer, C.J. (2004). *Siren Song: Chilean Water Law as a Model for International Reform*. Washington, DC: Resources for the Future.

Baviskar, A. (1999). *In the Belly of the River: Tribal Conflicts over Development in the Narmada Valley*. Oxford, UK: Oxford University Press.

Baviskar, A. (2019). Nation's Body, River's Pulse: Narratives of Anti-dam Politics in India. *Thesis Eleven* 150 (1): 26–41.

Bayliss, K. (2017). Material Cultures of Water Financialisation in England and Wales. *New Political Economy* 22 (4): 383–397.

Beaty, Z., Harper, L, and Saner, E. (2022). "People Say, 'You Started it All!'": The Lionesses Who Paved the Way for England's Historic Win. *The Guardian* (1 August). https://www.theguardian.com/football/2022/aug/01/people-say-you-started-it-all-the-lionesses-who-paved-the-way-for-englands-historic-win (accessed 2 August 2022).

Bennett, V., Dávila-Poblete, S., and Nieves Rico, M. (2008). Water and Gender: The Unexpected Connection that Really Matters. *Journal of International Affairs* 61 (2): 107–126.

Bhan, M. (2014). Morality and Martyrdom: Dams, "Dharma", and the Cultural Politics of Work in Indian-Occupied Kashmir. *Biography* 37 (1): 191–224.

Bieler, A. (2019). Against the Trend: Structure and Agency in the Struggle for Public Water in Europe. In: *Water Politics: Governance, Justice and the Right to Water* (ed. F. Sultana and A. Loftus), 129–142. Abingdon, UK: Routledge.

Bieler, A. (2021). *Fighting for Water: Resisting Privatization in Europe*. London: Bloomsbury.

Binoy, P. (2021). Pollution Governance in the Time of Disasters: Testimonials of Caste/d Women and the Politics of Knowledge in Kathikudam, Kerala. *Geoforum* 124: 175–184.

Birkenholtz, T. (2009a). Irrigated Landscapes, Produced Scarcity, and Adaptive Social Institutions in Rajasthan, India. *Annals of the Association of American Geographers* 99 (1): 118–137.

Birkenholtz, T. (2009b). Groundwater Governmentality: Hegemony and Technologies of Resistance in Rajasthan's (India) Groundwater Governance. *The Geographical Journal* 175 (3): 208–220.

Birkenholtz, T. (2013). "On the Network, Off the Map": Developing Intervillage and Intragender Differentiation in Rural Water Supply. *Environment and Planning D: Society and Space* 31 (2): 354–371.

Birkenholtz, T. (2015). Recentralizing Groundwater Governmentality: Rendering Groundwater and Its Users Visible and Governable. *WIREs Water* 2 (1): 21–30.

Birkenholtz, T. (2016). Dispossessing Irrigators: Water Grabbing, Supply-Side Growth and Farmer Resistance in India. *Geoforum* 69: 94–105.

Birkett, T.M. and Montoya, T. (2019). For Standing Rock: A Moving Dialogue. In: *Standing with Standing Rock: Voices from the #NoDAPL Movement* (ed. N. Estes and J. Dhillon), 261–280. Minneapolis, MN: University of Minnesota Press.

Blake, D. and Barney, K. (2018). Structural Injustice, Slow Violence? The Political Ecology of a "Best Practice" Hydropower Dam in Lao PDR. *Journal of Contemporary Asia* 48 (5): 808–834. https://doi.org/10.1080/00472336.20114825 608.1482560.

Bledsoe, A. (2017). Marronage as a Past and Present Geography in the Americas. *Southeastern Geographer* 57 (1): 30–50.

Blomley, N. (1994). *Law, Space and the Geographies of Power*. New York: Guilford.

Blomquist, W. (2020). Beneath the Surface: Complexities and Groundwater Policy-Making. *Oxford Review of Economic Policy* 36 (1): 154–170.

Blomquist, W., Schlager, E., and Heikkila, T. (2004a). *Common Waters, Diverging Streams: Linking Institutions and Water Management in Arizona, California, and Colorado*. Washington, DC: Resources for the Future.

Blomquist, W., Schlager, E., and Heikkila, T. (2004b). Building the Agenda for Institutional Research in Water Management Research. *Journal of the American Water Resources Association* 40 (4): 925–936.

Boelens, R. (2014). Cultural Politics and the Hydrosocial Cycle: Water, Power and Identity in the Andean Highlands. *Geoforum* 57 (1): 234–247.

Boelens, R. and Doornbos, B. (2001). The Battlefield of Water Rights: Rule Making Amidst Conflicting Normative Frameworks in the Ecuadorian Highlands. *Human Organization* 60: 343–355.

Boelens, R. and Vos, J. (2012). The Danger of Naturalizing Water Policy Concepts: Water Productivity and Efficiency Discourse from Field Irrigation to Virtual Water Trade. *Agricultural Water Management* 108: 16–26.

Boelens, R. and Vos, J. (2014). Legal Pluralism, Hydraulic Property Creation and Sustainability: The Materialized Nature of Water Rights in User-Managed Systems. *Current Opinion in Environmental Sustainability* 11: 55–62.

Boelens, R., Getches, D., and Guevara-Gil, A. ed. (2010). *Out of the Mainstream: Water Rights, Politics and Identity*. London: Earthscan.

Boelens, R., Perreault, T., and Vos, J. ed. (2018). *Water Justice*. Cambridge, UK: Cambridge University Press.

Bonds, A. and Inwood, J. (2016). Beyond White Privilege: Geographies of White Supremacy and Settler Colonialism. *Progress in Human Geography* 40 (6): 715–733.

Borgomeo, E., Jägerskog, A., Talbi, A., Wijnen, M., Hejazi, M., and Miralles-Wilhelm, F. (2018). *The Water-Energy-Food Nexus in the Middle East and North Africa: Scenarios for a Sustainable Future*. Washington, DC: World Bank.

Borras Jr., S.M. and Franco, J.C. (2013). Global Land Grabbing and Political Reactions 'From Below.' *Third World Quarterly* 34 (9): 1723–1747. https://doi.org/10.1080/01436597.2013.843845.

Bosworth, K. (2021). "They're treating us like Indians!": Political Ecologies of Property and Race in North American Pipeline Populism. *Antipode* 53 (3): 665–685.

Braverman, I. (2020). Silent Springs: The Nature of Water and Israel's Military Occupation. *Environment and Planning E: Nature and Space* 3 (2): 527–551.

Braverman, I., Blomley, N.K., Delaney, D., and Kedar, A. (2014). *The Expanding Spaces of Law: A Timely Legal Geography*. Palo Alto, CA: Stanford Law Books.

Bréthaut, C. and Pflieger, G. (2020). *Governance of a Transboundary River: The Rhône*. Cham, Switzerland: Palgrave Macmillan.

Brewis, A. and Wutich, A. (2022). *Lazy, Crazy, and Disgusting: Stigma and the Undoing of Global Health*. Baltimore, MD: Johns Hopkins University Press.

Brewis, A., Choudhary, N., and Wutich, A. (2019b). Household Water Insecurity May Influence Common Mental Disorders Directly and Indirectly through Multiple Pathways: Evidence from Haiti. *Social Science & Medicine* 238: 112520.

Brewis, A., Wutich, A., du Bray, M.V., Maupin, J., Schuster, R.C., and Gervais, M.M. (2019a). Community Hygiene Norm Violators are Consistently Stigmatized: Evidence from Four Global Sites and Implications for Sanitation Interventions. *Social Science & Medicine* 220 (1): 12–21.

Brite, E.B. (2018). The Hydrosocial Empire: The Karakum River and the Soviet Conquest of Central Asia in the 20th Century. *Journal of Anthropological Archaeology* 52: 123–136.

Bruno, T. and Jepson, W. (2018). Marketisation of Environmental Justice: US EPA Environmental Justice Showcase Communities Project in Port Arthur, *Texas. Local Environment* 23 (3): 276–292.

Bruns, B.R. and Meinzen-Dick, R. (2005). Framework for Water Rights: An Overview of Institutional Options. In: *Water Rights Reform: Lessons for Institutional Design* (ed. B. Randolph Bruns, C. Ringler, and R. Meinzen-Dick), 3–26. Washington, DC: International Food Policy Research Institute.

Bruns, A., Meisch, S., Ahmed, A., Meissner, R., and Romero-Lankao, P. (2022). Nexus Disrupted: Lived Realities and the Water-Energy-Food Nexus from an Infrastructure Perspective. *Geoforum* 133 (1): 79–88. https://doi.org/10.1016/j.geoforum.2022.05.007.

Budds, J. (2004). Power, Nature and Neoliberalism: The Political Ecology of Water in Chile. *Singapore Journal of Tropical Geography* 25 (3): 322–342.

Budds, J. (2009a). Contested H_2O: Science, Policy, and Politics in Water Resources Management in Chile. *Geoforum* 40 (3): 418–430.

Budds, J. (2009b). The 1981 Water Code: The Impacts of Private Tradable Water Rights on Peasant and Indigenous Communities in Northern Chile. In: *Lost in the Long Transition: Struggles for Social Justice in Neoliberal Chile* (ed. W.L. Alexander), 41–60. Latham, MD: Lexington Books.

Budds, J. (2013). Water, Power, and the Production of Neoliberalism in Chile, 1973–2005. *Environment and Planning D: Society and Space* 31 (2): 301–318.

Budds, J. and McGranahan, G. (2003). Are the Debates on Water Privatization Missing the Point? Experiences from Africa, Asia and Latin America. *Environment and Urbanization* 15 (2): 87–114.

Burbano M., Shin, S., Nguyen, K., and Pokhrela, Y. (2020). Hydrologic Changes, Dam Construction, and the Shift in Dietary Protein in the Lower Mekong River Basin. *Journal of Hydrology* 581: 124454. https://doi.org/10.1016/j.jhydrol.2019.124454.

Burchi, S. (2018). Legal Frameworks for the Governance of International Transboundary Aquifers: Pre-and Post-ISARM Experience. *Journal of Hydrology: Regional Studies* 20: 15–20. https://doi.org/10.1016/j.ejrh.2018.04.007.

Bustamante, R., Crespo, C., and Walnycki A. (2012). Seeing Through the Concept of Water as a Human Right in Bolivia. In: *The Right to Water: Politics, Governance and Social Struggles* (ed. F. Sultana and A. Loftus), 223–240. London: Earthscan.

Campbell, I.C., Poole, C., Giesen, W., and Valbo-Jorgensen, J. (2006). Species Diversity and Ecology of Tonle Sap Great Lake, Cambodia. *Aquatic Sciences* 68: 355–373. https://doi.org/10.1007/s00027-006-0855-0.

Cantor, A. (2016). The Public Trust Doctrine and Critical Legal Geographies of Water in California. *Geoforum* 72 (1): 49–57.

Cantor, A. (2017). Material, Political, and Biopolitical Dimensions of "Waste" in California Water Law. *Antipode* 49 (5): 1204–1222.

Cantor, A. and Emel, J. (2018). *New Water Regimes: An Editorial. Resources* 7 (2): 25.

Cantor, A., Kay, K., and Knudson, C. (2020). Legal Geographies and Political Ecologies of Water Allocation in Maui, *Hawai'i. Geoforum* 110 (1): 168–179.

Caretta, M.A., Mukherji, A., Arfanuzzaman, M., Betts, R.A., Gelfan, A., Hirabayashi, Y., Lissner, T.K., Lopez Gunn, E., Liu, J., Morgan, R., et al. ed. (2022). *Climate Change 2022: Impacts, Adaptation and Vulnerability*. Contribution of Working Group II to the Sixth Assessment Report of the Intergovernmental Panel on Climate Change. Cambridge UK: Cambridge University Press.

Carey, M. (2010). *In the Shadow of Melting Glaciers: Climate Change and Andean Society*. Oxford, UK and New York: Oxford University Press.

Carey, M., Molden, O.C., Rasmussen, M.B., Jackson, M., Nolin, A.W., and Mark, B.G. (2017). Impacts of Glacier Recession and Declining Meltwater on Mountain Societies. *Annals of the American Association of Geographers* 107 (2): 350–359.

Carmi, N., Alsayegh, M., and Zoubi, M. (2019). Empowering Women in Water Diplomacy: A Basic Mapping of the Challenges in Palestine, Lebanon and Jordan. *Journal of Hydrology* 569: 330–346.

Carmody, S. (2019). Five Years After Flint's Crisis Began, Is the Water Safe? *National Public Radio* (25 April). https://www.npr.org/2019/04/25/717104335/5-years-after-flints-crisis-began-is-the-water-safe (accessed 1 August 2022).

Cascão, A.E. and Nicol, A. (2016). GERD: New Norms of Cooperation in the Nile Basin? *Water International*, 41 (4): 550–573. https://doi.org/10.1080/02508060.2016.1180763.

Castro, J.E. (2006). *Water, Power and Citizenship: Social Struggle in the Basin of Mexico*. Basingstoke, UK and New York: Palgrave Macmillan.

Castro, J.E. (2008). Neoliberal Water and Sanitation Policies as a Failed Development Strategy: Lessons from Developing Countries. *Progress in Development Studies* 8 (1): 63–83.

Celio, M., Scott, C.A., and Giordano, M. (2010). Urban–Agricultural Water Appropriation: The Hyderabad, India Case. *Geographical Journal* 176: 39–57.

Cereijido, A. (2017). Valley of Contrasts. Episode #1734, *Latino USA* (18 August). https://www.latinousa.org/episode/valley-of-contrasts/ (accessed 19 April 2022).

Chen, D., Zhang, D., Luo, Z., Webber, M., and Rogers, S. (2019). Water–Energy Nexus of the Eastern Route of China's South-to-North Water Transfer Project. *Water Policy* 21 (5): 945–963.

Chenoweth, J., Hadjikakou, M., and Zoumidesm, C. (2014). Quantifying the Human Impact on Water Resources: A Critical Review of the Water Footprint Concept. *Hydrology and Earth System Sciences* 18: 2325–2342.

Chester, M.V. and Allenby, B. (2019). Toward Adaptive Infrastructure: Flexibility and Agility in a Non-stationarity Age. *Sustainable and Resilient Infrastructure* 4 (4): 173–191. https://doi.org/10.1080/23789689.2017.1416846.

Chester, M.V., Markolf, S., and Allenby, B. (2019). Infrastructure and the Environment in the Anthropocene. *Journal of Industrial Ecology* 23 (5): 1006–1015.

Chiarelli, D.D., D'Odorico, P., Davis, K.F., Rosso, R., and Rulli, M.C. (2021). Large-Scale Land Acquisition as a Potential Driver of Slope Instability. *Land Degradation and Development* 2021 (32): 1773–1785. https://doi.org/10.1002/ldr.3826.

Chief, K. (2018). Emerging Voices of Tribal Perspectives in Water Resources. *Journal of Contemporary Water Research & Education* 163 (1): 1–5.

Chief, K., Meadow, A., and Whyte, K. (2016). Engaging Southwestern Tribes in Sustainable Water Resources Topics and Management. *Water* 8 (8): 1–21.

Christophers, B. (2015). The Limits to Financialization. *Dialogues in Human Geography* 5 (2): 183–200.

Clark, C. (2012). The Centrality of Community Participation to the Realization of the Right to Water: The Illustrative Case of South Africa. In: *The Right to Water: Politics, Governance and Social Struggles* (ed. F. Sultana and A. Loftus), 174–189. London: Earthscan.

Clark, C. (2019). Race, Austerity, and Water Justice in the United States: Fighting for the Human Right to Water in Detroit and Flint, Michigan. In: *Water Politics: Governance, Justice, and the Right to Water* (ed. F. Sultana and A. Loftus), 175–188. Abingdon, UK: Routledge.

Clark, G. (2000). *Pension Fund Capitalism.* Oxford, UK: Oxford University Press.

Clarvis, M.H., Allan, A., and Hannah, D.M. (2014). Water, Resilience and the Law: From General Concepts and Governance Design Principles to Actionable Mechanisms. *Environmental Science & Policy* 43 (1): 98–110.

Cleaver, F. (1998a). Choice, Complexity, and Change: Gendered Livelihoods and the Management of Water. *Agriculture and Human Values* 15 (4): 293–299.

Cleaver, F. (1998b). Incentives and Informal Institutions: Gender and the Management of Water. *Agriculture and Human Values* 15 (4): 347–360.

Cohen-Sacham, E., Walters, G., Janzen, C., and Maginnis, S. ed. (2016). *Nature-Based Solutions to Address Global Societal Challenges.* Gland, Switzerland: IUCN.

Commons, J.R. (1957). *Legal Foundations of Capitalism.* Madison, WI: University of Wisconsin Press.

Community and School Gardens Program (CSGP). (2022). *The Sonoran Desert School Gardener's Almanac.* https://schoolgardens.arizona.edu/curriculum/sonoran-desert-school-gardeners-almanac (accessed 18 April 2022).

Conca, K. (2005). *Governing Water: Contentious Transnational Politics and Global Institution Building.* Cambridge, MA: MIT Press.

Cosens, B. and Gunderson, L. (2021). Adaptive Governance in North American Water Systems: A Legal Perspective on Resilience and Reconciliation. In: *Water Resilience* (ed. J. Baird and R. Plummer), pp. 171–192. New York: Springer International Publishing.

Cotula, L. (2012). The International Political Economy of the Global Land Rush: A Critical Appraisal of Trends, Scale, Geography, and Drivers. *The Journal of Peasant Studies* 39 (3–4): 649–680.

Cronon, W. (1996). The Trouble with Wilderness: Or, Getting Back to the Wrong Nature. *Environmental History* 1 (1): 7–28.

Crow, B. and Sultana, F. (2002). Gender, Class, and Access to Water: Three Cases in a Poor and Crowded Delta. *Society and Natural Resources* 15 (8): 709–724.

Crow, B. and Singh, N. (2003). Floods and International Relations in South Asia: An Assessment of Multi-track Diplomacy. https://people.ucsc.edu/~boxjenk/floods_crow&singh_D3.pdf (accessed 13 June 2022).

Crow-Miller, B. (2015). Discourses of Deflection: The Politics of Framing China's South-North Water Transfer Project. *Water Alternatives* 8 (2): 173–192.

Cumbers, A. and Paul, F. (2022). Remunicipalisation, Mutating Neoliberalism, and the Conjuncture. *Antipode* 54 (1): 197–217.

Curley, A. (2019a). "Our Winters' Rights": Challenging Colonial Water Laws. *Global Environmental Politics* 19 (3): 57–76.

Curley, A. (2019b). Beyond Environmentalism: #NoDAPL as Assertion of Tribal Sovereignty. In: *Standing with Standing Rock: Voices from the #NoDAPL Movement* (ed. N. Estes and J. Dhillon), 158–168. Minneapolis, MN: University of Minnesota Press.

Curley, A. (2021). Unsettling Indian Water Settlements: The Little Colorado River, the San Juan River, and Colonial Enclosures. *Antipode* 53 (3): 705–723.

Curley, A. and Smith, S. (2020). Against Colonial Grounds: Geography on Indigenous Lands. *Dialogues in Human Geography* 10 (1): 37–40.

Curley, A. (2021a). Infrastructures as Colonial Beachheads: The Central Arizona Project and the Taking of Navajo Resources. *Environment and Planning D: Society and Space* 39 (3): 387–404.

Curley, A. (2021b). Unsettling Indian Water Settlements: The Little Colorado River, the San Juan River, and Colonial Enclosures. *Antipode* 53 (3): 705–723.

Daigle, M. (2018). Resurging through Kishiichiwan: The Spatial Politics of Indigenous Water Relations. *Decolonization: Indigeneity, Education & Society* 7 (1): 159–172.

De Albuquerque, C. (2014). *Realising the Human Rights to Water and Sanitation: A Handbook by the UN Special Rapporteur Catarina De Albuquerque.* https://unhabitat.org/realising-the-human-rights-to-water-and-sanitation-a-handbook-by-the-un-special-rapporteur-catarina-de-albuquerque (accessed 14 October 2022).

De Chaisemartin, M. (2020). Measuring Transboundary Water Cooperation within the Framework of Agenda 2030: A Proposal for a Revision of SDG Indicator 6.5.2. *Water International* 45 (1): 60–78. https://doi.org/10.1080/02508060.2019.1708659.

Deedat, H. and Cottle, E. (2002). Cost Recovery and Prepaid Water Meters and the Cholera Outbreak in Kwa-Zulu Natal. In: *Cost Recovery and the Crisis of Service Delivery in South Africa* (ed. D.A. McDonald and J. Pape), pp. 81–100. London: Zed Press.

Degefu, D.M., Weijun, H., Zaiyi, L., Yuan, L., Huang, Z., and An, M. (2018). Mapping Monthly Water Scarcity in Global Transboundary Basins at Country-Basin Mesh Based Spatial Resolution. *Scientific Reports* 8 (2144). https://doi.org/10.1038/s41598-018-20032-w.

Deininger, K., Byerlee, D., Lindsay, J., Norton, A., Selod, H., and Stickler, M. (2011). *Rising Global Interest in Farmland: Can it Yield Sustainable and Equitable Benefits?* Washington, DC: World Bank.

Deitz, S. and Meehan, K. (2019). Plumbing Poverty: Mapping Hot Spots of Racial and Geographic Inequality in U.S. Household Water Insecurity. *Annals of the American Association of Geographers* 109 (4): 1092–1109.

Del Bene, D., Scheidel, A., and Temper, L. (2018). More Dams, More Violence? A Global Analysis on Resistances and Repression Around Conflictive Dams Through Co-produced Knowledge. *Sustainability Science* 13 (3): 617–633.

DeMyers, C., Warpinski, C., and Wutich, A. (2017). Urban Water Insecurity: A Case Study of Homelessness in Phoenix, Arizona. *Environmental Justice* 10 (3): 72–80.

Desai, A. (2001). *We Are the Poors: Community Struggles in Post-Apartheid South Africa.* New York: Monthly Review Press.

de Vos, H., Boelens, R., and Bustamante, R. (2006). Formal Law and Local Water Control in the Andean Region: A Fiercely Contested Field. *International Journal of Water Resources Development* 22: 37–48.

Diamond, L. and McDonald, J.W. (1996). *Multi-track Diplomacy: A Systems Approach.* Sterling, VA: Kumarian Press.

Dickin, S. and Caretta, M.A. (2022). Examining Water and Gender Narratives and Realities. *Wiley Interdisciplinary Reviews (WIREs): Water* 9 (5): e1602. https://doi.org/10.1002/wat2.1602.

D'Odorico, P., Carr, J., Dalin, C., Dell'Angelo, J., Konar, M., Laio, F., Ridolfi, L., Rosa, L., Suweis, S., and Tamea, S. (2019). Global Virtual Water Trade and the Hydrological Cycle: Patterns, Drivers, and Socio-environmental Impacts. *Environmental Research Letters* 14: 053001. https://doi.org/10.1088/1748-9326/ab05f4.

D'Odorico, P., Davis, K.F., Rosa, L., Carr, J.A., Chiarelli, D., Dell'Angelo, J., Gephart, J., MacDonald, G.K., Seekell, D.A., Suweis, S., et al. (2018). The Global Food-Energy-Water Nexus. *Reviews of Geophysics* 56 (3): 456–531. https://doi.org/10.1029/2017RG000591.

Domosh, M. (2015). Practising Development at Home: Race, Gender, and the "Development" of the American South. *Antipode* 47 (4): 915–941.

Doshi, S. (2014). Imperial Water, Urban Crisis: A Political Ecology of Colonial State Formation in Bombay, 1850–1890. *Review (Fernand Braudel Center)* 37 (3–4): 173–218.

Doyle, M. (2012). America's Rivers and the American Experiment. *Journal of the American Water Resources Association* 48 (4): 820–837.

Doyle, M.W., Stanley, E.H., and Harbor, J.M. (2003). Channel Adjustments Following Two Dam Removals in Wisconsin. *Water Resources Research* 39 (1) https://doi.org/10.1029/2002WR001714.

Doyle, M.W., Stanley, E.H., Havlick, D.G., Kaiser, M.J., Steinbach, G., Graf, W.L., Galloway, G.E., and Riggsbee, J.A. (2008). Aging Infrastructure and Ecosystem Restoration. *Science* 319 (5861): 286–287.

Drost, S., de Wilde, J., and Drennen, Z. (2017). Bunge: Key Position in Cerrado State Puts Zero-Deforestation Commitment at Risk. Chain Reaction Research, Washington DC. https://chainreactionresearch.com/wp-content/uploads/2017/12/bunge-report-191217.pdf (accessed 18 October 2022).

D'Souza, R. (2006). *Drowned and Dammed: Colonial Capitalism and Flood Control in Eastern India.* New Delhi: Oxford University Press.

Dunbar-Ortiz, R. (2014). *An Indigenous Peoples' History of the United States.* Boston, MA: Beacon Press.

Earle, A., Cascao, A., Hansson, S., Jägerskog, A., Swain, A., and Öjendal, J. (2015). *Transboundary Water Management and the Climate Change Debate.* Abingdon, UK: Routledge.

Eaton, R. (1993). *Rise of Islam and the Bengal Frontier, 1204–1760.* Berkeley, CA: University of California Press.

Estes, N. (2019). *Our History is the Future: Standing Rock Versus the Dakota Access Pipeline and the Long Tradition of Indigenous Resistance.* London: Verso.

Estes, N. and Dhillon, J. ed. (2019). *Standing with Standing Rock: Voices from the #NoDAPL Movement.* Minneapolis, MN: University of Minnesota Press.

Eyler, B., Basist, A., Kwan, R., Weatherby, C., and Williams, C. (2022). Mekong Dam Monitor at One Year: What Have We Learned? Report. Mekong Fish Network. www.stimson.org/2022/mdm-one-year-findings/ (accessed 13 June 2022).

Falkenmark, M. (2020). Water Resilience and Human Life Support – Global Outlook for the Next Half Century. *International Journal of Water Resources Development* 36 (2–3): 377–396. https://doi.org/10.1080/07900627.2019.1693983.

Falkenmark, M., Wang-Erlandsson, L., and Rockström, J. (2019). Understanding of Water Resilience in the Anthropocene. *Journal of Hydrology X* 2: 100009.

Fanon, F. (1961). *The Wretched of the Earth*. New York: Grove Press.

Ferretti, F. (2019). Decolonizing the Northeast: Brazilian Subalterns, Non-European Heritages, and Radical Geography in Pernambuco. *Annals of the American Association of Geographers* 109 (5): 1632–1650.

Fine, B. (2013). Financialization from a Marxist Perspective. *International Journal of Political Economy* 42 (4): 47–66.

Fletcher, R. and Breitling, J. (2012). Market Mechanism or Subsidy in Disguise? Governing Payment for Environmental Services in Costa Rica. *Geoforum* 43 (3): 402–411.

Flyvbjerg, B. (2014). What You Should Know about Megaprojects and Why: An Overview. *Project Management Journal* 45 (2): 6–19.

Food First Information and Action Network (FIAN) International, Rede Social de Justiça e Direitos Humanos and Comissão Pastoral da Terra (CPT). (2018). *The Human and Environmental Cost of Land Business: The Case of MATOPIBA, Brazil*. FIAN International, Rede Social de Justiça e Direitos Humanos and Comissão Pastoral da Terra (CPT).

Folke, C. (2003). Freshwater for Resilience: A Shift in Thinking. *Philosophical Transactions of the Royal Society B: Biological Sciences* 358 (1440): 2027–2036.

Ford, J. and Plimmer, G. (2018). Pioneering Britain Has a Rethink on Privatization. *Financial Times* (22 January). https://www.ft.com/content/b7e28a58-f7ba-11e7-88f7-5465a6ce1a00 (accessed 14 October 2022).

Foucault, M. (2020, 1975). *Discipline and Punish: The Birth of the Prison*. London: Penguin Random House UK.

Fox, C.A. and Sneddon, C.S. (2019). Political Borders, Epistemological Boundaries, and Contested Knowledges: Constructing Dams and Narratives in the Mekong River Basin. *Water* 11 (3): 413.

Fox, C.A., Magilligan, F.J., and Sneddon, C.S. (2016). "You Kill the Dam, You are Killing a Part of Me:" Dam Removal and the Environmental Politics of River Restoration. *Geoforum* 70 (1): 93–104.

Fox, C.A., Reo, N.J., Fessell, B., and Dituri, F. (2022). Native American Tribes and Dam Removal: Restoring the Ottaway, Penobscot and Elwha Rivers. *Water Alternatives* 15 (1): 31–55.

Fragkou, M.C., and McEvoy, J. (2016). Trust Matters: Why Augmenting Water Supplies via Desalination May Not Overcome Perceptual Water Scarcity. *Desalination* 397: 1–8.

François, B., Schlef, K.E., Wi, S., and Brown, C.M. (2019). Design Considerations for Riverine Floods in a Changing Climate: A Review. *Journal of Hydrology* 574: 557–573.

Furlong, K. (2014). STS Beyond the "Modern Infrastructure Ideal": Extending Theory by Engaging with Infrastructure Challenges in the South. *Technology in Society* 38: 139–147.

Furlong, K. (2016). *Leaky Governance: Alternative Service Delivery and the Myth of Water Utility Independence.* Vancouver, BC: UBC Press.

Furlong, K. (2021). Full-Cost Recovery = Debt Recovery: How Infrastructure Financing Models Lead to Overcapacity, Debt, and Disconnection. *WIREs Water* 8 (2): e1503.

Gaber, N. (2021). Blue Lines and Blues Infrastructures: Notes on Water, Race, and Space. *Environment and Planning D: Society and Space* 39 (6): 1073–1091.

Galvin, M. (2015). Talking Shit: Is Community-Led Total Sanitation a Radical and Revolutionary Approach to Sanitation? *WIREs: Water* 2 (1): 9–20.

Gandy, M. (2002). *Concrete and Clay: Reworking Nature in New York City.* Cambridge, MA: MIT Press.

Garrick, D., De Stefano, L., Yu, W., Jorgensen, I., O'Donnell, E., Turley, L., Aguilar-Barajas, I., Dai, X., de Souza Leão, R., Punjabi, B., et al. (2019). Rural Water for Thirsty Cities: A Systematic Review of Water Reallocation from Rural to Urban Regions. *Environmental Research. Letters* 14 (4): 043003.

Garrick, D., Hanemann, M., and Hepburn, C. (2020). Rethinking the Economics of Water: An Assessment. *Oxford Review of Economic Policy* 36 (1): 1–23.

Garrick, D., Siebentritt, M.A., Aylward, B., Bauer, C.J., and Purkey, A. (2009). Water Markets and Freshwater Ecosystem Services: Policy Reform and Implementation in the Columbia and Murray-Darling Basins. *Ecological Economics* 69 (2): 366–379.

Gassert, F., Reig, P., Luo, T., and Maddocks, A. (2013). *Aqueduct Country and River Basin Rankings: A Weighted Aggregation of Spatially Distinct Hydrological Indicators.* Working paper. Washington, DC: World Resources Institute. https://www.wri.org/research/aqueduct-country-and-river-basin-rankings (accessed 10 June 2022).

Gasson, C. (2010). The Human Right to a National Water Plan. *Global Water Intelligence* (28 July). https://www.globalwaterintel.com/news/2010/30/the-human-right-to-a-national-water-plan (accessed 18 June 2022).

Gasteyer, S., Isaac, J., Hillal, J., and Hodali, K. (2012). Water Grabbing in Colonial Perspective: Land and Water in Israel/Palestine. *Water Alternatives* 5 (2): 450–468.

Gergan, M.D. (2017). Living with Earthquakes and Angry Deities at the Himalayan Borderlands. *Annals of the American Association of Geographers* 107 (2): 490–498.

Gergan, M.D. (2020). Disastrous Hydropower, Uneven Regional Development, and Decolonization in India's Eastern Himalayan Borderlands. *Political Geography* 80 (1): 102175.

Gergan, M.D. and McCreary, T. (2022). Disrupting Infrastructure of Colonial Hydro-Modernity: Lepcha and Dakelh Struggles against Temporal and Territorial Displacements. *Annals of the American Association of Geographers* 112 (3): 789–798.

Gijsbers, P.J.A. and Loucks, D.P. (1999). Libya's Choices: Desalination or the Great Man-Made River Project. *Physics and Chemistry of the Earth, Part B: Hydrology, Oceans and Atmosphere* 24 (4): 385–389.

Gilmont, M. (2014). Decoupling Dependence on Natural Water: Reflexivity in the Regulation and Allocation of Water in Israel. *Water Policy* 16 (1): 79–101.

Gilmore, R.W. (2006). *Golden Gulag: Prisons, Surplus, Crisis, and Opposition in Globalizing California*. Berkeley, CA: University of California Press.

Giordano, M., Drieschova, A., Duncan, J.A., Sayama, Y., De Stefano, L., and Wolf, A.T. (2014). A Review of the Evolution and State of Transboundary Freshwater Treaties. *International Environmental Agreements: Politics, Law and Economics* 14 (3): 245–264. https://doi.org/10.1007/s10784-013-9211-8.

Goddard, J.J., Ray, I., and Balazs, C. (2021). How Should Water Affordability be Measured in the United States? A Critical Review. *Wiley Interdisciplinary Reviews (WIREs): Water* 9: e1573.

Gondo, R., Kolawole, O.D., and Mbaiwa, J.E. (2019a). Institutions and Water Governance in the Okavango Delta, Botswana. *Chinese Journal of Population Resources and Environment* 17 (1): 67–78.

Gondo, R., Kolawole, O.D., and Mbaiwa, J.E. (2019b). Dissonance in Customary and Statutory Water Management Institutions: Issues of Cultural Diversity in the Management of Water Resources in the Okavango Delta, Botswana. *Environment, Development and Sustainability* 21 (3): 1091–1109.

Gondo, R., Kolawole, O.D., Mbaiwa, J.E., and Motsholapheko, M.R. (2019c). Stakeholders' Perceptions on Water Resources Management in the Okavango Delta, Botswana. *Transactions of the Royal Society of South Africa* 74 (3): 283–296.

Gosling, L. and Tobin, K. (2020). Ten Years of the Human Rights to Water and Sanitation: Our Contributions and the Road Ahead. https://washmatters. wateraid.org/blog/ten-years-human-rights-water-sanitation (14 October 2022).

Gosnell, H. and Kelly, E.G. (2010). Peace on the River? Socio-ecological Restoration and Large Dam Removal in the Klamath Basin, USA. *Water Alternatives* 3 (2): 361–383.

Grafton, Q.R., Colloff, M.J., Marshall, V., and Williams, J. (2020). Confronting a 'Post-Truth Water World' in the Murray-Darling Basin, Australia. *Water Alternatives* 13 (1): 1–28.

Graham, S. and Marvin, S. (2001). *Splintering Urbanism: Networked Infrastructures, Technological Mobilities, and the Urban Condition*. Abingdon, UK: Routledge.

GRAIN (2018). *Failed Farmland Deals: A Growing Legacy of Disaster and Pain*. GRAIN. Barcelona. https://grain.org/en/article/5958-failed-farmland-deals-a-growing-legacy-of-disaster-and-pain (accessed 10 June 2022).

Grandin, G. (2006). *Empire's Workshop: Latin America, the United States, and the Rise of the New Imperialism*. New York: Metropolitan Books, Henry Holt and Company.

Greene, J. (2018). Bottled Water in Mexico: The Rise of a New Access to Water Paradigm. *WIREs Water* 5 (4): e1286.

Grill, G., Lehner, B., Thieme, M., Geenen, B., Tickner, D., Antonelli, F., Babu, S., Borrelli, P., Cheng, L., Crochetiere, H., et al. (2019). Mapping the World's Free-Flowing Rivers. *Nature* 569: 215–221. https://doi.org/10.1038/s41586-019-1111-9.

Goldman, F. (2012). Camila Vallejo, the World's Most Glamorous Revolutionary. *The New York Times Magazine* (5 April). https://www.nytimes.com/2012/04/08/magazine/camila-vallejo-the-worlds-most-glamorous-revolutionary.html (accessed 1 June 2022).

Goldman, M. (2007). How "Water for All!" Policy Became Hegemonic: The Power of the World Bank and its Transnational Policy Networks." *Geoforum* 38 (1): 786–800.

Grafton, R.Q., Williams, J., Perry, C.J., Molle, F., Ringler, C., Steduto, P., Udall, B., Wheeler, S.A., Wang, Y., Garrick, D., et al. (2018). The Paradox of Irrigation Efficiency. *Science* 80 (361): 748–750.

Green, O., Cosens, B., and Garmestani, A. (2013). Resilience in Transboundary Water Governance: The Okavango River Basin. *Ecology and Society* 18 (2): 23. https://doi.org/10.5751/ES-05453-180223.

The Guardian, ed. (1960). Five Million Families in Britain Living in Houses Without Baths, *The Guardian* (29 November). https://www.theguardian.com/global-development/2018/nov/29/five-million-families-in-britain-living-in-houses-without-baths-archive-1960 (accessed 20 June 2022).

Gupta, A. (2012). *Red Tape: Bureaucracy, Structural Violence, and Poverty in India*. Durham, NC: Duke University Press.

Haasnoot, M., van Aalst, M., Rozenberg, J., Dominique, K., Matthews, J., Bouwer, L.M., Kind, J., and Poff, N.L. (2020). Investments under Non-stationarity: Economic Evaluation of Adaptation Pathways. *Climatic Change* 161 (3): 451–463.

Haines, D. (2017). *Rivers Divided: Indus Basin Waters in the Making of India and Pakistan*. London, UK: Hurst & Company.

Hallegatte, S. (2009). Strategies to Adapt to an Uncertain Climate Change. *Global Environmental Change* 19 (2): 240–247.

Halls, A.S., Paxton, B.R., Hall, N., Peng Bun, N., Lieng, S., Pengby, N., and So, N. (2013). *The Stationary Trawl (Dai) Fishery of the Tonle Sap-Great Lake, Cambodia*. MRC Technical Paper No. 32, Mekong River Commission, Phnom Penh, Cambodia.

Hamandawana, H. and Chanda, R. (2010). Natural and Human Dimensions of Environmental Change in the Proximal Reaches of Botswana's Okavango Delta. *Geographical Journal* 176 (1): 58–76.

Han, X. and Webber, M. (2020a). From Chinese Dam Building in Africa to the Belt and Road Initiative: Assembling Infrastructure Projects and Their Linkages. *Political Geography* 77: 102102.

Han, X. and Webber, M. (2020b). Extending the China Water Machine: Constructing a Dam Export Industry. *Geoforum* 112: 63–72.

Hanemann, M. and Young, M. (2020). Water Rights Reform and Water Marketing: Australia vs the US West. *Oxford Review of Economic Policy* 36 (1): 108–131.

Haraway, D. (1988). Situated Knowledges: The Science Question in Feminism and the Privilege of Partial Perspective. *Feminist Studies* 14 (3): 575–599.

Hardin, G. (1968). The Tragedy of the Commons. *Science* 162 (3859): 1243–1248.

Harris, L. (2008). Water Rich, Water Poor: Intersections of Gender, Poverty, and Vulnerability in Newly Irrigated Areas of Southeastern Turkey. *World Development* 36 (12): 2643–2662.

Harris, L.M. (2014). Imaginative Geographies of Green: Difference, Postcoloniality, and Affect in Environmental Narratives in Contemporary Turkey. *Annals of the Association of American Geographers* 104 (4): 801–815.

Harris, L., Kleiber, D., Goldin, J., Darkwah, A., and Morinville, C. (2017). Intersections of Gender and Water: Comparative Approaches to Everyday Gendered Negotiations of Water Access in Underserved Areas of Accra, Ghana and Cape Town, South Africa. *Journal of Gender Studies* 26 (5): 561–582.

Harrower, M. (2009). Is the Hydraulic Hypothesis Dead Yet? Irrigation and Social Change in Ancient Yemen. *World Archaeology* 41 (1): 58–72.

Harvey, D. (1974). Population, Resources, and the Ideology of Science. *Economic Geography* 50 (3): 256–277.

Harvey, D. (1982). *The Limits to Capital*. Oxford, UK: Blackwell.

Harvey, D. (2000). *Spaces of Hope*. Berkeley, CA: University of California Press.

Harvey, D. (2003). *The New Imperialism*. Oxford, UK: Oxford University Press.

Hawkins, G. (2017). The Impacts of Bottled Water: An Analysis of Bottled Water Markets and Their Interactions with Tap Water Provision. *Wiley Interdisciplinary Reviews (WIREs): Water* 4: e1203.

Heller, L. (2022). *The Human Rights to Water and Sanitation*. Cambridge, UK: Cambridge University Press.

Herrera, V. (2017). *Water and Politics: Clientelism and Reform in Urban Mexico*. Ann Arbor, MI: University of Michigan Press.

Heynen, N., Kaika, M., and Swyngedouw, E. ed. (2006). *In the Nature of Cities: Urban Political Ecology and the Politics of Urban Metabolism*. Abingdon, UK: Routledge.

Hoang Thi Ha and Seth, F.N. (2021). The Mekong River Ecosystem in Crisis: ASEAN Cannot Be a Bystander. *ISEAS Perspective* 69: 1–9.

Hoang, L.P., Van Vliet, M.T.H., Kummu, M., Lauri, H., Koponen, J., Supit, I., Leemans, R., Kabat, P., and Ludwig F. (2019). The Mekong's Future Flows under Multiple Drivers: How Climate Change, Hydropower Developments, and Irrigation Expansions Drive Hydrological Changes. *Science of the Total Environment* 649: 601–609.

Hobbes, T. (2017, 1651). *Leviathan*. London: Penguin.

Hoekstra, A.Y. and Mekonnen, M.M. (2012). The Water Footprint of Humanity. *PNAS* 109 (9): 3232–3237.

Hoekstra, A.Y., Mekonnen, M.M., Chapagain, A.K., Mathews, R.E., and Richter, B.D. (2012). Global Monthly Water Scarcity: Blue Water Footprints versus Blue Water Availability. *PloS One* 7 (2): e32688.

Hoff, H. (2011). Understanding the Nexus: Background Paper for the Bonn 2011 Conference. The Water, Energy and Food Security Nexus. Stockholm Environment Institute, Stockholm.

Holden, J. and Pagel, M. (2013). Transnational Land Acquisitions: What are the Drivers, Levels, and Destinations, of Recent Transnational Land Acquisitions? EPS Peaks. https://assets.publishing.service.gov.uk/media/57a08a48ed915d3cfd0006ac/Transnational_land_acquisitions_10.pdf (accessed 10 June 2022).

Hommes, L., Boelens R., Bleeker, S., Duarte-Abadía, B., Stoltenborg, D., and Vos, J. (2020). Water Governmentalities: The Shaping of Hydrosocial Territories, Water Transfers and Rural–Urban Subjects in Latin America. *Environment and Planning E: Nature and Space* 3 (2): 399–422.

Hoogesteger, J. (2018). The Ostrich Politics of Groundwater Development and Neoliberal Regulation in Mexico. *Water Alternatives* 11 (3): 552–571.

Hooper, V. (2015). The Importance of the "Urban" in Agricultural-to-Urban Water Transfers. Insights from Comparative resEarch in India and China. Unpublished PhD thesis. Norwich, UK: University of East Anglia.

Household Water Insecurity (HWISE)-Research Coordination Network (RCN). (2022). Research network includes many resources and scholars who work in this area, funded by the National Science Foundation: https://hwise-rcn.org/ (accessed 23 June 2022).

Huang, Y., Lin, W., Li, S., and Ning, Y. (2018). Social Impacts of Dam-Induced Displacement and Resettlement: A Comparative Case Study in China. *Sustainability* 10 (11): 4018.

Huber, A. and Joshi, D. (2015). Hydropower, Anti-Politics, and the Opening of New Political Spaces in the Eastern Himalayas. *World Development* 76 (1): 13–25.

Huitema, D. and Meijerink, S. (2014). *The Politics of River Basin Organizations: Coalitions, Institutional Design Choices and Consequences.* Cheltenham, UK: Edward Elgar.

Hunt, S. (2015). *Large-Scale Land Acquisitions.* Christian Aid Ireland.

Ide, T. (2019). The Impact of Environmental Cooperation on Peacemaking: Definitions, Mechanisms, and Empirical Evidence. *International Studies Review* 21 (3): 327–346. https://doi.org/10.1093/isr/viy014.

Ide, T. (2020). The Dark Side of Environmental Peacebuilding. *World Development* 127. https://doi.org/10.1016/j.worlddev.2019.104777.

Immerzeel, W.W., Lutz, A.F., Andrade, M., Bahl, A., Biemans, H., Bolch, T., Hyde, S., Brumby, S., Davies, B.J., Elmore, A.C., et al. (2020). Importance and Vulnerability of the World's Water Towers. *Nature* 577 (7790): 364–369. https://doi.org/10.1038/s41586-019-1822-y.

Inclusive Development International (2020). Laos: Pursuing Accountability for Deadly Dam Collapse – Xe-Pian Xe-Namnoy Hydropower Dam. Inclusive Development International. https://www.inclusivedevelopment.net/cases/laos-xe-pian-xe-namnoy-dam-collapse/ (accessed 13 June 2022).

Ingram, H. and White, D. (1993). International Boundary and Water Commission: An Institutional Mismatch for Resolving Transboundary Water Problems. *Natural Resources Journal* 33 (1): 153–175.

Inclusive Development International and International Rivers (2019). Reckless Endangerment: Assessing Responsibility for the Xe Pian-Xe Namnoy Dam Collapse. Report. Asheville, NC: Inclusive Development International; Oakland, CA: International Rivers.

Intergovernmental Science-Policy Platform on Biodiversity and Ecosystem Services (IPBES). (2018). The IPBES Assessment Report on Land Degradation and Restoration (ed. L. Montanarella, R. Scholes, and A. Brainich). *Secretariat of the Intergovernmental Science-Policy Platform on Biodiversity and Ecosystem Services*, Bonn, Germany.

International Commission on Large Dams (ICLD). (n.d.). World Register of Dams. https://www.icold-cigb.org/GB/world_register/world_register_of_dams.asp (accessed 18 October 2022).

Islam, S. and Smith, K. (2020). *Interdisciplinary Collaboration for Water Diplomacy: A Principled and Pragmatic Approach*. New York: Routledge.

Jansen, R.B. (2003). Dams, Dikes, and Levees. In: *Encyclopedia of Physical Science and Technology, Third Edition* (ed. R.A. Meyers), 171–190. Elsevier. https://www.sciencedirect.com/referencework/9780122274107/encyclopedia-of-physical-science-and-technology (accessed 18 October 2022).

Jasanoff, S. ed. (2004). *States of Knowledge: The Co-Production of Science and the Social Order*. London: Routledge.

Jepson, W. (2012). Claiming Space, Claiming Water: Contested Legal Geographies of Water in South Texas. *Annals of the Association of American Geographers* 102 (3): 614–631.

Jepson, W. (2014). Measuring "No-Win" Waterscapes: Experience-Based Scales and Classification Approaches to Assess Household Water Security in Colonias on the US-Mexico Border. *Geoforum* 51 (1): 107–120.

Jepson, W., Budds, J., Eichelberger, L., Harris, L., Norman, E., O'Reilly, K., Pearson, A., Shah, S., Shinn, J., Staddon, C., et al. (2017). Advancing Human Capabilities for Water Security: A Relational Approach. *Water Security* 1: 46–52.

Johansson, E.L., Fader, M., Seaquist, J.W., and Nicholas, K.A. (2016). Water Demand from Land Acquisitions in Africa. *PNAS* 113 (41): 11471–11476.

Joshi, D. (2011). Caste, Gender and the Rhetoric of Reform in India's Drinking Water Sector. *Economic and Political Weekly* 46 (18): 56–63.

Joslin, A.J. and Jepson, W.E. (2018). Territory and Authority of Water Fund Payments for Ecosystem Services in Ecuador's Andes. *Geoforum* 91: 10–20.

Kaika, M. (2004). Interrogating the Geographies of the Familiar: Domesticating Nature and Constructing the Autonomy of the Modern Home. *International Journal of Urban and Regional Research* 28 (2): 265–286.

Kaika, M. (2005). *City of Flows: Modernity, Nature, and the City*. London: Routledge.

Kallio, M. and Kummu, M. (2021). Comment on 'Changes of Inundation Area and Water Turbidity of Tonle Sap Lake: Responses to Climate Changes or Upstream Dam Construction?' *Environmental Research Letters* 16 (5): 058001. https://doi.org/10.1088/1748-9326/abf3da.

Katz, C. (2001). Vagabond Capitalism and the Necessity of Social Reproduction. *Antipode* 33 (4): 709–728.

Katz, C. (2008). Bad Elements: Katrina and the Scoured Landscape of Social Reproduction. *Gender, Place & Culture* 15 (1): 15–29.

Kelley, S.H. and Valdés Negroni, J.M. (2021). Tracing Institutional Surprises in the Water-Energy Nexus: Stalled Projects of Chile's Small Hydropower Boom. *Environment and Planning E: Nature and Space* 4 (3): 1171–1195.

Kenney, D.S. (2005). Prior Appropriation and Water Rights Reform in the Western United States. In: *Water Rights Reform: Lessons for Institutional Design* (ed. B. Randolph Bruns, C. Ringler, and R. Meinzen-Dick), 167–182. Washington, DC: International Food Policy Research Institute.

Kerf, M. (1998). *Concessions for Infrastructure: A Guide to their Design and Award.* Washington, DC: World Bank.

Keulertz, M. and Woertz, E. (2016). States as Actors in International Agro-Investments. In: *Large-Scale Land Acquisitions: Focus on South-East Asia* (ed. C. Gironde, C. Golay, and P. Messerli), 30–52. Leiden, Boston: Brill Nijhoff.

King, P., Gaddis, E., Grellier, J., Grobicki, A.M., Hay, R., Mirumachi, N., Mudd, G., Mukhtarov, F., and Rast, W. (2019). Freshwater Policy. In: *Global Environment Outlook – GEO-6: Healthy Planet, Healthy People*. Report. United Nations Environment Programme. Nairobi. https://www.unep.org/resources/global-environment-outlook-6 (accessed 13 June 2022).

Kishimoto, S., Lobina, E., and Petitjean, O. (2015). *Our Public Water Future: The Global Experience with Remunicipalisation*. Amsterdam: Transnational Institute (TNI)/Public Services International Research Unit (PSIRU)/Multinationals Observatory/Municipal Services Project (MSP)/European Federation of Public Service Unions (EPSU).

Kittikhoun, A. and Schmeier, S. (2021). *River Basin Organizations in Water Diplomacy*, Abingdon, UK: Routledge.

Klein, N. (2007). *The Shock Doctrine: The Rise of Disaster Capitalism*. Toronto, ON: Canada Knopf.

Klenk, N., Fiume, A., Meehan, K., and Gibbes, C. (2017). Local Knowledge in Climate Adaption Research: Moving Knowledge Frameworks from Extraction to Co-production. *WIREs Climate Change* 8 (5): e475.

Klingensmith, D. (2003). Building India's 'Modern Temples': Indians and Americans in the Damodar Valley Corporation, 1945–60. In: *Regional Modernities: The Cultural Politics of Development in India* (ed. K. Sivaramakrishnan and A. Agarwal), 122–142. Redwood City, CA: Stanford University Press.

Koch, N. (2021). The Political Lives of Deserts. *Annals of the American Association of Geographers* 111 (1): 87–104.

Komakech, H.C., van der Zaag, P., and van Koppen, B. (2012). The Last Will Be First: Water Transfers from Agriculture to Cities in the Pangani River Basin, Tanzania. *Water Alternatives* 5 (3): 700–720.

Kooy, M. (2014). Developing Informality: The Production of Jakarta's Urban Waterscapes. *Water Alternatives* 7 (1): 35–53.

Kooy, M. and Bakker, K. (2008). Splintered Networks: The Colonial and Contemporary Waters of Jakarta. *Geoforum* 39 (6): 1843–1858.

Kooy, M., and Furlong, K. (2020). Explanations and Implications of Bottled Water Trends across Disciplines: Politics, Piped Water and Consumer Preferences. *WIREs Water* 7 (4): e1385.

Kotsila, P., Angueloveski, I., Baró, F., Langemeyer, J., Sekulova, F., and Connolly, J. (2021). Nature-Based Solutions as Discursive Tools and Contested Practices in Urban Nature's Neoliberalisation Processes. *Environment and Planning E: Nature and Space* 4 (2); 252–274.

Kummu, M., Tes, S., Yin, S., Adamson, P., Józsa, J., Koponen, J., Richey, J., and Sarkkula, J. (2014). Water Balance Analysis for the Tonle Sap Lake–Floodplain System. *Hydrological Processes* 28: 1722–1733.

Lakhani, N. (2020). Detroit Suspends Water Shutoffs over Covid-19 Fears. *The Guardian* (12 March). https://www.theguardian.com/us-news/2020/mar/12/detroit-water-shutoffs-unpaid-bills-coronavirus (accessed 20 June 2022).

Land Matrix (2021). Deals. Public Database. https://landmatrix.org/list/deals (accessed 10 June 2022).

Langford, M. and Russell, A.F. ed. (2017). *The Human Right to Water: Theory, Practice and Prospects*. Cambridge, UK: Cambridge University Press.

Lankford, B. and Hepworth, N. (2010). The Cathedral and the Bazaar: Monocentric and Polycentric River Basin Management. *Water Alternatives* 3 (1): 82–101.

Lansing, S. (2009). *Priests and Programmers: Technologies of Power in the Engineered Landscape of Bali*. Princeton, NJ: Princeton University Press.

Lapavitsas, C. (2013). *Profiting without Producing: How Finance Exploits Us All*. London: Verso Books.

Larder, N. (2015). Space for Pluralism? Examining the Malibya Land Grab. *Journal of Peasant Studies* 42 (3–4): 839–858. https://doi.org/10.1080/03066150.2015.1029461.

Latour, B. (1993). *We Have Never Been Modern*. Cambridge, MA: Harvard University Press.

Lave, R. (2012). *Fields and Streams: Stream Restoration, Neoliberalism, and the Future of Environmental Science*. Athens, GA: University of Georgia Press.

Lave, R., Doyle, M., and Robertson, M. (2010). Privatizing Stream Restoration in the US. *Social Studies of Science* 40 (5): 677–703.

Lave, R., Robertson, M.M., and Doyle, M.W. (2008). Why You Should Pay Attention to Stream Mitigation Banking. *Ecological Restoration* 4: 287–289.

Lawrence, G., Sippel, S.R., and Burch, D. (2015). The Financialization of Food and Farming. In: *Handbook on the Globalisation of Agriculture* (ed. G.M. Robinson and D.A. Carson), 309–327. Cheltenham, UK: Edward Elgar.

Leb, C. (2020). *Data Innovations for Transboundary Freshwater Resources Management: Are Obligations Related to Information Exchange Still Needed?* Brill Research Perspectives in International Law 4 (4). Leiden: Brill.

Lebel, L., Anderies, J.M., Campbell, B., Folke, C., Hatfield-Dodds, S., Hughes, T.P., and Wilson, J. (2006). Governance and the Capacity to Manage Resilience in Regional Social-Ecological Systems. *Ecology and Society* 11 (1): 19.

Lebel, L. and Lebel, B. (2018). Nexus Narratives and Resource Insecurities in the Mekong Region. *Environmental Science & Policy* 90: 164–172. https://doi.org/10.1016/j.envsci.2017.08.015.

Lee, C. (2017). *The Specter of Global China: Politics, Labor, and Foreign Investment in Africa*. Chicago, IL: University of Chicago Press.

Lewis, E.W., Staddon, C., and Sirunda, J. (2019). Urban Water Management Challenges and Achievements in Windhoek, Namibia. *Water Practice and Technology* 14 (3): 703–713.

Lilienthal, D.E. (1944). *TVA: Democracy on the March*. London: Harper and Brothers.

Lin, G.C.S. (2017). Water, Technology, Society and the Environment: Interpreting the Technopolitics of China's South–North Water Transfer Project. *Regional Studies* 51 (3): 383–388.

Linton, J. (2008). Is the Hydrologic Cycle Sustainable? A Historical-Geographical Critique of a Modern Concept. *Annals of the Association of American Geographers* 98 (3): 630–649.

Linton, J. (2010). *What Is Water?: The History of a Modern Abstraction*. Vancouver, BC: UBC Press.

Linton, J. and Brooks, D.B. (2011). Governance of Transboundary Aquifers: New Challenges and New Opportunities. *Water International* 36 (5): 606–618.

Linton, J. and Budds, J. (2014). The Hydrosocial Cycle: Defining and Mobilizing a Relational-Dialectical Approach to Water. *Geoforum* 57 (1): 170–180.

Lipton, G. (2020). What Are – and Aren't – Nature-Based Solutions? Multinationals are Increasingly Harnessing Nature to Reach Green Goals, But It Can Be a Tricky Business. *Landscape News* https://news.globallandscapesforum.org/48171/what-are-and-arent-nature-based-solutions/ (accessed 8 August 2022).

Liptrot, T. and Hussein, H. (2020). Between Regulation and Targeted Expropriation: Rural-to-urban Groundwater Reallocation in Jordan. *Water Alternatives* 13 (3): 864–885.

Loftus, A. (2006). Reification and the Dictatorship of the Water Meter. *Antipode* 38 (5): 1023–1045.

Loftus, A. (2007). Working the Socio-Natural Relations of the Urban Waterscape in South Africa. *International Journal of Urban and Regional Research* 31 (1): 41–59.

Loftus, A. (2009). Rethinking the Political Ecologies of Water. *Third World Quarterly* 30 (5): 953–968.

Loftus, A. (2015). Water (In)security: Securing the Right to Water. *The Geographical Journal* 181 (4): 350–356.

Loftus, A., and March, H. (2016). Financializing Desalination: Rethinking the Returns of Big Infrastructure. *International Journal of Urban and Regional Research* 40 (1): 46–61.

Loftus, A. and March, H. (2016). Financializing Desalination: Rethinking the Returns of Big Infrastructure. *International Journal of Urban and Regional Research* 40 (1): 46–61.

Loftus, A. and March, H. (2019). Integrating What and for Whom? Financialisation and the Thames Tideway Tunnel. *Urban Studies* 56 (11): 2280–2296.

Loftus, A., March, H., and Nash, F. (2016). Water Infrastructure and the Making of Financial Subjects. *Water Alternatives* 9 (2): 319–335.

Loftus, A.J. and McDonald, D.A. (2001). Of Liquid Dreams: A Political Ecology of Water Privatization in Buenos Aires. *Environment and Urbanization* 13 (2): 179–199.

Loftus, A., March, H., and Purcell, T.F. (2019). The Political Economy of Water Infrastructure: An Introduction to Financialization. *WIREs Water* 6 (1): e1326.

Lohr, A.M., Bell, M.L., Coulter, K., Marston, S., Thompson, M., Carvajal, S.C., Wilkinson-Lee, A.M., Gerald, L.B., and Korchmaros, J. (2022). The Association between Duration of School Garden Exposure and Self-Reported Learning and School Connectedness. *Health Education & Behavior.* https://doi.org/10.1177/10901981221084266.

Lord, A. (2016). Citizens of a Hydropower Nation: Territory and Agency at the Frontiers of Hydropower Development in Nepal. *Economic Anthropology* 3 (1): 145–160.

Lord, A., Drew, G., and Gergan, M.D. (2020). Timescapes of Himalayan Hydropower: Promises, Project Life Cycles, and Precarities. *Wiley Interdisciplinary Reviews: Water* 7 (6): e1469.

Lowe, L. (2019). *No Place to Go: How Public Toilets Fail Our Private Needs.* Brooklyn, NY and London: Melville House.

Ludden, D. (2005). Development Regimes in South Asia: History and the Governance Conundrum. *Economic and Political Weekly* 40 (37): 4042–4051.

McAfee, K. (1999). Selling Nature to Save It? Biodiversity and Green Developmentalism. *Environment and Planning D: Society and Space* 17 (2): 133–154.

McCaffrey, Stephen C. (2019). *The Law of International Watercourses*, 3rd edition. Oxford, UK: Oxford University Press.

McCracken, M. and Meyer, C. (2018). Monitoring of Transboundary Water Cooperation: Review of Sustainable Development Goal Indicator 6.5.2 Methodology. *Journal of Hydrology* 563: 1–12.

McCully, P. (1996). *Silenced Rivers: The Ecology and Politics of Large Dams.* London: Zed Books.

McDonald, D.A. (2018). Remunicipalization: The Future of Water Services? *Geoforum* 91: 47–56.

McDonald, D.A. and Ruiters, G. (2005). *The Age of Commodity: Water Privatization in Southern Africa.* Abingdon: Routledge.

McDonald-Wilmsen, B. and Webber, M. (2010). Dams and Displacement: Raising the Standards and Broadening the Research Agenda. *Water Alternatives* 3 (2): 142–161.

McEvoy, J. (2014). Desalination and Water Security: The Promise and Perils of a Technological Fix to the Water Crisis in Baja California Sur, Mexico. *Water Alternatives* 7 (3): 518–541.

McEvoy, J. and Wilder, M. (2012). Discourse and Desalination: Potential Impacts of Proposed Climate Change Adaptation Interventions in the Arizona-Sonora Border Region. *Global Environmental Change* 22 (2): 353–363.

McGowan. C. (2019). Chile Protests: What Prompted the Unrest? *Al Jazeera* (30 October). https://www.aljazeera.com/news/2019/10/30/chile-protests-what-prompted-the-unrest (accessed 1 June 2022).

McGregor, D. (2014). Traditional Knowledge and Water Governance: The Ethic of Responsibility. *AlterNative: An International Journal of Indigenous Peoples* 10 (5): 493–507.

Magilligan, F., Graber, B.E., Nislow, K.H., Chipman, J.W., Sneddon, C.S., and Fox, C.A. (2016). River Restoration by Dam Removal: Enhancing Connectivity at Watershed Scales. *Elementa: Science of the Anthropocene.* 4: 1–14. https://doi.org/10.12952/journal.elementa.000108.

Maiklem, L. (2019). *Mudlarking: Lost and Found on the Thames River.* London: Bloomsbury.

Malthus, T.R. (1992). *An Essay on the Principle of Population* (selected and introduced by D. Winch). Cambridge: Cambridge University Press.

March, H. (2015). Taming, Controlling and Metabolizing Flows: Water and the Urbanization Process of Barcelona and Madrid (1850–2012). *European Urban and Regional Studies* 22 (4): 350–367.

March, H. and Purcell, T. (2014). The Muddy Waters of Financialisation and New Accumulation Strategies in the Global Water Industry: The Case of AGBAR. *Geoforum* 53 (1): 11–20.

March, H., Saurí, D., and Rico-Amorós, A.M. (2014). The End of Scarcity? Water Desalination as the New Cornucopia for Mediterranean Spain. *Journal of Hydrology* 519: 2642–2651.

Marshall, T.H. (1992, 1949). *Citizenship and Social Class.* London: Pluto Press.

Marshall, V. (2017). *Overturning Aqua Nullius: Securing Aboriginal Water Rights.* Canberra, ACT: Aboriginal Studies Press.

Masamba, W. and Motsholapheko, M. (2017). Water Resources Management and Irrigation Development in the Okavango Delta, Botswana. In: *Partnerships and Power Games: Natural Resources Governance and Management in the Okavango Delta* (ed. L. Magole and A. Delaney), 143–164. Zurich: Lit Verlag.

Mascarenhas, M. (2012). *Where the Waters Divide: Neoliberalism, White Privilege, and Environmental Racism in Canada.* Washington, DC: Lexington Books.

Mason, M. (2022). Infrastructure under Pressure: Water Management and State-Making in Southern Iraq. *Geoforum* 132: 52–61. https://doi.org/10.1016/j.geoforum.2022.04.006

Mason, M., Zeitoun, M., and El Sheikh, R. (2011). Conflict and Social Vulnerability to Climate Change: Lessons from Gaza. *Climate and Development* 3 (4): 285–297.

Massey, D. (2005). *For Space*. Thousand Oaks, CA: SAGE Publications.

Mazer, K., Danyluk. M., Hunchuck. E., and Cowen, D. (2019). Mapping a Many-Headed Hydra: Transnational Infrastructures of Extraction and Resistance. In: *Standing with Standing Rock: Voices from the #NoDAPL Movement* (ed. N. Estes and J. Dhillon), 354–381. Minneapolis, MN: University of Minnesota Press.

Mazzucato, M. (2011). The Entrepreneurial State. *Soundings* 49: 131–142.

Mdee, A. and Harrison, E. (2019). Critical Governance Problems for Farmer-Led Irrigation: Isomorphic Mimicry and Capability Traps. *Water Alternatives* 12 (1): 30–45.

Meehan, K. (2010). *Greywater and the Grid: Explaining Informal Water Use in Tijuana*. PhD dissertation. Tucson, AZ: The University of Arizona.

Meehan, K. (2013). Disciplining De Facto Development: Water Theft and Hydrosocial Order in Tijuana. *Environment and Planning D: Society and Space* 31 (2): 319–336.

Meehan, K. (2014). Tool-Power: Water Infrastructure as Wellsprings of State Power. *Geoforum* 57 (1): 215–224.

Meehan, K. (2019). Making Space for Practical Authority: Policy Formalization and the Right to Water in Mexico. In: *Water Politics: Governance, Justice and the Right to Water* (ed. F. Sultana and A. Loftus), 28–41. London and New York: Routledge.

Meehan, K. and Moore, A. (2014). Downspout Politics, Upstream Conflict: Formalizing Rainwater Harvesting in the United States. *Water International* 39 (4): 417–430.

Meehan, K. and Strauss, K. ed. (2015). *Precarious Worlds: New and Contested Geographies of Social Reproduction*. Athens, GA: University of Georgia Press.

Meehan, K., Jepson, W., Harris, L.M., Wutich, A., Beresford, M., Fencel, A., London, J., Pierce, G., Radonic, L., Wells, C., et al. (2020b). Exposing the Myths of Household Water Insecurity in the Global North: A Critical Review. *Wiley Interdisciplinary Reviews Water* 7 (6): e1486.

Meehan, K., Jurjevich, J.R., Chun, N.M.J.W., and Sherrill, J. (2020a). Geographies of Insecure Water Access and the Housing-Water Nexus in US Cities. *Proceedings of the National Academy of Sciences* 117 (46): 28700–28707.

Meehan, K., Jurjevich, J.R., Griswold, A., Chun, N.M.J.W., and Sherrill. J. (2021). *Plumbing Poverty in U.S. Cities: A Report on Gaps and Trends in Household Water Access, 2000 to 2017*. London: King's College London. https://doi.org/10.18742/pub01-061.

Meehan, K., Odetola, M., and Griswold, A. (2022). *Homelessness, Water Insecurity, and the Human Rights to Water and Sanitation*. London: King's College London. https://doi.org/10.18742/pub01-085.

Meehan, K., Ormerod, K.J., and Moore, S.A. (2013). Remaking Waste as Water: The Governance of Recycled Effluent for Potable Water Supply. *Water Alternatives* 6 (1): 67–85.

Mégret, F. and Alston, P. ed. (2020). *The United Nations and Human Rights: A Critical Appraisal*. Oxford, UK: Oxford University Press.

Mehta, L., Veldwisch, G.J., and Franco, J. (2012). Introduction to the Special Issue: Water Grabbing? Focus on the (Re) appropriation of Finite Water Resources. *Water Alternatives* 5 (2): 193–207.

Meinzen-Dick, R. and Zwarteveen, M. (1998). Gendered Participation in Water Management: Issues and Illustrations from Water Users' Associations in South Asia. *Agriculture and Human Values* 15 (4): 337–345.

Mekong River Commission (MRC). (2010). *State of the Basin Report 2010*. Vientiane: MRC.

Mekonnenn, M.M. and Gerbens-Leenes, W. (2020). The Water Footprint of Global Food Production. *Water* 12 (10): 2696. https://doi.org/10.3390/w12102696.

Mekonnen, M.M. and Hoekstra, A.Y. (2020). Sustainability of the Blue Water Footprint of Crops. *Advances in Water Resources* 143: 103679.

Melosi, M.V. (2011). *Precious Commodity: Providing Water for America's Cities*. Pittsburgh, PA: University of Pittsburgh Press.

Menga, F. (2015). Building a Nation Through a Dam: The Case of Rogun in Tajikistan. *Nationalities Papers* 43 (3): 479–494.

Menga, F. and Swyngedouw, E. ed. (2018). *Water, Technology, and the Nation-State*. Oxford, UK: Routledge.

Merme, V., Ahlers, R., and Gupta, J. (2014). Private Equity, Public Affair: Hydropower Financing in the Mekong Basin. *Global Environmental Change* 24: 20–29.

Merrifield, A. (2006). *Henri Lefebvre: A Critical Introduction*. New York: Taylor & Francis.

Merryman, J.H. and Pérez-Perdomo, R. (2018). *The Civil Law Tradition: An Introduction to the Legal Systems of Europe and Latin America*. Palo Alto, CA: Stanford University Press.

Mies, M. (2014). *Patriarchy and Accumulation on a World Scale: Women in the International Division of Labour*. London: Zed Books.

Mighty Earth (2022). Mighty Earth's New Monitoring Data Reveals Deforestation Connected to Soy Trader and Meatpackers in Brazil More than Doubled over Two-Year Period. *Soy Report*. https://www.mightyearth.org/2021/04/28/mighty-earths-new-monitoring-data-reveals-deforestation-connected-to-soy-trader-and-meatpackers-in-brazil-more-than-doubled-over-two-year-period/ (accessed 10 June 2022).

Mirumachi, N. (2007). The Politics of Water Transfer between South Africa and Lesotho: Bilateral Cooperation in the Lesotho Highlands Water Project. *Water International* 32 (4): 558–570.

Mirumachi, N. (2015). *Transboundary Water Politics in the Developing World*. Abingdon, UK: Routledge.

Mirumachi, N. (2020). Informal Water Diplomacy and Power: A Case of Seeking Water Security in the Mekong River Basin. *Environmental Science and Policy* 114 (1): 86–95.

Mirumachi, N., Duda, A., Gregulska, J., and Smetek, J. (2021). The Human Right to Drinking Water: Impact of Large-Scale Agriculture and Industry: In-Depth

Analysis. Publications Office. European Union, Policy Department, Directorate-General for External Policies. https://doi.org/10.2861/714431.

Mogomotsi, G.E.J., Mogomotsi, P.K., and Mosepele, K (2020). Legal Aspects of Transboundary Water Management: An Analysis of the Intergovernmental Institutional Arrangements in the Okavango River Basin. *Leiden Journal of International Law* 33 (2): 391–408.

Molle, F. (2008). Nirvana Concepts, Narratives and Policy Models: Insights from the Water Sector. *Water Alternatives* 1 (1): 131–156.

Molle, F. and Berkoff, J. (2009). Cities vs. Agriculture: A Review of Intersectoral Water Re-allocation. *Natural Resources Forum* 33: 6–18.

Molle, F., Wester, F., and Hirsch, P. (2010). River Basin Closure: Processes, Implications and Responses. *Agricultural Water Management* 97: 569–577.

Molle, F. Wester, F., Hirsch, P., Jensen, J.R., Murray-Rust, H., Paranjpye, V., Pollard, S., and van der Zaag, P. (2007). River Basin Development and Management. In: *Water for Food, Water for Life: A Comprehensive Assessment of Water Management in Agriculture* (ed. D. Molden), 585–620. London: Earthscan.

Mollinga, P.P., Meinzen-Dick, R.S., and Merrey, D.J. (2007). Politics, Plurality, and Problemsheds: A Strategic Approach for Reform of Agricultural Water Resources Management. *Development Policy Review* 25: 699–719.

Molnar, K., Cuppari, R., Schmeier, S., and Demuth, S. (2017). *Preventing Conflicts, Fostering Cooperation: The Many Roles of Water Diplomacy*. Stockholm, Sweden: UNESCO's International Centre for Water Cooperation (ICWC) at SIWI; Koblenz, Germany: UNESCO's International Centre for Water Resources and Global Change (ICWRGC).

Morales, M. del C., Harris, L, and Öberg, G. (2015). Citizenshit: The Right to Flush and the Urban Sanitation Imaginary. *Environment and Planning A: Economy and Space*. 46 (12): 2816–2833.

Morrison, K. (2010). Dharmic Projects, Imperial Reservoirs, and New Temples of India: An Historical Perspective on Dams in India. *Conservation and Society* 8 (3): 182–195.

Mossallam, A. (2014). We Are the Ones Who Made This Dam "High!" A Builders' History of the Aswan High Dam. *Water History* 6 (4): 297–314.

Muehlmann, S. (2013). *Where the River Ends: Contested Indigeneity in the Mexican Colorado Delta*. Durham, NC: Duke University Press.

Mukuyu, P., Lautze, J., Rieu-Clarke, A., Saruchera, D., and McCartney, M. (2020). The Devil's in the Details: Data Exchange in Transboundary Waters, *Water International* 45 (7–8): 884–900. https://doi.org/10.1080/02508060.2020.1850026.

Müller, A., Mirumachi, N., Tickner, D., Louw, D., and Weston, D. (2022). Stalemate of the Hydrological Master Variable? The Challenge of Implementing Environmental Flows in the Orange-Senqu Basin. *Water International* 47 (3): 458–479.

Müller-Mahn, D., Gebreyes, M., Allouche, J., and Debarry, A. (2022). The Water-Energy-Food Nexus beyond "Technical Quick Fix": The Case of Hydro-

Development in the Blue Nile Basin, Ethiopia, *Frontiers in Water* 4: 787589. https://doi.org/10.3389/frwa.2022.787589.

Murray-Hudson, M. and Dauteuil, O. (2019). Botswana's Okavango Delta is Created by a Delicate Balance, but for How Much Longer? *The Conversation* (November 14). https://theconversation.com/botswanas-okavango-delta-is-created-by-a-delicate-balance-but-for-how-much-longer-125323 (accessed 6 August 2022).

National Audit Office (2015). The Economic Regulation of the Water Sector. https://www.nao.org.uk/wp-content/uploads/2014/07/The-economic-regulation-of-the-water-sector.pdf (accessed 13 June 2022).

Native American Advancement Foundation (NAAF). (2021). June: News from GuVo District, Tohono O'odham Nation. https://www.naafnow.org/so/cfNc5PT-v?languageTag=en#/main (accessed 17 May 2022).

Naz, F. (2015). Water, Water Lords, and Caste: A Village Study from Gujarat, *India*. *Capitalism Nature Socialism* 26 (3): 89–101.

Neves-Silva, P., Martins, G.I., and Heller, L. (2018). "We Only Have Access as a Favour, Don't We?" The Perception of Homeless Population on the Human Rights to Water and Sanitation. *Cadernos de Saúde Pública* 34 (3): e00024017.

Oliveira, G. de L.T., Murton, G., Rippa, A., Harlan, T., and Yang, Y. (2020). China's Belt and Road Initiative: Views from the Ground. *Political Geography* 82: 102225. https://doi.org/10.1016/j.polgeo.2020.102225.

O'Reilly, K, Halvorson, S., Sultana, F., and Laurie, N. (2009). Introduction: Global Perspectives on Gender-Water Geographies. *Gender, Place & Culture* 16 (4): 381–385.

Ormerod, K.J. (2016). Illuminating Elimination: Public Perception and the Production of Potable Water Reuse. *WIREs Water* 3 (4): 537–547.

Ormerod, K.J. (2019). Toilet Power: Potable Water Reuse and the Situated Meaning of Sustainability in the Southwestern United States. *Journal of Political Ecology* 26 (1): 633–651.

Ormerod, K.J. and Scott, C.A. (2013). Drinking Wastewater: Public Trust in Potable Reuse. *Science, Technology, & Human Values* 38 (3): 351–373.

Ornaghi, C. and Tonic, M. (2021). The Effects of the Universal Metering Programme on Water Consumption, Welfare and Equity. *Oxford Economic Papers* 73 (1): 399–422.

Orr, S., Pittock, J., Chapagain, A., and Dumaresq, D. (2012). Dams on the Mekong River: Lost Fish Protein and the Implications for Land and Water Resources. *Global Environmental Change* 22 (4): 925–932.

Osaka, S., Bellamy, R., and Castree, N. (2021). Framing "Nature-Based" Solutions to Climate Change. *WIREs Climate Change* 12 (5): e729.

Ostrom, E. (1990). *Governing the Commons: The Evolution of Institutions for Collective Action*. Cambridge, UK: Cambridge University Press.

O'Toole, F. (2014). The Irish Rebellion over Water. *New York Times* (19 December). https://www.nytimes.com/2014/12/20/opinion/fintan-otoole-the-irish-rebellion-over-water.html (last accessed 11 October 2022).

Ottinger, G., Barandiarán, J., and Kimura, A.H. (2016). Environmental Justice: Knowledge, Technology, and Expertise. In: *The Handbook of Science and Technology Studies* (ed. U. Felt, C.A. Miller, R. Fouché, and L. Smith-Doerr), 1029–1058. Cambridge, MA: The MIT Press.

Owen, D., Cantor, A., Nylen, N.G., Harter, T., and Kiparsky, M. (2019). California Groundwater Management, Science-Policy Interfaces, and the Legacies of Artificial Legal Distinctions. *Environment Research Letters* 14 (4): 045016.

Pahl-Wostl, C., Lebel, L., Knieper, C., and Nikitina, E. (2012). From Applying Panaceas to Mastering Complexity: Toward Adaptive Water Governance in River Basins. *Environmental Science & Policy*, 23: 24–34. https://doi.org/10.1016/j.envsci.2012.07.014.

Parks, N. and Grant, G. (2009). A Ravenous River Reclaims Its True Course: The Tale of Marmot Dam's Demise. *Science Findings* 111. https://www.fs.usda.gov/research/treesearch/32018 (accessed 30 July 2022).

Pattnaik, I. and Lahiri-Dutt, K. (2021). Do Women Like to Farm? Evidence of Growing Burdens of Farming on Women in Rural India. *The Journal of Peasant Studies* 49 (3): 629–651. https://doi.org/10.1080/03066150.2020.1867540.

Pauli, B.J. (2019). *Flint Fights Back: Environmental Justice and Democracy in the Flint Water Crisis*. Cambridge, MA: MIT Press.

Pauli, B.J. (2020). The Flint Water Crisis. *Wiley Interdisciplinary Reviews: Water* 7 (3): e1420.

Pechey, L., White, C., Rowcroft, P., and Smith, S. (2013). *The Role of Payments for Ecosystem Services in Climate Change Adaptation*. London: Defra.

Peet, R. (1985). Introduction to the Life and Thought of Karl Wittfogel. *Antipode* 17 (1): 3–21.

Peoplepower21 (2019). Allegation Letter on Xe-Pian Xe-Namnoy Hydropower Dam Collapse. People's Solidarity for Participatory Democracy. https://www.peoplepower21.org/English/1609025 (accessed 13 June 2022).

The Permanent Okavango River Basin Water Commission (OKACOM). (2020). Realising the Benefits of Transboundary Water Cooperation in the Cubango-Okavango River Basin. Gaborone, Botswana: OKACOM. https://unece.org/fileadmin/DAM/env/water/activities/Benefits_cooperation/OKACOM_Policy_Document_June_2020.pdf (accessed 5 August 2022).

Perramond, E.P. (2012). The Politics of Scaling Water Governance and Adjudication in New Mexico. *Water Alternatives* 5 (1): 62–82.

Perramond, E.P. (2013). Water Governance in New Mexico: Adjudication, Law, and Geography. *Geoforum* 45 (1): 83–93.

Perramond, E.P. (2016). Adjudicating Hydrosocial Territory in New Mexico. *Water International* 41 (1): 173–188.

Perreault, T. (2006). From the Guerra Del Agua to the Guerra Del Gas: Resource Governance, Neoliberalism and Popular Protest in Bolivia. *Antipode* 38 (1): 150–172.

Perreault, T. (2008). Custom and Contradiction: Rural Water Governance and the Politics of *Usos y Costumbres* in Bolivia's Irrigators' Movement. *Annals of the Association of American Geographers* 98 (4): 834–854.

Perreault, T. (2014). What Kind of Governance for What Kind of Equity? Towards a Theorization of Justice in Water Governance. *Water International* 39 (2): 233–245.

Pierce, G., El-Khattabi, A.R., Gmoser-Daskalakis, K., and Chow, N. (2021). Solutions to the Problem of Drinking Water Service Affordability: A Review of the Evidence. *Wiley Interdisciplinary Reviews (WIREs): Water* 8: e1522.

Pitta, F. and Mendonça, M.L. (2022). Industrial Soy: Expansion in Brazil. Financialization, Deforestation, and Dispossession in the Birthplace of Waters. Rede Social de Justiça e Direitos Humanos (Network for Social Justice and Human Rights) and Friends of the Earth United States.

Pitta, F., Mendonça, M.L., Stefano, D., and Conant, J. (2022). Red-Handed Deforestation and Bunge's Silent Conquest: How Land-Grabbers and Soy Speculators Enable the Destruction of Brazil's Cerrado. Rede Social de Justiça e Direitos Humanos (Network for Social Justice and Human Rights) and Friends of the Earth United States.

Poelina, A. and McDuffie, M. (2017). *Mardoowarra's Right to Life*. Broome, Australia: Madjulla Association. https://vimeo.com/205996720 (accessed 10 July 2022, password: Kimberley).

Poelina, A. (producer) and McDuffie, M. (director). (2021). *A Voice for Martuwarra*. Madjulla, Inc. Broome, Australia. https://vimeo.com/424782302 (accessed 12 July 2022).

Poelina, A., Taylor, K.S., and Perdristat, I. (2019). Martuwarra Fitzroy River Council: An Indigenous Cultural Approach to Collaborative Water Governance. *Australasian Journal of Environmental Management* 26 (3): 236–254.

Poff, N.L. and Matthews, J.H. (2013). Environmental Flows in the Anthropocene: Past Progress and Future Prospects. *Current Opinion in Environmental Sustainability* 5 (6): 667–675.

Poff, N.L., Allan, J.D., Bain, M.B., Karr, J.R., Prestegaard, K.L., Richter, B.D., Sparks, R.E., and Stromberg, J.C. (1997). The Natural Flow Regime. *BioScience*, 47 (11): 769–784. https://doi.org/10.2307/1313099.

Pohl, B., Carius, A., Conca, K., Dabelko, G.D., Kramer, A., Michel, D., Schmeier, S., Swain, A., and Wolf, A. (2014). *The Rise of Hydro-Diplomacy: Strengthening Foreign Policy for Transboundary Waters*. Berlin: Adelphi.

Pohlner, H. (2016). Institutional Change and the Political Economy of Water Megaprojects: China's South-North Water Transfer. *Global Environmental Change* 38: 205–216.

Postel, S. and Richter, B. (2012). *Rivers for Life: Managing Water for People and Nature*. Washington, DC: Island Press.

Prasetiawan, T., Nastily, A., and Muntalif, B.S. (2017). 'Bad' Piped Water and Other Perceptual Drivers of Bottled Water Consumption in Indonesia. *WIREs Water* 4 (4): e1219.

Prieto, M. (2015). Privatizing Water in the Chilean Andes: The Case of Las Vegas de Chiu-Chiu. *Mountain Research and Development* 35: 220–229.

Prieto, M. (2021). Indigenous Resurgence, Identity Politics, and the Anticommodification of Nature: The Chilean Water Market and the Atacameño People. *Annals of the*

American Association of Geographers 112 (2): 487–504. https://doi.org/10.1080/2469 4452.2021.1937036.

Prieto, M. and Bauer, C. (2012). Hydroelectric Power Generation in Chile: An Institutional Critique of the Neutrality of Market Mechanisms. *Water International* 37 (2): 131–146.

Prieto, M., Calderón, M., and Fragkou, M.C. (2020). Water Policy and Management in Chile. In: *Encyclopedia of Water: Science, Technology, and Society* (ed. P.A. Maurice), 2589–2600. Oxford, UK: Wiley-Blackwell.

Pulido, L. (2015). Geographies of Race and Ethnicity I: White Supremacy vs White Privilege in Environmental Racism Research. *Progress in Human Geography* 39 (6): 809–817.

Pulido, L. (2016). Flint, Environmental Racism, and Racial Capitalism. *Capitalism Nature Socialism* 27 (3): 1–16.

Pulido, L. (2017). Geographies of Race and Ethnicity II: Environmental Racism, Racial Capitalism, and State-Sanctioned Violence. *Progress in Human Geography* 41 (4): 524–533.

Pulido, L. (2018). Geographies of Race and Ethnicity III: Settler Colonialism and Nonnative People of Color. *Progress in Human Geography* 42 (2): 309–318.

Pulido, L., Kohl, E., and Cotton, N.M. (2016). State Regulation and Environmental Justice: The Need for Strategy Reassessment. *Capitalism Nature Socialism* 27 (2): 12–31.

Purcell, T.F., Loftus, A., and March, H. (2020). Value–Rent–Finance. *Progress in Human Geography* 44 (3): 437–456.

Purvis, K. (2016). "Water Is a Human Right . . . But It Can Have a Price." *The Guardian* (20 October). https://www.theguardian.com/global-development-professionals-network/2016/oct/20/water-human-right-price-united-nations (accessed 18 June 2022).

Prudham, S. (2004). Poisoning the Well: Neoliberalism and the Contamination of Municipal Water in Walkerton, Ontario. *Geoforum* 35 (3): 343–359.

Quick, T. and Woodhouse, P. (2014). The Impact of Large Scale Land Acquisitions on Water Resources – A Background Note. Department for International Development, Evidence on Demand. http://dx.doi.org/10.12774/eod_hd.october2014.quicketal.

Radcliffe, S. (2012). Development for the Postneoliberal Era? *Sumak kawsay*, Living Well, and the Limits to Decolonization in Ecuador. *Geoforum* 43: 240–249.

Ramsay, R. (2020). Water Wars: The Mexican State Where Farmers and Authorities Are in Combat over a Diminishing Resource. *Sky News*. https://news.sky.com/story/water-wars-the-mexican-state-where-farmers-and-authorities-are-in-combat-over-a-diminishing-resource-12111191 (accessed 13 June 2022).

Ranganathan, M. (2014a). Paying for Pipes, Claiming Citizenship: Political Agency and Water Reforms at the Urban Periphery. *International Journal of Urban and Regional Research* 38 (2): 590–608.

Ranganathan, M. (2014b). 'Mafias' in the Waterscape: Urban Informality and Everyday Public Authority in Bangalore. *Water Alternatives* 7 (1): 89–105.

Ranganathan, M. (2016). Thinking with Flint: Racial Liberalism and Roots of an American Water Tragedy. *Capitalism Nature Socialism* 27 (3): 17–33.

Ranganathan, M. (2018). Rule by Difference: Empire, Liberalism, and the Legacies of Urban "Improvement." *Environment and Planning A: Economy and Space* 50 (7): 1386–1406.

Ranganathan, M. and Balazs, C. (2015). Water Marginalization at the Urban Fringe: Environmental Justice and Urban Political Ecology across the North-South Divide. *Urban Geography* 36 (3): 403–423.

Ranganathan, M. and Bonds, A. (2022). Racial Regimes of Property: Introduction to the Special Issue. *Environment and Planning D: Society and Space* 40 (2): 197–207.

Reisner, M. (1993). *Cadillac Desert: The American West and its Disappearing Water*. New York: Penguin.

Republic of South Africa Government Gazette (RSA). (2001). Regulations Relating to Compulsory National Standards and Measures to Conserve Water. http://ec. europa.eu/echo/files/evaluation/watsan2005/annex_files/SA/SA1%20-%20 South%20Africa%20Watsan%20Regulations.pdf (accessed 18 June 2022).

Richey, A.S., Thomas, B.F., Lo, M.-H., Reager, J.T., Famiglietti, J.S., Voss, K., Swenson, S., and Rodell, M. (2015). Quantifying Renewable Groundwater Stress with GRACE. *Water Resources Research* 51: 5217–5238. https://doi.org/10.1002/2015WR017349.

Rieu-Clarke, A., Moynihan, R., and Magsig, B-O. (2012). United Nations Watercourses Convention: User's Guide. IHP-HELP Centre for Water Law, Policy, and Science (under the auspices of UNESCO), *United Kingdom*. https://unece.org/fileadmin/DAM/env/ water/meetings/Water_Convention/2016/10Oct_From_Practitioner_to_Practitioner/ UN_Watercourses_Convention_-_User_s_Guide.pdf (accessed 13 June 2022).

Rizvi, M. (2020). *The Ethics of Staying: Social Movements and Land Rights Politics in Pakistan*. Stanford, CA: Stanford University Press.

Robbins, P. (2019). *Political Ecology: A Critical Introduction* (3rd edition). Oxford, UK: Wiley.

Robbins, P., and Smith, S.H. (2017). Baby Bust: Towards Political Demography. *Progress in Human Geography* 41 (2): 199–219.

Robertson, M. (2004). The Neoliberalization of Ecosystem Services: Wetland Mitigation Banking and Problems in Environmental Governance. *Geoforum* 35 (3): 361–373.

Robertson, M. (2006). The Nature That Capital Can See: Science, State, and Market in the Commodification of Ecosystem Services. *Environment and Planning D: Society and Space* 24 (3): 367–387.

Robertson, M. (2012). Measurement and Alienation: Making a World of Ecosystem Services. *Transactions of the Institute of British Geographers* 37 (3): 386–401.

Robinson, C. (2000). *Black Marxism: The Making of the Black Radical Tradition*. Chapel Hill, NC: University of North Carolina.

Rockowitz, D., Askew-Merwin, C., Sahai, M., Markley, K., Kay, C., and Reames, T. (2018). *Household Water Security in Metropolitan Detroit: Measuring the Affordability Gap*. Ann Arbor, MI: University of Michigan Poverty Solutions.

Rockström, J., Falkenmark, M., Allan, T., Folke, C., Gordon, L., Jägerskog, A., Kummu, M., Lannerstad, M., Meybeck, M., Molden, D., et al. (2014). The Unfolding Water Drama in the Anthropocene: Towards a Resilience-Based Perspective on Water for Global Sustainability. *Ecohydrology* 7: 1249–1261.

Rodina, L. (2018). Defining "Water Resilience": Debates, Concepts, Approaches, and Gaps. *WIREs Water* 6 (2): e1334.

Rodina, L. and Chan, K.M.A. (2019). Expert Views on Strategies to Increase Water Resilience: Evidence from a Global Survey. *Ecology and Society* 24 (4): art28. https://doi.org/10.5751/ES-11302-240428.

Rodríguez-de-Francisco, J.C., Duarte-Abadía, B., and Boelens, R. (2019). Payment for Ecosystem Services and the Water-Energy-Food Nexus: Securing Resource Flows for the Affluent? *Water* 11 (6): 1143.

Rogers, S., Chen, D., Jiang, H., Rutherfurd, I., Wang, M., Webber, M., Crow-Miller, B., Barnett, J., Finlayson, B., Jiang, M., et al. (2020). An Integrated Assessment of China's South-North Water Transfer Project. *Geographical Research* 58 (1): 49–63.

Root, R.L. (2020). Exclusive: Inside the Battle over a UN Report on Water Privatization. *Devex* (26 October). https://www.devex.com/news/exclusive-inside-the-battle-over-a-un-report-on-water-privatization-98385 (accessed 18 June 2022).

Rosinger, A.Y. (2020). Biobehavioral Variation in Human Water Needs: How Adaptations, Early Life Environments, and the Life Course Affect Body Water Homeostatis. *American Journal of Human Biology* 32 (1): e23338.

Rosinger, A.Y. and Brewis, A. (2020). Life and Death: Toward a Human Biology of Water. *American Journal of Human Biology* 32 (1): e23361.

Rosinger, A.Y. and Young, S.L. (2020). The Toll of Household Water Insecurity on Health and Human Biology: Current Understandings and Future Directions. *Wiley Interdisciplinary Reviews (WIREs): Water* 7 (6): e1468.

Roth, D., Boelens, R., and Zwarteveen, M. ed. (2005). *Liquid Relations: Contested Water Rights and Legal Complexity*. New Brunswick, NJ: Rutgers University Press.

Rosa, L., Chiarelli, D.D., Rulli, M.C., Dell'Angelo, J., and D'Odorico P. (2020). Global Agricultural Economic Water Scarcity. *Science Advances* 6 (18): eaaz6031. https://doi.org/10.1126/sciadv.aaz6031.

Rostow, W. (1960). *The Stages of Economic Growth: A Non-communist Manifesto*. Cambridge, UK: Cambridge University Press.

Rudolph, M. and Kurian, R. (2022). Hydro-Hegemony, Water Governance, and Water Security: Palestinians under Israeli Occupation in the Jordan Valley, West Bank. *Water Alternatives* 15 (1): 73–92.

Rulli, M.C. and D'Odorico, P. (2013). The Water Footprint of Land Grabbing. *Geophysical Research Letters* 40 (23): 6130–6135.

Rulli, M.C. and D'Odorico, P. (2014). Food Appropriation through Large Scale Land Acquisition. *Environmental Research Letters* 9: 064030.

Salmond, A., Tadaki, M., and Gregory, T. (2014). Enacting New Freshwater Geographies. *New Zealand Geographer* 70: 47–55. https://doi.org/10.1111/nzg.12039.

Salzman, J., Bennett, G., Carroll, N., Goldstein, A., and Jenkins, M. (2018). The Global Status and Trends of Payments for Ecosystem Services. *Nature Sustainability* 1 (3): 136–144.

Saruchera, D. and Lautze, J. (2016). Transboundary River Basin Organizations in Africa: Assessing the Secretariat. *Water Policy* 18 (5): 1053–1069.

Savenije, H.H. and van der Zaag, P. (2002). Water as an Economic Good and Demand Management: Paradigms with Pitfalls. *Water International* 27 (1): 9.

Schlager, E. and Ostrom, E. (1992). Common Property and Natural Resources: A Conceptual Analysis. *Land Economics* 68 (3): 249–252.

Schmidt, J.J. (2014). Historicising the Hydrosocial Cycle. *Water Alternatives* 7 (1): 220–234.

Schwartz, K., Luque, M.T., Rusca, M., and Ahlers, R. (2015). (In)formality: The Meshwork of Water Service Provisioning. *Wiley Interdisciplinary Reviews (WIREs): Water* 2 (1): 31–36.

Scott, C.A., Meza, F.J., Varady, R.G., Tiessen, H., McEvoy, J., Garfin, G.M., Wilder, M., Farfán, L.M., Pablos, N.P., and Montaña, E. (2013). Water Security and Adaptive Management in the Arid Americas. *Annals of the Association of American Geographers* 103 (2): 280–289.

Scudder, T. (2012). Resettlement Outcomes of Large Dams. In: *Impacts of Large Dams: A Global Assessment* (ed. C. Tortajada, D. Altinbilek, A. and Biswas), 37–67. Berlin, Heidelberg: Springer.

Seemann, M. (2016). *Water Security, Justice and the Politics of Water Rights in Peru and Bolivia*. New York: Palgrave Macmillan.

Sengupta, S. (2022). Chile Writes a New Constitution, Confronting Climate Change Head On. *The New York Times* (28 December 2021; updated 6 January 2022). https://www.nytimes.com/2021/12/28/climate/chile-constitution-climate-change.html (accessed 1 June 2022).

Shapiro, A. (2018). What's Changed and What Hasn't When It Comes to the Flint Water Crisis. *National Public Radio* (26 October). https://www.npr.org/2018/10/26/661136990/whats-changed-and-what-hasnt-when-it-comes-to-the-flint-water-crisis (accessed 1 August 2022).

Sheng, J. and Webber, M. (2019). Governance Rescaling and Neoliberalization of China's Water Governance: The Case of China's South-North Water Transfer Project. *Environment and Planning A: Economy and Space* 51 (8): 1644–1664.

Shinn, J.E. (2016). Adaptative Environmental Governance of Changing Socio-ecological Systems: Empirical Insights from the Okavango Delta, Botswana. *Global Environmental Change* 40: 50–59.

Shinn, J.E., King, B., Young, K.R., and Crews, K.A. (2014). Variable Adaptations: Micro-politics of Environmental Displacement in the Okavango Delta, Botswana. *Geoforum* 57: 21–29.

Shove, E. (2002). Sustainability, System Innovation, and the Laundry. Department of Sociology, Lancaster University, Lancaster LA1 4YN, UK. https://www.lancaster.

ac.uk/fass/resources/sociology-online-papers/papers/shove-sustainability-system-innovation.pdf (accessed 5 August 2022).

Sidaway, J.D. and Woon, C.Y. (2017). Chinese Narratives on "One Belt, One Road" (一带一路) in Geopolitical and Imperial Contexts. *The Professional Geographer* 69 (4): 591–603.

Smakhtin, V., Revenga, C., and Döll, P. (2004). A Pilot Global Assessment of Environmental Water Requirements and Scarcity. *Water International* 29 (3): 307–317. https://doi.org/10.1080/02508060408691785.

Smith, D.M. (1982). *Living Under Apartheid: Aspects of Urbanization and Social Change in South Africa*. London: George Allen & Unwin.

Smith, S. (2020). *Political Geography: A Critical Introduction*. Hoboken, NJ: Wiley Blackwell.

Sneddon, C. (2015). *Concrete Revolution: Large Dams, Cold War Geopolitics, and the US Bureau of Reclamation*. Chicago, IL: The University of Chicago Press.

Sneddon, C. and Fox, C. (2006). Rethinking Transboundary Waters: A Critical Hydropolitics of the Mekong Basin. *Political Geography* 35 (12): 181–202. https://doi.org/10.1016/j.polgeo.2005.11.002.

Sneddon, C. and Fox, C. (2011). The Cold War, the US Bureau of Reclamation, and the Technopolitics of River Basin Development, 1950–1970. *Political Geography* 30 (8): 450–460.

Sneddon, C., Barraud, R., and Germaine, M.A. (2017). Dam Removals and River Restoration in International Perspective. *Water Alternatives* 10 (3): 648.

Sneddon, C., Magilligan, F.J., and Fox, C.A. (2017). Science of the Dammed: Expertise and Knowledge Claims in Contested Dam Removals. *Water Alternatives* 10 (3): 677.

Sneddon, C., Magilligan, F.J., and Fox, C.A. (2021). Peopling the Environmental State: River Restoration and State Power. *Annals of the American Association of Geographers* 112 (1): 1–18.

Sneddon, N., Chausson, A., Berry, P., Girardin, C.A.J., Smith, A., and Turner, B. (2020). Understanding the Value and Limits of Nature-Based Solutions to Climate Change and Other Global Challenges. *Philosophical Transactions of the Royal Society B: Biological Sciences* 375 (1794): 20190120.

Sojamo, S., Keulertz, M., Warner, J., and Allan, J.A. (2012). Virtual Water Hegemony: The Role of Agribusiness in Global Water Governance. *Water International* 37 (2): 169–182.

Spagnuolo, F. (2017). Global Land Rush, Water Grabbing and the Human Right to Water. In: *Agricultural Law: Current Issues from a Global Perspective* (ed. M. Alabrese, M. Brunori, S. Rolandi, and A. Saba), 293–304. Cham, Switzerland: Springer.

Stiglitz, J.E. (2002). *Globalization and Its Discontents*. New York: Norton.

Strategic Foresight (2021). Water Cooperation Quotient. https://www.strategicforesight.com/inner-keyinitiatives.php?id=6 (accessed 18 October 2022).

Strauss, K. (2013). Unfree Again: Social Reproduction, Flexible Labor Markets and the Resurgence of Gang Labour in the UK. *Antipode* 45 (1): 180–197.

Sugg, Z., Ziaja, S., and Schlager, E. (2016). Conjunctive Groundwater Management as a Response to Socio-ecological Disturbances: A Comparison of Four Western US States. *Texas Water Journal* 7 (1): 1–24.

Sultana, F. (2009). Fluid Lives: Subjectivities, Gender and Water in Rural Bangladesh. *Gender, Place & Culture* 16 (4): 427–444.

Sultana, F. (2011). Suffering for Water, Suffering from Water: Emotional Geographies of Resource Access, Control, and Conflict. *Geoforum* 4: 163–172.

Sultana, F. (2012). Producing Contaminated Citizens: Toward a Nature–Society Geography of Health and Well-Being. *Annals of the Association of American Geographers* 102 (5): 1165–1172. https://doi.org/10.1080/00045608.2012.671127

Sultana, F. (2018). Water Justice: Why It Matters and How to Achieve It. *Water International* 43 (4): 483–493.

Sultana, F. (2020). Embodied Intersectionalities of Urban Citizenship: Water, Infrastructure, and Gender in the Global South. *Annals of the American Association of Geographers* 110 (5): 1407–1424.

Sultana, F. and Loftus, A. ed. (2012). *The Right to Water: Politics, Governance and Social Struggles*. London: Earthscan.

Sultana, F. and Loftus A. ed. (2019). *Water Politics: Governance, Justice and the Right to Water*. Abingdon, UK: Routledge.

Sum, N.-L. (2019). The Intertwined Geopolitics and Geoeconomics of Hopes/Fears: China's Triple Economic Bubbles and the "One Belt One Road" Imaginary. *Territory, Politics, Governance* 7 (4): 528–552. https://doi.org/10.1080/21622671.2018.1523746.

Swain, M., McKinney, E., and Susskind, L. (2020). Water Shutoffs in Older American Cities: Causes, Extent, and Remedies. *Journal of Planning Education and Research*. https://doi.org/10.1177/0739456X20904431.

Swyngedouw, E. (2004). *Social Power and the Urbanization of Water: Flows of Power*. New York and Oxford, UK: Oxford University Press.

Swyngedouw, E. (2009). The Political Economy and Political Ecology of the Hydro-social Cycle. *Journal of Contemporary Water Research & Education* 142 (1): 56–60.

Swyngedouw, E. (2013). Into the Sea: Desalination as Hydro-Social Fix in Spain. *Annals of the Association of American Geographers* 103 (2): 261–270.

Swyngedouw, E. and Williams, J. (2016). From Spain's Hydro-Deadlock to the Desalination Fix. *Water International* 41 (1): 54–73.

Taylor, K.S., Moggridge, B.J., and Poelina, A. (2016). Australian Indigenous Water Policy and the Impacts of the Ever-Changing Political Cycle. *Australasian Journal of Water Resources* 20 (2): 132–147.

Tecklin, D., Bauer, C., and Prieto, M. (2011). Making Environmental Law for the Market: The Emergence, Character, and Implications of Chile's Environmental Regime. *Environmental Politics* 20 (6): 879–898.

Teodoro, M.P. (2019). Water and Sewer Affordability in the United States. *AWWA Water Science* 1 (2): e1129.

Teodoro, M.P. and Saywitz, R.R. (2020). A Snapshot of Water and Sewer Affordability in the United States, 2019. *Journal of the American Water Works Association* 112 (8): 10–19.

Thiel, A., Blomquist, W.A., and Garrick, D.E. ed. (2019). *Governing Complexity: Analysing and Supplying Polycentricity*. Cambridge, UK: Cambridge University Press.

Thomas, K. (2021). International Rivers as Border Infrastructures: En/forcing Borders in South Asia. *Political Geography* 89: 102448.

Tignino, M. and Bréthaut, C. (2020). The Role of International Case Law in Implementing the Obligation Not to Cause Significant Harm. *International Environmental Agreements: Politics, Law and Economics* 20: 631–648. https://doi.org/10.1007/s10784-020-09503-6.

Tola, M. (2018). Between Pachamama and Mother Earth: Gender, Political Ontology and the Rights of Nature in Contemporary Bolivia. *Feminist Review* 118 (1): 25–40. https://doi.org/10.1057/s41305-018-0100-4.

Transnational Institute (2020). Over 100 Civil Society Organizations Stand Behind UN Special Rapporteur Léo Heller, Denounce Private Water Industry Interference. Declaration. (21 October). https://www.tni.org/en/article/over-100-civil-society-organizations-stand-behind-un-special-rapporteur-leo-heller-denounce (accessed 18 June 2022).

Truelove, Y. (2011). (Re-)Conceptualizing Water Inequality in Dehli, India through a Feminist Political Ecology Framework. *Geoforum* 42 (2): 143–152.

Truelove, Y. (2019a). Rethinking Water Insecurity, Inequality and Infrastructure through an Embodied Urban Political Ecology. *Wiley Interdisciplinary Reviews (WIREs): Water* 6 (3): e1342.

Truelove, Y. (2019b). Gray Zones: The Everyday Practices and Governance of Water beyond the Network. *Annals of the American Association of Geographers* 109 (6): 1758–1774.

Truelove, Y. and Ruszczyk, H. (2022). Bodies as Urban Infrastructure: Gender, Intimate Infrastructures and Slow Infrastructural Violence. *Political Geography* 92. https://doi.org/10.1016/j.polgeo.2021.102492.

Tuck, E. and Yang, K.W. (2012). Decolonization Is Not a Metaphor. *Decolonization: Indigeneity, Education & Society* 1 (1): 1–40.

UK Ministry of Housing, Communities & Local Government (2017). *50 Years of the English Housing Survey, 1967–2017*. London, UK. https://www.gov.uk/government/publications/50-years-of-the-english-housing-survey (accessed 20 June 2022).

United Nations Convention to Combat Desertification (UNCCD). (2017). The Global Land Outlook, First Edition. Secretariat of the United Nations Convention to Combat Desertification. Bonn, Germany. https://www.unccd.int/sites/default/files/documents/2017-09/GLO_Full_Report_low_res.pdf (accessed 10 June 2022).

United Nations Economic Commission for Europe (UNECE). (2015). *Policy Guidance Note on the Benefits of Transboundary Water Cooperation: Identification, Assessment and Communication*. New York, Geneva: United Nations Economic

Commission for Europe. https://unece.org/fileadmin/DAM/env/water/publications/ WAT_Benefits_of_Transboundary_Cooperation/ECE_MP.WAT_47_Policy GuidanceNote_BenefitsCooperation_1522750_E_pdf_web.pdf (accessed 13 June 2022).

United Nations Environment Programme-DHI (UNEP-DHI). (2016). *Transboundary River Basins: Status and Trends.* Nairobi: United Nations Environment Programme. http://www.geftwap.org/publications/river-basins-spm (accessed 12 June 2022).

United Nations Office of the High Commissioner for Human Rights (UN OHCHR). (2016). Flint: Fundamentally about Human Rights – UN Experts Underline (3 March). https://www.ohchr.org/EN/NewsEvents/Pages/DisplayNews.aspx?NewsID=17139& (accessed 18 June 2022).

United Nations Office of the High Commissioner for Human Rights (UN OHCHR). (2020). Privatization and the Human Rights to Water and Sanitation: Report A/75/208. (20 July). https://www.ohchr.org/EN/Issues/WaterAndSanitation/SRWater/Pages/PrivateSectorParticipation.aspx (accessed 18 June 2022).

United Nations Water (UN-Water). (2018a). *Progress on Transboundary Water Cooperation: Global Baseline for SDG Indicator 6.5.2.* Paris: UNESCO.

United Nations Water (UN-Water). (2018b). *United Nations World Water Development Report 2018: Nature-Based Solutions for Water.* Paris: UNESCO.

United Nations Water (UN-Water). (2019). *The United Nations World Water Development Report 2019: Leaving No One Behind.* Paris: UNESCO.

United Nations Water (UN-Water). (2020). *Summary Progress Update 2021 – SDG 6 – Water and Sanitation for All.* Version: 1 March 2021. Geneva, Switzerland.

United Nations Water (UN-Water). (2021). *Progress on Transboundary Water Cooperation Global Status of SDG Indicator 6.5.2 and Acceleration Needs.* Paris, United Nations and UNESCO.

United Nations Water (UN-Water). (2022). *The United Nations World Water Development Report 2022: Groundwater: Making the Invisible Visible.* Paris: UNESCO.

Usher, M. (2022). Territory, Hydraulics, Biopolitics: Internal Colonization through Urban Catchment Management in Singapore. *Territory, Politics, Governance* https://doi.org/10.1080/21622671.2022.2056503.

Vaishnava, S. and Baka. J. (2021). Unruly Mountains: Hydropower Assemblages and Geological Surprises in the Indian Himalayas. *Environment and Planning E: Nature and Space* 5 (3). https://doi.org/10.1177/25148486211050780.

Van den Berge J., Boelens, R., and Vos, J. (2019). Citizen Mobilization for Water: The Case of Thessaloniki, Greece. In: *Water Politics: Governance, Justice and the Right to Water* (ed. F. Sultana and A. Loftus). Abingdon, UK: Routledge. https://journals.sagepub.com/doi/full/10.1177/25148486211050780.

Velásquez, T.A. (2018). Tracing the Political Life of Kimsacocha: Conflicts over Water and Mining in Ecuador's Southern Andes. *Latin American Perspectives* 222 45 (5): 154–169.

Veracini, L. (2010). *Settler Colonialism: A Theoretical Overview*. New York: Palgrave Macmillan.

Villar, P.C. (2016). International Cooperation on Transboundary Aquifers in South America and the Guarani Aquifer Case. *Revista Brasileira de Politica Internacional* 59 (01). https://doi.org/10.1590/0034-7329201600107.

Wainwright, M. (2008). Leeds Students Ban Bottled Water. *The Guardian* (16 December). https://www.theguardian.com/education/2008/dec/16/students-waste (accessed 2 August 2022).

Wani, S.P., Rockström, J., and Oweis, T.Y. (2009). *Rainfed Agriculture: Unlocking the Potential*. Wallingford, UK: CAB International.

Warner, J., Mirumachi, N., Farnum, R., Grandi, M., Menga, F., and Zeitoun, M. (2017). Transboundary 'Hydro-hegemony': 10 Years Later. *WIREs Water* 4 (6): e1242.

Water Footprint Network (2022). What Is a Water Footprint. https://waterfootprint.org/en/water-footprint/what-is-water-footprint/ (accessed 25 July 2022).

Water Land Ecosystems Mekong (WLE Mekong). (2015). Mapping the Dams of the Mekong. WLE-Mekong. https://wle-mekong.cgiar.org/wp-content/uploads/unnamed-11.jpg (accessed 12 June 2022).

Waughray, D. ed. (2011). *Water Security: The Water-Energy-Food-Climate Nexus*. Washington, DC and London: Island Press.

Weaver, E.M. and Drapalski, H.J. (2020). The Navigable Waters Protection Rule: Streamlining Waters of the United States. *American Bar Association* (26 May). https://www.americanbar.org/groups/litigation/committees/environmental-energy/practice/2020/navigable-waters-protection-rule-streamlining-waters-of-the-united-states/ (accessed 7 June 2022).

Weaver, M. (2022). Emergency Water Plant in London Unusable Despite Drought Risk. *The Guardian* (4 August). https://www.theguardian.com/environment/2022/aug/04/emergency-water-plant-london-unusable-despite-drought-risk (accessed 5 August 2022).

Webber, M., Crow-Miller, B., and Rogers, S. (2017). The South–North Water Transfer Project: Remaking the Geography of China. *Regional Studies* 51 (3): 370–382.

Wescoat, J. (2000). Wittfogel East and West: Changing Perspectives on Water Development in South Asia and the US, 1670–2000. In: *Cultural Encounters with the Environment: Enduring and Evolving Geographic Themes* (ed. V. Haarmann), 109–132. Lanham, MD: Rowman and Littlefield.

Wescoat, J.L. and Halvorson, S.J. (2012). Emerging Regional Perspectives on Water Research and Management: An Introductory Comment. *Eurasian Geography and Economics* 53 (1): 87–94.

Wester, P., Rap, E., and Vargas-Velázquez, S. (2009). The Hydraulic Mission and the Mexican Hydrocracy: Regulating and Reforming the Flows of Water and Power. *Water Alternatives* 2 (3): 395–415.

Wheeler, K.G., Hall, J.W., Abdo, G.M., Dadson, S.J., Kasprzyk, J.R., Smith, R., and Zagona, E.A. (2018). Exploring Cooperative Transboundary River Management Strategies for the Eastern Nile Basin. *Water Resources Research* 54 (11): 9224–9254.

White, R. (1996). *The Organic Machine: The Remaking of the Columbia River.* New York: Macmillan.

Whittington, D. and Thomas, D.A. (2020). Policy Note: Invited Opinion Interview with Professor Tony Allan: "Water Scarcity, Food Production, and Virtual Water" (Part 1). *Water Economics and Policy* 6 (1): 1971003.

Williamson, F. (2020). Responding to Extremes: Managing Urban Water Scarcity in the Late Nineteenth-Century Straits Settlements. *Water History* 12 (3): 251–263.

Williamson, J. (1993). Democracy and the "Washington Consensus." *World Development* 21 (8): 1329–1336.

Wilmsen, B. (2018). Is Land-Based Resettlement Still Appropriate for Rural People in China? A Longitudinal Study of Displacement at the Three Gorges Dam. *Development and Change* 49 (1): 170–198.

Wilson, N.J. (2014). Indigenous Water Governance: Insights from the Hydrosocial Relations of the Koyukon Athabascan Village of Ruby, *Alaska. Geoforum* 57 (1): 1–11.

Wilson, N.J. and Inkster, J. (2018). Respecting Water: Indigenous Water Governance, Ontologies, and the Politics of Kinship on the Ground. *Environment and Planning E: Nature and Space* 1 (4): 516–538.

Wilson, N.J., Montoya, T., Arsenault, R., and Curley, A. (2021). Governing Water Insecurity: Navigating Indigenous Water Rights and Regulatory Politics in Settler Colonial States. *Water International* 46 (6): 783–801.

Winston, C. (2021). Maroon Geographies. *Annals of the American Association of Geographers* 111 (7): 2185–2199.

Wittfogel, K. A. (1957). *Oriental Despotism: A Comparative Study of Total Power.* New Haven, CT: Yale University Press.

Wolf, A., Yoffe, S.B., and Giordano, M. (2003). International Waters: Identifying Basins at Risk. *Water Policy* 5: 29–60.

Wolfe, P. (2006). Settler Colonialism and the Elimination of the Native. *Journal of Genocide Research* 8 (4): 387–409.

Wood, W.W. and Cherry, J.A. (2021). Food Security and Inaccurate Quantification of Groundwater Irrigation Use. *Groundwater* 59 (6): 782–783.

World Bank (2016). *High and Dry: Climate Change, Water, and the Economy.* Washington, DC: World Bank. https://www.worldbank.org/en/topic/water/publication/high-and-dry-climate-change-water-and-the-economy (accessed 5 August 2022).

World Bank (2019). *The Cubango-Okavango River Basin Multi-Sector Investment Opportunities Analysis: Summary Report.* Washington, DC: World Bank.

World Commission on Dams (WCD). (2000). *Dams and Development: A New Framework for Decision-Making: The Report of the World Commission on Dams.* London: Earthscan.

Worster, D. (1992). *Rivers of Empire: Water, Aridity, and the Growth of the American West.* New York and Oxford, UK: Oxford University Press.

Worster, D. (2011). The Flow of Empire: Comparing Water Control in China and the United States. *Rachel Carson Center Perspectives* 5: 1–23.

Wutich, A. (2009). Intrahousehold Disparities in Women and Men's Experiences of Water Insecurity and Emotional Distress in Urban Bolivia. *Medical Anthropology Quarterly* 23 (4): 436–454.

Wutich, A. (2010). After the Cochabamba Water War of 2000: A Common Pool Resource Institution in the Urban Andes. In: *Human Ecology* (ed. D. Bates and J. Tucker), 297–316. Boston, MA: Springer. https://doi.org/10.1007/978-1-4419-5701-6_19.

Wutich, A. (2020). Water Insecurity: An Agenda for Research and Call to Action for Human Biology. *American Journal of Human Biology* 32 (1): e23345.

Wutich, A. and Ragsdale, K. (2008). Water Insecurity and Emotional Distress: Coping with Supply, Access, and Seasonal Variability of Water in a Bolivian Squatter Settlement. *Social Science & Medicine* 67 (12): 2116–2125.

Wutich, A., Brewis, A., and Tsai, A. (2020). Water and Mental Health. *WIREs Water* 7 (5): e1461.

Wutich, A., Budds, J., Eichelberger, L., Geere, J., Harris, L.M., Horney, J.A., Jepson, W., Norman, E., O'Reilly, K., Pearson, A.L., et al. (2017). Advancing Methods for Research on Household Water Insecurity: Studying Entitlements and Capabilities, Socio-cultural Dynamics, and Political Processes, Institutions, and Governance. *Water Security* 2: 1–10.

Yihdego, Z., Rieu-Clarke, A., and Cascão, A.E. ed. (2018). *The Grand Ethiopian Renaissance Dam and the Nile Basin: Implications for Transboundary Water Cooperation*. New York: Routledge.

Yoffe, S., Wolf, A.T., and Giordano, M. (2003). Conflict and Cooperation over International Freshwater Resources: Indicators of Basins at Risk. *Journal of the American Water Resources Association*, 39 (5): 1109–1126. https://doi.org/10.1111/j.1752-1688.2003.tb03696.x.

Yoshida, Y., Lee, H.S., Trung, B.H., Tran. H-D., Lall, M.K., Kakar, K., and Xuan, T.D. (2020). Impacts of Mainstream Hydropower Dams on Fisheries and Agriculture in Lower Mekong Basin. *Sustainability* 12 (6): 2408. https://doi.org/10.3390/su12062408.

Zarfl, C., Lumsdon, A. E., Berlekamp, J., Tydecks, L., and Tockner, K. (2015). A Global Boom in Hydropower Dam Construction. *Aquatic Sciences* 77 (1): 161–170.

Zeitoun, M. and Mirumachi, N. (2008). Transboundary Water Interaction I: Reconsidering Conflict and Cooperation. *International Environmental Agreements: Politics, Law and Economics* 8 (4): 297–316. https://doi.org/10.1007/s10784-008-9083-5.

Zeitoun, M., Cascao, A., Warner, J., Mirumachi, N., Matthews, N., Menga, F., and Farnum, R. (2017). Transboundary Water Interaction III: Contesting Hegemonic Arrangements. *International Environmental Agreements* 17 (2): 271–294.

Zeitoun, M., Mirumachi, N., and Warner, J. (2011). Transboundary Water Interaction II: The Influence of "Soft" Power. *International Environmental Agreements: Politics, Law and Economics* 11: 159–178. http://dx.doi.org/10.1007/s10784-010-9134-6.

Zeitoun, M., Mirumachi, N., and Warner, J. (2020). *Water Conflicts: Analysis for Transformation*. Oxford, UK: Oxford University Press.

Zeitoun, M. and Warner, J. (2006). Hydro-hegemony: A Framework for Analysis of Trans-boundary Water Conflicts. *Water Policy* 8 (5): 435–460.

Zikhali-Nyoni, T. (2021). The Role of SADC in Transboundary Water Interactions: The Case of the Incomati International River Basin. *Journal of Southern African Studies*, 47 (4): 703–718.

Zwarteveen, M.Z. (1997). From Basic Need to Commodity: A Discussion on Gender and Water Rights in the Context of Irrigation. *World Development* 25 (8): 1335–1349.

Zwarteveen, M.Z. (2010). A Masculine Water World: The Politics of Gender and Identity in Irrigation Expert Thinking. In: *Out of the Mainstream: Water Rights, Politics, and Identity* (ed. R. Boelens, D. Getches, and A. Guevera Gil), 75–89. London: Routledge.

Zwarteveen, M.Z., and Boelens, R. (2014). Defining, Researching and Struggling for Water Justice: Some Conceptual Building Blocks for Research and Action. *Water International* 39 (2): 143–158.

Index

Page numbers in italics refer to figures; page numbers in bold refer to tables; page numbers followed by "n" refer to footnotes.

Printed and bound by CPI Group (UK) Ltd, Croydon, CR0 4YY

27/10/2024

14580204-0002